T0314656

J.D. Bernal: A Life in Science and Politics

An eminent molecular physicist and path-breaking crystallographer, an eloquent and prescient writer on the social implications of science, an early foe of pseudo-scientific racism and an indefatigable campaigner for peace and civil rights: as a scientist and a Communist intellectual, J.D. Bernal was caught up in many of the central dramas of the twentieth century.

As Eric Hobsbawm describes here, Bernal played a major role in the dynamic 'red science' movement of the 1930s, whose ideas on the links between science and society are only now being accorded their full significance. Bernal's *The Social Function of Science* remains a classic analysis of the way in which wider social relations may determine the boundaries of both scientific understanding and practice.

Impressed by Bernal's relentless questioning of received ideas, Mountbatten recruited him to the brilliant scientific team of his 'Department of Wild Talents' during World War Two, to help in planning the Normandy landings. After the war, Bernal strove to combine running the Department of Physics at Birkbeck College, London, with travelling and campaigning throughout six continents against the nuclear threat of the Cold War. In a field notorious for its misogynism, Bernal's laboratories at Birkbeck were a haven for many of the leading women scientists of the day, among them Rosalind Franklin and the Nobel laureate Dorothy Crowfoot Hodgkin. And, as James Watson has acknowledged, Bernal's X-ray photographs of molecular structures formed a vital piece of evidence on the path leading to the discovery of DNA.

In this wide-ranging collection of essays, different facets of Bernal's life and work are recounted and assessed by Eric Hobsbawm, Peter Trent, Ritchie Calder, Hilary Rose, Steven Rose, Chris Freeman, Fred Steward, Ivor Montagu, Peter Mason, Chris Whittaker, Roy Johnston, Ann Synge and Earl Mountbatten of Burma.

Brenda Swann and **Francis Aprahamian** worked at different periods with J.D. Bernal at Birkbeck College, London, on the research for several of his most important works, including *The Social Function of Science, Science in History, A Prospect of Peace* and *The Extension of Man*.

J. D. Bernal

A Life in Science and Politics

Edited by
BRENDA SWANN and
FRANCIS APRAHAMIAN

VERSO
London • New York

First published by Verso 1999
© individual contributors 1999
All rights reserved

The moral rights of the authors and the editors have been asserted

Verso
UK: 6 Meard Street, London W1F 0EG
US: 20 Jay Street, Suite 1010, Brooklyn, NY 11201

Verso is the imprint of New Left Books

ISBN 978-1-78663-759-8

British Library Cataloguing in Publication Data
A catalogue record for this book is available from the British Library

Library of Congress Cataloging-in-Publication Data
A catalog record for this book is available from the Library of Congress

Typeset by SetSystems Ltd, Saffron Walden, Essex
Printed in the United States

Contents

Initials and Acronyms   vii

Preface   ix
Eric Hobsbawm

Introduction   xxi
Brenda Swann

1 Early Years and Influences   1
Ann Synge

2 Irish Roots   17
Roy Johnston

3 Political Formation   37
Fred Steward

4 The Scientist   78
Peter Trent

5 The Social Function of Science   101
Chris Freeman

6 Red Scientist: Two Strands from a Life in Three Colours   132
Hilary Rose and Steven Rose

7 Bernal at War   160
Ritchie Calder

8 Memories of Desmond Bernal   191
Earl Mountbatten of Burma

9 D-Day Diaries   196
J. D. Bernal

10   The Peacemonger   212
     Ivor Montagu

11   Bernal at Birkbeck   235
     Eric Hobsbawm

12   Science in History   255
     Peter Mason

13   Building Tomorrow   268
     Chris Whittaker

     Obiter Dicta   295

     Books by J. D. Bernal   302

     Calendar of Events   303

     Notes on Editors and Contributors   307

     Index   310

Initials and Acronyms

ARP	Air Raid Precautions
ASLIB	Association of Special Libraries and Information Bureaux
AScW	Association of Scientific Workers
AUT	Association of University Teachers
BA	British Association for the Advancement of Science
BRS	Building Research Station
CLASP	Consortium of Local Authorities' Schools Programme
CND	Campaign for Nuclear Disarmament
COSSAC	Chief of Staff, Supreme Allied Command
CSAWG	Cambridge Scientists' Anti-War Group
CSIRO	Commonwealth Scientific Industrial Research Organization
CUSS	Cambridge University Socialist Society
DNA	Deoxyribonucleic Acid
DSIR	Department of Scientific and Industrial Research
FIL	For Intellectual Liberty
GNP	Gross National Product
ICI	Imperial Chemical Industries
IUC	International Union of Crystallography
LRD	Labour Research Department
MARS	Modern Architecture Research Group
NALGO	National Association of Local Government Officers
NUSW	National Union of Scientific Workers
NUT	National Union of Teachers
R&D	Research and Development
RI	Royal Institution
RIBA	Royal Institute of British Architects
RNA	Ribonucleic Acid
RS	Royal Society
SCR	Society for Cultural Relations with the Soviet Union

STS   Scientific and Technical Services
UDC   Union of Democratic Control
WFSW   World Federation of Scientific Workers
WPC   World Peace Council

Preface

Eric Hobsbawm

There are two reasons for publishing this book about John Desmond Bernal (1901–71), written largely by authors who knew him. The first is personal, the second historical.

As a person, and indeed as a scientist, Bernal was fascinating, memorable and extraordinarily impressive. All who knew him were struck by the combination of intellectual brilliance, curiosity, bohemian culture and an erudition that leaped across the borders between science, history and the arts. Most of those who knew him thought he was a genius, or at all events a man uniquely gifted, to quote an admiring friend, in 'shooting an arrow of original thought' into any target presented to him. As a scientist, though it was patently well within his capacities, he did not win the accepted accolade of a Nobel Prize, probably because he spread himself too thin, and never focused his energies and effort long enough on any of the numerous fields that attracted his interest. Perhaps more to the point, though he was a molecular biologist before the term was invented, and much of the field rested on his pioneering work in crystallography, he remained on the fringes of the greatest triumph of molecular biology in his time, the discovery of the 'double helix'.

His contribution to the sciences is described elsewhere in this volume, and I am not qualified to assess it. However, one thing is patent: 'Bernal inspired others. That was his role.'[1] At least three of his pupils and protégés became Nobel laureates (Dorothy Crowfoot Hodgkin, Max Perutz, Maurice Wilkins); at least two others Nobel laureates (Francis Crick and Aaron Klug – not to mention Rosalind Franklin, who died too soon for the Nobel to which she had a serious claim) went out of their way to work in his establishment; and at a memorial meeting for him another two (P. M. S. Blackett and J. C. Kendrew) considered that the range and power of his mind and his scientific originality were superior to theirs. That is an impressive score of peer assessment.

Inspiring others is not necessarily a good recipe for posthumous fame, especially as word-of-mouth reputation tends to die with those who were its bearers. All those who spoke at the memorial meeting organized by his London college are now themselves dead: the Nobel physicist Blackett, the Nobel chemist Kendrew, Solly Zuckerman, later the British government's chief scientific adviser, who, with Bernal and Geoffrey Pyke, had formed Mountbatten's 'Department of Wild Talents' as they helped to plan the 1944 invasion of France, and C. P. Snow, who made Bernal into the hero of a novel. (Are he and J. B. S. Haldane, who appears thinly disguised in one of Aldous Huxley's novels, the only British scientists who have broken into fiction?) The causes to which he devoted so much of his time and energy are gone. The Soviet-sponsored World Peace Council and World Federation of Scientific Workers are now reduced to topics of potential PhD theses. The Soviet Union itself, child of the October Revolution to which, like so many who became Communists in its wake, Bernal retained a troubled loyalty, is now past history.

The years since his death have eroded his memory. Those who knew this short, abrupt, bushy-haired, reedy-voiced man with piercing eyes and a clumsy nautical walk, lover of science, women and the arts (except for a notorious deafness to music, unusual for a scientist), are today a diminishing band of senior citizens. Those who knew him in his prime, that is to say before the Cold War and his refusal to distance himself from the USSR marginalized him, must be an even older and smaller body. Scientists now in their prime were born after the date of the first in the series of strokes that put an end to Bernal's career in his mid-sixties. Most of them are unlikely to have a clear idea of his contribution to science in the first half of the twentieth century, a golden age for science in Britain. Controversy, both political and personal, has since his death stood in the way of the major biography that is undoubtedly his due. No doubt he will get it in the end, but when it is written it will have to be exclusively, or almost entirely, on the basis of the written sources. The present work cannot replace it. However, it can help to rescue an astonishing figure from the present mist of semi-oblivion.

The second reason for the present book is Bernal's central role in the political radicalization and mobilization of British scientists in the 1930s, which was also, and not fortuitously, one of the rare periods when Marxism made a significant impact on the natural sciences.

Bernal was a central figure in this, as it turned out, temporary era of what has been called 'red science' in three ways. He was at the heart of the political mobilization of scientists on the left – against war both before and after World War Two, and for the organization of scientific workers. He was, without doubt, the earliest as well as the most prominent and long-lasting convert to Communism among the significant scientists in Britain, and as such he was crucial in giving Marxism its public presence in the sciences in the 1930s. (After the war his – at least in public – uncritical loyalty to the Soviet Union had the opposite effect.) Last, but far from least, he was the most influential prophet of the unlimited potential of science for progress, and therefore of the transformation of science, and scientists, into a recognized, publicly structured and funded force of production (and, as it turned out, destruction): an ambiguous and problematic legacy.

For, as we can now see, this was a momentous era in the social and public history of twentieth-century science, not least because it laid the foundations for the postwar public recognition that the activities of scientists are the indispensable foundations for technological and economic development, not to mention the winning of wars. Hence the astonishing explosion that followed both in the number of scientists and in the amount of resources devoted by governments, public and – to a lesser extent – private institutions to the natural sciences. In the 1930s Bernal became perhaps the chief prophet of this transformation through his remarkably influential book *The Social Function of Science* (London, 1939), the impact of which extended far beyond scientists on the political left.

Though a small group of scientists with left-wing opinions was on the scene in the 1920s – Bernal among them – there is little doubt that, in Britain at least, in the 1930s there was a substantial radicalization of the younger scientists. It was most noticeable in the sciences at the cutting edge of progress. In Cambridge, which virtually monopolized top-level British scientific achievement in the first half of the twentieth century, young left-wing scientists were most prominent at the Cavendish Laboratory (Physics) and the Biochemistry Laboratory. These were the strongholds of the Cambridge Scientists' Anti-War Group, formed by eighty researchers in 1934, an active campaigning organization. This does not mean that the bulk of young scientists became political militants, let alone Marxists. What happened was, rather, that an active nucleus of the left set the political tone, reducing the apolitical or politically moderate majority of scientists to a marginal minority. C. P. Snow, a perceptive and well-informed observer of science and public life, and with not very successful experience of both, thought that in 1936 a poll of the 200 brightest physicists under the age of forty would have revealed about 15 as Communists or fellow-travellers, about 50 as politically on the left, another 100 as passively sympathetic to the left, and the remainder neutral, apart from five or six on the eccentric wings of the right. That is to say, more than three quarters would have considered themselves left of the political centre. What needs explaining is this hegemony of the radical minority within the community of scientists at this period.

It is hardly surprising that Western intellectuals, scientists among them, were radicalized in the 1930s, although the question remains to be answered, why those attracted to Communism contained what the historian of inter-war Cambridge called 'quite so many intellectually gifted people'.[2] Nevertheless, there were specific reasons for the new political concerns of scientists.

The first, at least in Britain, is perhaps best indicated by the title of a book by Julian Huxley, a progressive but far from revolutionary member of a family distinguished in science since Darwin: *Scientific Research and Social Need*. The mere conjunction of the words 'research' and 'social need' was a political manifesto against the orthodoxy of 'high science', the 'pure' pursuit of truth, for itself alone and for no other purpose, which had until then dominated the scientific community. Even in 1933, just after Hitler had come to power, A. V. Hill, a Nobel laureate and pillar of the Royal Society, could still announce that 'the Society frowns on inducting members who meddle in politics' – that is to say, in other kinds of politics than those considered suitable for respectable professors at Oxford and Cambridge.

And yet, in a period of economic depressions and mass unemployment, not merely was it likely that scientists would be more aware than hitherto of social needs but the evident incapacity of the economic and political system to deal with these problems leaped to the eye. That is why, in the USA of the Great Depression, the short-lived phenomenon of 'technocracy' mobilized men from a normally apolitical or conventionally conservative group of successful experts in their own fields, such as engineers, to demand political power. As they watched the fumblings of politicians and businessmen misled by the orthodoxy of self-adjusting free markets, scientists, confident in the knowledge that they were part of what the great physicist Rutherford called 'the heroic period' of their subjects, might well feel that they had something to contribute to the solution of their society's problems. From the Cavendish Laboratory even the great Rutherford, not a man given to radicalism or unconventionality in intellectual and political matters, supported the project put forward by a dissident Conservative, Harold Macmillan, for planned economic policy, a National Development Board and a far greater measure of social justice. *Nature*, the most prestigious journal in British science, under its editor R. A. Gregory, gave a voice to those who pressed for greater recognition of the social implications of science, greater scientific planning, and a greater role for science in political decision-making.[3] The 'red' scientists, whose movement, like the Marxist student movement, also emerged in the early 1930s, were thus part of a wider 'reformist' current in science, and at one moment something like a scientific 'popular front' actually appeared, when the Marxists united with the *Nature* group to secure the creation of a Division for the Social and International Relations of Science within the British Association for the Advancement of Science (1938–45).

The second reason for the radicalization of scientists in the 1930s was the triumph of Hitler. Few scientists in the democratic countries could remain neutral about a political system based on a crude form of racism, which drove many of Germany's most brilliant scientific talents into exile; no Western scientist could fail to have personal contact with some refugee colleague or student who had first-hand knowledge of the consequences for science of Hitler and, after 1938, of Italian fascism. Moreover, Nazi ideology was expressly hostile to the principles of rational enquiry which most scientists believed to be the foundations of their activity. From 1933 on, antifascism was politics directly relevant to scientists.

However, the element that was likely to move them to political action was not simply National Socialism or fascism which, after all, were very unlikely to come to power under their own steam in Britain, but that, in the contemporary phrase, 'fascism meant war'. Anti-war sentiment was already strong in the 1920s, especially on the left. It was greatly reinforced at the end of the 1920s by the wave of memoirs, novels and plays from writers who had fought in the first of modern total wars, and who at last found themselves able to give shape to the experience of 1914–18: Erich Maria Remarque's *All Quiet on the Western Front* (1929) was probably the most internationally influential of such works. From 1933 on, it was evident to all that the danger of a second world war came entirely from the systematic policy of expansion and military aggression of Hitler's Germany and its allies, fascist Italy and militarist Japan.

As it happened, after the Great War which initiated modern aerial and chemical warfare, war was a matter on which scientists were expected to have opinions, even those who were slow to be moved to action by dislike of fascist or Nazi regimes. The aggressions of Nazi Germany made it possible to merge the fight against fascism with the fight against war.

In any such mobilization, at least in Britain, even a small left-wing minority was likely to play a disproportionate part, if only because, in the absence of other initiatives and under governments dedicated to the 'appeasement' of Hitler and Mussolini, it found itself acting as the most effective voice and often the only active organizer of an increasingly widely supported movement of resistance to the international aggressors.

The peculiarity of the scientists' political movement was that it mobilized not merely their opinions but also their special expertise as scientists. Who but nuclear physicists could know that a German discovery in 1938 showed atomic bombs to be feasible, and that Nazi Germany possessed the expertise to construct them? Who but anti-Nazi scientists would immediately think of warning governments? Or, indeed, how could an atom bomb be built except by means of a vast and unprecedented allocation of resources and scientific manpower, which incidentally laid the foundations for the 'big science' of the second half of the twentieth century? This was the tragic paradox of the radical era in the laboratories: when it turned out that Nazi Germany had not made nuclear weapons, the justification for making them disappeared. Those without whom the atom bomb neither could nor would have been built were left protesting and powerless when it was actually dropped.

Even before the nuclear effort, however, the left had initiated the process that was to lead to the systematic mobilization of scientists by government. In doing so, it demonstrated the crucial role of science in society, of which Bernal had made himself the major spokesman, and transformed its social role. In Britain the scientists' anti-war effort had initially been entirely defensive. The major activity of anti-war scientists in the 1930s was to warn the public against the dangers of aerial, chemical and biological warfare, which they rightly recognized would be directed primarily against noncombatants. Criticizing the deficiencies of government policy, they explored more effective ways to protect the civilian population. Bernal was first drawn directly into the government on the eve of the war as an expert on protection against air raids whose criticism impressed the authorities, and later as a more general analyst of the effects of bombing. Once mobilized, he and

other scientists could and did bring their methods, specific and generalized ('operational research'), to bear on every aspect of the national
war effort.

All this might have happened in any case. Nevertheless it is significant, though not surprising, that the initiative in this transformation
was taken by scientists of, or associated with, the left: believers in the
unlimited potential of science for humanity, champions of its mobilization – in effect, by sympathetic governments – for social purposes,
and, not least, believers in the superiority of reason, theory and
experiment over ignorance and obscurantism.

This is true not only of Britain, for in France it was the Popular Front
government of 1936 which – urged on by the socialist physicist Jean
Perrin and the subsequently Communist physicist Paul Langevin –
made the encouragement of science an important plank in its programme. It established both a government undersecretaryship for
scientific research (under the physicist Irène Joliot-Curie, who was to
become a Communist together with her husband Frédéric) and the
general framework for a nationally funded research agency. The Centre
National de la Recherche Scientifique began to function in 1939, and
is still in operation.

Political radicalization between the First and Second World Wars
need not necessarily have increased the influence of Marxism among
scientists, although in the early 1930s the USSR with its immunity to
the mass unemployment that ravaged the West, and its enthusiastic
dedication to and – it seemed – material support for science made its
ideology ('scientific socialism') seem worth serious consideration.[4] It is
easy to discover socialist and Communist scientists in the 1930s who
took no interest in, and even some who disagreed with, Marxist theory.
Not that anyone at this time doubted that Marxist theory was relevant
to the natural sciences. On the contrary, what an opponent of Marxism
like Bertrand Russell found hard to understand in 1938 was why
'Britain's leading holders of scientific chairs were adherents to philosophical quackery'.[5] Probably, with historians, young scientists were the
branch of intellectuals most strongly influenced by Marxism in 1930s
Britain.

Nevertheless, though Marx had taken an interest in the natural
sciences – two fellows of the Royal Society were among the few who
attended his funeral in 1883 – and Engels had written extensively on
scientific matters, there seemed to be far less of professional relevance
in the writing of the classics for natural scientists than for social
scientists and historians. Indeed, asked by the Social Democratic Party

of Germany after Engels's death for an opinion on Engels's scientific
writings, the only academic scientist in the party, Dr Leo Arons (to
exclude whom from academic teaching the German Empire passed a
special law) advised against publication. After the October Revolution,
the question of publishing Engels's *Dialectics of Nature* arose again, and
Einstein himself was consulted in 1924; he wrote that the manuscript
was of no special interest from the point of view of either the history of
physics or twentieth-century physics. However, he added tactfully, since
Engels was an important historical figure a case might be made for
publishing it as an interesting contribution to the study of his intellectual development. And indeed Ryazanov, the editor of the proposed
Gesamtausgabe of the works of Marx and Engels, while trying to explain
away these assessments, chose to publish it, not in the main body of
works of Marx and Engels but in a specialized periodical reprinting,
among other things, lesser material from the Marxian archives.[6] (Only
under Stalin did it become a canonical text – one of the few works
of Marx and Engels to be specifically cited in the ultimate com-
pendium of Communist orthodoxy, the 1939 *History of the CPSU (b):
Short Course.*) Again, it was plain that developments in physics since
Planck and Einstein were difficult to fit into the framework of Lenin's
Materialism and Empiriocriticism which was not to be criticized in the new
USSR.[7]

Yet there are intellectual reasons, especially in the British context,
why a number of scientists between the wars nevertheless found inspiration in the Engels-formulated Marxism that came out of the USSR. As
has been correctly observed,[8] they read Engels not for the details of his
scientific knowledge, but for the light his ideas might throw on the
problems on which as scientists they were engaged. Thus we find both
Bernal and Joseph Needham at different times quoting the same text
by Engels on the significance of the three decisive discoveries of
nineteenth-century science: the unification of the various 'forces' in
nature – mechanical force, heat, light, electricity, magnetism, chemical
bonding, et cetera – which he traced back to the thermodynamic
discoveries; the discovery of the organic cell; and the theory of evolution. What remained was to 'explain the origin of life from inorganic
nature',[9] which Engels believed to be within the reach of biochemistry.
Leaving details aside, scientists of the 1920s and 1930s would have been
struck by an approach that had (a) anticipated and welcomed the
rupture of the framework of classical physics, (b) recognized that the
discovery of a single basic unit of life made possible the analysis of
living organisms and systems of increasing complexity at a series of

levels, and (c) recognized that diachronicity, that is, history, inevitably entered the sciences with the theory of evolution.[10] Historians of science must be left to trace the precise ways in which Engels's dialectical materialism seemed to fit the specific work on which the scientific converts to Marxism were engaged, but it clearly appeared to have some relevance to them – though perhaps more directly to J. B. S. Haldane and Needham than to Bernal.

First, its perspective was essentially historical, and its concern was with change and transformation. This appealed particularly to scientists engaging with essentially historical questions such as the origins and development of life on earth (opened up by the famous 1924 paper of the Soviet scientist Oparin), or with essentially nonstatic phenomena. Hence in this period the influence of Marxism was particularly powerful among workers in and on the life sciences: J. B. S. Haldane, Joseph Needham, C. H. Waddington and Bernal himself.

Incidentally, the Marxist potential for revolutionizing the history of science was demonstrated by the striking impact of B. Hessen's paper on 'The Social and Economic Roots of Newton's *Principia*' at the famous London International Congress on the History of Science and Technology of 1931, to which the USSR unexpectedly sent a powerful delegation not yet reduced to Stalinist unanimity and whose reports were available in English.[11] It inspired Joseph Needham to adapt and republish the first volume of his vast *Chemical Embryology* under the title of *A History of Embryology* in 1934, and it had repercussions in sociology and even in non-Marxist history.[12]

Second, Marxism appealed – once again, most obviously to those in the life sciences – because of the necessary complexity of their simultaneously multi-level subject matters, by definition both stable and unstable, self-reproducing and self-changing. Since the vitalists who believed in the essential autonomy of life and its irreducibility to nonlife were clearly losing their battle, the weakness of their old antagonists, the reductionist (materialist) mechanists, was now evident, and this attracted attention to Marxism, antivitalist but deeply antimechanical and antireductionist. Next to Hessen's paper on Newton, and Bukharin's attack on the separation of 'pure' and 'applied' or useful science, the paper that made the greatest impression at the 1931 Congress was by the Soviet biologist B. M. Zavadovsky, who argued that dialectical materialism had superseded the debate between vitalists and mechanists.[13]

In short, 'dialectical materialism' justified the efforts of such biologists as J.B.S. Haldane and Joseph Needham – perhaps the most interesting and neglected thinker among the 'red' scientists – to apply experiment and suitable mathematical modelling to the analysis of complex and essentially nonmechanical organisms and systems, the feedbacks within them and their often brusque transformations. For them Marxism was a natural development of approaches already established in biology and in organicist thinking.[14]

Third, Marxism seemed to provide a vision of the totality of the phenomena in nature that allowed both for its unity and its limitless diversity: a 'theory of everything' that was not at the same time mechanically reductionist. This may well have proved particularly attractive to adventurous young talents who took all knowledge and experience for their province – and it is presumably no accident that Marxism appealed to minds that refused to accept disciplinary boundaries and limitations: Bernal, whose range in science and art was even more encyclopaedic than Haldane's; Needham, brought up in a household of science and art, who also wrote about the seventeenth-century English revolution and made himself the world authority on the history of science in China as well as on many other matters; or the young casualty of the Spanish Civil War Christopher Caudwell, who left behind writings on aesthetics, psychoanalysis, cultural criticism and the 'Crisis in Physics'.

Moreover, to the essentially interdisciplinary or transdisciplinary Young Turks of interwar science the cosmic ambitions of Marxist theory might well have seemed to provide a welcome counterweight to the deliberate specialization of orthodox science departments. In any case, it anticipated what is today a common, if as yet unrealized, ambition in science: to produce an all-embracing analysis of the structure, dynamics and historical evolution of our universe from the original big bang to (at least) the present state of the earth's biosphere and the behaviour of its inhabitants.

However politically radicalized, scientists working purely in the inorganic fields like physics seem to have been largely indifferent to dialectical materialism. This is at first sight surprising, since the nature of the new physics seemed almost to call for the use of terms familiar in Marxist thought, like 'contradiction' and 'dialectic'. Such terms are, indeed, found in Niels Bohr, though in reference to Hegel and Kierkegaard rather than to Marx and Engels.[15] But a few hints at a descriptive vocabulary did not give much practical help to physicists. Unlike in the life sciences, dialectical materialism could make no plausible contribution to their research methods, or models of theory formation. Even for the greatest minds, like Einstein, their problem was epistemological, external to the laboratory or systems of equations: how to square the findings of post-Planck/Einstein physics with the philosophical foundations on which science had hitherto been based. What could orthodox 1930s Marxism contribute to this debate, committed as it was to a pre-Einsteinian Leninist 'materialism' which, had this been possible, would have preferred not even to accept the findings of the new physics? In some ways, this was also the difficulty of the relations of Marxism to the philosophers in Britain who, though often distinctly on the political left, and sometimes philo-Soviet – Wittgenstein briefly considered actual emigration to the USSR – remained overwhelmingly immune to Marxist influence.

The marriage between Marxism and the natural sciences proved to be short-lived, although it left behind a heritage that can still be traced in the life sciences (for example in Stephen Jay Gould, who still in the 1970s regretted the neglect of Engels by scientists[16]). It is easy to ascribe its break-up to Soviet Stalinism, especially after 1947, when Stalinism obviously and almost provocatively alienated Western scientists and forced even most Communist ones in such fields as genetics to choose between science and defending the scientifically indefensible. The Cold War did the rest. Nevertheless, neither the specific context of the 1930s nor a purely political explanation can fully explain the rise and fall of the union between Marxism and the natural sciences. It cannot, for instance, explain why the revival of Marxism in the 1960s and 1970s left the natural sciences totally to one side, or even argued the nonapplicability of Marx's thought (as distinct from that of Engels, which was regarded as separable and different) to the field of the natural sciences. Nor can it help us judge whether the bond between Marxian thought and the natural sciences has been snapped for good or whether, and under what circumstances or where, it may be re-established. Was the phase of 'red science' a historic freak or not?

If such questions are to be answered – and perhaps the time has come to confront them – the era of 'red science', and indeed the general history of twentieth-century science, must be seen in a wider historical and cultural perspective. This cannot yet be done, if only because there has as yet been no adequate study of the politics of science, or of its different disciplinary fields, in its major centres – or at least no adequate comparative study. In the meantime it is important to show that the questions raised by the era of 'red science', in which J. D. Bernal and his work played such a crucial role, go far beyond the study of a small group of scientists in one country between the world wars. They are major questions, which it is no longer possible to neglect. It is to be hoped that the present book may help to make such neglect more difficult.

Notes

1. Maurice Goldsmith, *Sage: a Life of J. D. Bernal*, London, 1980, p. 175.
2. T. E. B. Howarth, *Cambridge Between Two Wars*, London, 1978, p. 209.
3. See *Nature*, 26 September 1931, and more especially the editorial 'The planning of research', 28 July 1934.
4. The illusion of a Soviet Union dedicated to science in society and a model of how to organize scientific research soon faded; even among Communists it could hardly survive the imposition of a mandatory official version of genetics and the destruction of the version disapproved by authority, with or without the physical elimination of its adherents.
5. Bertrand Russell in *New Statesman*, 12 February 1938, cited in Edwin A. Roberts, *The Anglo-Marxists: A Study in Ideology and Culture*, Lanham-Oxford, 1997, p. 89.
6. *Marx-Engels Archiv. Zeitschrift des Marx-Engels Instituts in Moskau.* Edited by D. Rjazanov. Vol. II, Frankfurt, 1927, 'Dialektik und Natur'. See especially pp. 140–41, 147–9.
7. For a recognition of these difficulties see J. B. S. Haldane *The Marxist Philosophy and the Sciences*, London, 1938, pp. 56, 60.
8. By Mauro Ceruti in 'Il materialismo dialettico e la scienza negli anni 1930', in E. J. Hobsbawm, G. Haupt, F. Marek, E. Ragionieri, V. Strada and C. Vivanti eds., *Storia del Marxismo*, Torino, 1981, pp. 493–550. This essay constitutes the best guide to the subject. See also Gary Werskey, *The Visible College*, London, 1978; and Roberts, especially pp. 143–208.
9. Karl Marx and Frederick Engels, *Collected Works*, Vol. 25 (London, 1987), pp. 476–8. (The passage was originally drafted for inclusion in *Ludwig Fuerbach and the End of Classical German Philosophy*.)
10. Ceruti, p. 515.
11. N. I. Bukharin, ed., *Science at the Crossroads*, London, 1931, and (with a foreword by Joseph Needham, London 1971). The delegation included the plant geneticist N. I. Vavilov, subsequently a victim of Stalin's later support of the scientific heterodoxies of Lysenko. In fact, one of the things that attracted J. B. S. Haldane to Marxism was that Soviet genetics in the 1920s was more receptive than British genetics to his own linkage between Darwinism and Mendelian genetics (see Roberts, p. 180).
12. For example R. K. Merton's 'Science, Technology and Society in Seventeenth Century England', *Osiris* IV, No 2, 1938, pp. 360–632; G. N. Clark, *Science and Social Welfare in the Age of Newton*, Oxford, 1937.
13. Roberts, p. 150. This paper brought Engels's recently published *Dialectics of Nature* before a wider public for the first time.
14. 'The dialectical materialism of British biologists of the 1930s ... originates in an explicit synthesis of the ideas of Woodger, Weiss, Whitehead's philosophy, stimulations from texts of Engels and Lenin and some contributions from contemporary Soviet scientists (Zavadovsky), with the object of treating problems derived not only from cell physiology and genetics, but also from embryology and the theory of evolution – for these often required a type of explanation different from the one which predominated in biology in the strict sense.' Ceruti, pp. 533–4.
15. Ceruti, pp. 517–30.
16. Stephen Jay Gould, *Ever Since Darwin*, New York, 1977, pp. 210–11.

Introduction

Brenda Swann

This collection has been a long time in the making. After the death of
Desmond Bernal in 1971, Francis Aprahamian was appointed his
literary executor. Francis had worked for Desmond after World War
Two and collaborated extensively with him on his books, particularly
on *Science in History*, and *World without War*. Indeed, Desmond had
written of him in the acknowledgements to *World without War* that the
book 'has been made possible by the effort and skill of Mr Francis
Aprahamian . . . It is almost entirely due to his labours that I have been
able to present something like a balanced picture of the economies of
the different parts of the world today.' From time to time suggestions
were made to Francis and to Eileen, Desmond's widow, of names of
people who might write a biography, but it did not seem possible to
find anyone who could cover all aspects of Desmond's life and work.
Finally it was suggested that the biography should take the form of a
collection of essays, and a meeting was called to discuss the idea. There
was a complication in that Eileen was being pressurized to allow a
journalist to write the official biography. But Desmond had been
approached before his death and had stated firmly that he did not
want a journalist to write it and, in particular, not that journalist.

The meeting was held on 17 November 1976 at Eileen's house in
Camden Town. Present were: Eileen; Margot Heinemann, who had
shared his life for the last twenty years; Francis; Margaret Gardiner;
Martin, son of Margaret and Desmond, and now a professor at Cornell;
Ann Synge, a friend of both Eileen and Desmond and a contributor to
this book; Anita Rimmel, who had been his secretary since 1940; and
myself, Brenda Swann. I worked for Desmond, from early 1938 until
1940, as his secretary and research assistant on *The Social Function of
Science*. Having just retired from working at the Public Record Office, I
had been asked by Birkbeck College to collect, arrange, and catalogue

Desmond's papers. The meeting soon agreed that the official biography should take the form of a collection of articles, each written by an expert in that particular field. The topics to be covered were agreed and names were suggested. Francis was then working for the Open University and in charge of scientific publications; he and I were asked to act as editors. (Sadly, Francis died in 1991, when this volume was still incomplete.)

No book could do justice to every aspect of Desmond's life. After graduating from Cambridge in 1923 he worked with Sir William Bragg at the Royal Institution, using X-ray techniques to investigate the arrangement of atoms in molecules. Working on minerals, he went on to draw up the Bernal charts and to invent vital apparatus. He was appointed Lecturer in Structural Crystallography at Cambridge in 1927. In 1928, at the request of Sir William Bragg, Desmond toured scientific centres in Europe. The following year, he helped to organize a conference on X-ray crystallography out of which three committees were established, on publications, nomenclature and abstracts; Desmond was secretary of all three.

It was in 1934 that Desmond took the first X-ray photograph of a protein that was sufficiently clear to be used in analysis – a moment from which the origins of modern molecular biology may be traced. In the same year he was also appointed Assistant Director of Research in Crystallography. In 1939 he published his seminal work *The Social Function of Science*, the first attempt to present a social analysis of what science does and what science could do, to examine how scientific research and the applications of science interact with the aims of society, and to formulate a coherent policy on science.

At the outbreak of war Desmond was seconded to work at the Ministry of Home Security on the protection of the public from aerial attack. He went on to become scientific adviser to Lord Mountbatten; one of his many duties was to study the suitability of the landing beaches to be used for the second front. Returning after the war to Birkbeck College, where he had been appointed Professor of Physics in 1937, Desmond devoted himself to building up the physics department and the laboratories, while concentrating his scientific attention on research into the liquid state, particularly that of water.

Even with thirteen contributions to this volume, all Desmond's activities have not been covered. For example, he did a lot of work for the Association of Special Libraries and Information Bureaux (ASLIB) as he was very concerned that scientists should be able to find what published information they needed and not be swamped by the vast expansion of printed material. He believed very strongly in the trade union organization of scientific workers; he had joined the National Union of Scientific Workers in 1924, and in the 1930s he helped to organize the Cambridge Branch of the Association of Scientific Workers, which the NUSW had become. He helped the association to expand during the war (when its membership increased from 1,100 to 17,000) and became its president in 1949. With the AScW and through friendship with Julian Huxley, he helped to put the S into UNESCO. He helped to form the World Federation of Scientific Workers and drew up its constitution.

Desmond had become aware of socialism at Cambridge in 1919 and his interest in politics continued and deepened for the rest of his life. In 1934 he was much influenced by Paul Langevin whom he met after the February riots when Langevin was organizing French intellectuals to resist fascism. He often quoted Langevin's words, 'The scientific work I do can be done . . . by others, . . . but unless the political work is done there will be no science at all.' He felt this very strongly after Hiroshima. During the rest of his life he spent as much time as he could on peace work, trying to bridge the gap between east and west.

Desmond's membership of the Communist Party was very important to him. It has not been found possible to determine exactly when he joined. He told Gary Werskey he joined the Communist Party shortly after moving to London in 1923. I know he was active in the Holborn Labour Party which until 1926 contained Labour members and Communists. I think on returning to Cambridge in 1927 he lapsed. A friend, Magda Phillips, told me she recruited him to the Communist Party in 1933. Werskey states that he 'lost' his card in 1933 but he remained attached to the party for the rest of his life. The fact that he did not carry a card was almost certainly decided by the party. He was certainly a member in 1938 when I started working for him. I was told one reason I was given the job was that I was considered to be a reliable member of the party. I remember him telling me that he had not always been a member or active in the party because at one time the party had been suspicious of intellectuals. During the war he maintained contact with Harry Pollitt, and on two or three occasions they met in my flat. After the war he was chairman of the Engels Society, a debating society for Communist Party scientists.

He travelled a lot in Britain and all over the world. Wherever he went he was in demand to speak to scientific societies, to students, to scientific workers, to peace meetings. His interests ranged over all human activities except that he had little interest in music or sports, though he enjoyed rock climbing. He was widely read in British, American, and European literature; Donne was his favourite poet, he admired Joyce and could read *Finnegans Wake* with ease. His knowledge of art and architecture and their history was extensive. He reckoned to be able to date any old building in London to within ten years, and in the provinces to within twenty years. On his first visit to Egypt during World War Two, he was able to correct a guide who had mixed up the tombs they were looking at.

Dr Alan Mackay, a colleague at Birkbeck, wrote the following tribute to him after his death:

> Bernal's prodigious memory, his constant travel, his wide range of friends and acquaintances, his energetic mode of life, enabled him to accumulate an immense stock of facts and anecdotes, which he marshalled creatively in his own mind into new patterns. He contributed significantly to a wide range of topics besides crystallography. He was a brilliant raconteur and a propagandist for science and its applications and could captivate his audience, whether a scientific meeting or a pretty woman, by his discourses on almost any topic. He gave his ideas away freely, never doubting that he had plenty more. He was generous, considerate, and bohemian in his private life (and indeed also in the administration of his department). Finally he wore himself down doing his work for world peace as well as for science.

Perhaps the best description of Desmond was written by C. P. Snow and can be found in *Science of Science*, published in 1964. He appears also as a character in an early Snow novel, *The Search*. An entertaining account of Desmond's activities during the 1920s and 1930s can be found in Gary Werskey's *The Visible College*. A fuller account of his scientific work, written by Dorothy Hodgkin, can be found in *Memoirs of the Fellows of the Royal Society*, Volume 26, December 1980. A recent biography of Dorothy, written by Georgina Ferry, sheds fascinating light on Desmond's life and work.

In this collection, Ann Synge describes Desmond's family background in Ireland, while Roy Johnston discusses the influences – social, political and scientific – of the Irish cultural milieu in which Desmond grew up. Fred Steward examines Desmond's political development in the context of the 1920s and 1930s, and his growing fascination with Communism, psychology and sexual liberation. Peter Trent draws out the major features of Desmond's contribution to the science of crystallography. Chris Freeman discusses the impact and importance of Desmond's great work, *The Social Function of Science*, and Hilary Rose and

Steven Rose explore the questions raised by the theory and practice of what has come to be known as red science.

Desmond's work during World War Two is described here by Ritchie Calder, by Earl Mountbatten of Burma and by Desmond himself, in his 'D-Day Diaries'. Ivor Montagu brings to life the ambience of the World Peace Council, to which Desmond devoted so much of his energy in the decades following the war. Eric Hobsbawm describes Desmond's struggle to run the Department of Physics and to establish a separate Department of Crystallography at Birkbeck College in an increasingly hostile Cold War environment. Peter Mason, in his discussion of *Science and History*, takes up Desmond's main theme that the role of science should be to fulfil human needs. Finally, Chris Whittaker writes on Desmond's interest in architecture and construction, the sphere where art and science meet.

A list of Desmond's books can be found at the end of this volume. His papers are at the Cambridge University Library and can be consulted on application to the Keeper of Manuscripts and University Archives. Papers relating to his work for the World Peace Council are held in the Bernal Peace Library at the Marx Memorial Library, Clerkenwell, London.

Finally, we would like to thank Jane Bernal and Martin Bernal for their generous assistance, and Birkbeck College for their donation of an award towards the publication of this book.

Early Years and Influences

Ann Synge

John Desmond Bernal, crystallographer, molecular physicist, social scientist, Communist visionary and peace campaigner, was born on the farm of Brookwatson near Nenagh in County Tipperary on 10 May 1901. Although he lived most of his life in England, Desmond always regarded himself as an Irishman, and in his young days he was a very anti-English Irish nationalist.

His descent was, however, by no means purely Irish. The Bernals are a Sephardic Jewish family and his ancestors had only reached Ireland in 1840 from Spain, via Amsterdam and London. Desmond's grand-father, John Bernal, was distantly related to the MP and notable collector of *objets d'art* Ralph Bernal. He himself was an auctioneer and furniture dealer in Limerick, as well as being a director of the local railway company. The business was established in 1841 and seems to have prospered, for John Bernal lived in an imposing house called Albert Lodge, described as being 'very Frenchy' in style.

John and Catherine Bernal had twelve children. Samuel, Desmond's father, was born in 1865 and, by the time he was nineteen, was the only surviving boy in a family of girls. Whether because of the anti-Jewish feeling that was growing in Limerick at this time or from a simple wish to see the world and make a fortune, Samuel emigrated to Australia in 1884 and worked on a sheep farm. Although he was not happy there, he did not return until 1898 when his father died. He then went to live with his elder sister, Mrs Riggs-Miller, some twenty miles from Limerick in the neighbourhood of Nenagh, and helped with the management of her farm. At the same time Samuel was buying his own farm at nearby Brookwatson: the place had been neglected for years and the house was largely in ruins.

Mrs Riggs-Miller was an inveterate traveller and it was while accompanying her on a trip to the Continent that Samuel visited

Blankenberghe, in Belgium, and met the woman who was to become his wife. Elizabeth (Bessie) Miller was an American, lively and educated, the daughter of a Presbyterian clergyman, the Reverend William Miller, from San José, California. Bessie had had her early education at Mme Bovet's Academy in New Orleans, where she became a fluent French speaker. Her diaries were indeed written mainly in French, interspersed with a little German, Italian and English. Bessie attended Stanford University for a while and also went to lectures at the Sorbonne. She had travelled widely in Europe, undertaking extensive cycling tours with her brother Jack and indefatigably visiting churches, museums, theatres, concerts and picturesque views, while writing a series of articles about her experiences for such papers as the *San Francisco Argonaut*.

The courtship of Sam and Bessie was both eventful and rapid; within a month they were engaged. The Bernal family in Ireland had adopted the Catholic faith and this meant that Bessie would have to convert to Catholicism for the marriage to take place. Her parents were not pleased but accepted her decision. It was arranged that Bessie's marriage portion of £1,600 should be used to pay the mortgage on Brookwatson, and Sam went back to Ireland to organize the refurbishing there while Bessie went first to Paris and then on to London, combining an extensive shopping trip with instruction from priests. She was received into the Catholic Church in time for the wedding.

Sam and Bessie were married on 9 January 1900; he was thirty-five and she was thirty-one. They went to Paris for their honeymoon, accompanied by Bessie's younger sister, Laetitia, and then back to Ireland, where they stayed for about two months with Mrs Riggs-Miller while Brookwatson was being refurbished. It was not the easiest of times. On the day they moved into the still-uncompleted farmhouse, Bessie received the news of her father's death. She was unwell and the couple were having money troubles, partly caused by delays in Bessie's marriage settlement. Life in County Tipperary was very different to what Bessie had been accustomed to, both at home and on the Continent. The neighbours were not, for the most part, interested in literature and art, whilst Bessie had little inclination to play tennis or bridge. However, there was plenty to do in getting the house in order, in managing the servants and, especially, in planning the garden, which gave her great pleasure in itself and also provided a common interest with many of the women she met socially. And after four months Laetitia visited Brookwatson and stayed for about two months, which also helped Bessie to acclimatize to her new life.

From the start, Sam had been looking forward to raising a large family. Bessie was not so keen but she discovered she was pregnant before Laetitia's stay was over. After a rather uncomfortable pregnancy, John Desmond was born on 10 May 1901. Bessie had some difficulty accustoming herself to motherhood but Desmond was a healthy child and she enjoyed watching his progress. Her brother Jack got married in the autumn and came to visit her with his wife on their wedding trip, which cheered her, but Bessie was still not truly reconciled to country life and was glad of a week in Dublin and a longer visit to France the following summer. By now, however, she was pregnant again and on 22 January 1903 another son, Kevin O'Carroll, was born. There was less than two years' difference in age between the two boys and they would spend most of their time together, both at home and at school, until Desmond went up to Cambridge.

Even though she now had two small sons, Bessie Bernal still enjoyed travelling and in November 1903 she went back to San José to visit her mother, taking Desmond with her. He seems to have been a very attractive child, pale, with golden hair and already very talkative. The trip was very eventful, with many parties and visits, and Desmond's precocity caused much amusement. A fellow passenger on the boat on which they crossed the Atlantic was amazed that he could talk to his mother in both French and English. In March they returned to Brookwatson and by May Bessie was writing in her diary, 'I feel as if I had never been away.' During the next few years two daughters were born into the Bernal family, Geraldine in 1906 and Fiona in 1908. Unfortunately, all the children had whooping cough shortly after Fiona was born and she caught it too, and died.

Desmond learned to read early and easily, and soon he was sent to a small Catholic school in Nenagh, run by nuns. He learnt his catechism well and was taught lettering and geometry. He seems to have been a docile child as well as a good scholar. Later both he and Kevin were transferred to the Protestant parish school in Barrack Street. This was said to be because they might get lice and other infections if they went to the school kept by the Christian Brothers, but possibly also the educational standards were higher in the Diocesan School, which had about twenty girls and boys in the senior school being prepared for Trinity College Dublin.

In 1906 Bessie's mother died and Laetitia came to live with the Bernals at Brookwatson, helping Bessie with the garden and the children, of whom she was very fond: they always knew her as Cuddie. She was a motherly woman, more so than Bessie. Her early life had been marred by tragedy. Her brother Jack writes: 'Laetitia was engaged to marry a Scotch gentleman, Mr McLaren, an attorney and member of the English parliament. She came home to make preparations for her wedding and received word of his sudden death. This was in the early nineties and from this blow she never recovered.'

Desmond continued to be a good student but was not always a good boy. 'We used to eat chocolate and get caned every day or oftener if we were bold.'[1] As he got older, he joined in the fights with stones and water pistols which took place in the school yard when the masters had gone home to their lunch. He also enjoyed his lessons and wrote, 'The first lesson in Euclid was wonderful, even though I thought "obvious" meant "impossible". Algebra was not quite so exciting.'[2]

Life was not all school though: there was plenty to occupy the Bernal children at home. Recalling this period Desmond later wrote, 'House and yard, garden, fields and river, that was my world. The smell of cows, the sound of milking in the morning, hot hours with the tram builders in the lawn field, watching the men bring the horses home in the evening, those were my days. To play in lofts, or hay sheds, to walk down to the Holy Well to see the rushing water of the weir, or see it smooth and deep by the sandy banks of the burrow, those were my joys.'[3] There were parties too, especially in the winter, and a great deal of good food, both from the farm and garden and in the form of gifts from the American relatives.

Family relationships were, on the whole, harmonious. Desmond also wrote: 'Daddy was a fine man, a great playboy in his time. . . . He was always planning, building, improving. He had no liking for reading or writing, though he was a great talker and his delight was to walk over the fields looking at the cattle and the grain and scratching the bullocks' backs with the end of his stick. He was a good Catholic and a good husband, ruling the house and beating his children because he knew it was his duty, though it did not fit his easy and kindly nature. . . . Mammy was tall and beautiful . . . she had spent years in Italy and France, so that I can never remember a time when French was not my other language. A language of gentleness. When I was scolded I would always beg, "say it in French." She would spend her days among flowers and fruit in the garden, or making the most delicious cakes or reading French fairy stories to us. Through her I realized the outside world of beauty in form and language.' Her sister Cuddie was a fierce supporter of work and cleanliness. 'Loathing the Irish for their laziness, their dirt and their dishonesty, but unable to escape the charm of their language and the ingenuity of their evasions . . . unfailingly generous, she devoted her life to her sister and her perverse and ungrateful nephews.'

There were many comings and goings of relatives over the years: Americans passed through on their way to or from the Continent and the Irish members of the family travelled considerably themselves. In 1908, for example, Sam's sister Mrs Riggs-Miller made a trip to the Orient via Russia, and in 1909 she started on a tour of Europe with Sam, Bessie and Laetitia. They had not gone far, however, before the children developed measles and they had to return. In the summer of 1911, Desmond and Kevin went to stay with friends at Royan in France. The children bickered as children so often do on such visits but Desmond recalled: 'The midnight of the full moon was the great pêche. All the peasants went out with wine in their bullock carts. The men waded in up to their necks at the head of the long net, the women to the top of their black skirts. They swept back and forwards, and gleaming writhing fish piled on to the moonlit sand. Great stingrays, turbot and the small fry that were slung into the waves again. The carts moved away again, the men and women sang songs and we went to sleep lying on the straw.'[4]

Even as a young boy, Desmond was aware of his family's anomalous social position. Writing about it in 1955, Desmond recalled: 'There were in Ireland, at that time, two peoples. The English, the Ascendancy or gentry as they were called, owned most of the land and held practically all the positions in government. They were distinguished from the rest of the people, as much as anything, by their religion. The Ascendancy were Protestants, the Catholics were the people. My father was a small farmer and a Catholic; my mother was an American who had been brought up as a Protestant. I could not help seeing both sides ... it appeared to me, even as a schoolboy, that all the trouble would be ended if only the English could be driven out. And yet I mixed with those English who were, in many ways, as Irish as anyone could be. It was not, I sensed, even as a small child, these individuals who were responsible for the injustice and the struggle. In my own town the four most imposing buildings were the law court, the jail and the military and police barracks.'[5]

In school Desmond continued to earn good reports from his teachers. 'I learned things well. Learning was so easy with formal things, theorems of Euclid, declensions and conjugations. Everything clear-cut, you have it or you haven't it, true or untrue, right or wrong. I learned because I knew that learning would give me understanding of things. It might be dull but it was a clear necessity. Life seemed to be in front of me like a wheel, so many turns to go before the liberty of being grown up. History at school meant very little but the history of Ireland that I read to myself at home, the long oppressions, the repeated failures, moved me to self-pitying resentment, a determination to be myself the instrument of delivery.[6]

'And then, one day, I conceived the idea that dominated ten years of life. I would use science and apply it to war to liberate Ireland. But why stop at Ireland? . . . I was a singularly closed-in child. . . . My phantasy occupied half my reflective thought. Everything I did, learning and growing, was with reference to it. Science, and engineering, school would give me; for the art of war I would join the British Army to betray it. It would be a hard and unpleasant life, dangerous but worth the reward. I never doubted the actual possibility of the scheme, except when sin made me feel that God might not favour my prospects on account of my unworthiness.'[7]

In November 1910 another son, Godfrey, was born to the Bernals. He was to be their last child. In 1911 a bigger upheaval occurred in the lives of Desmond and Kevin. Whether for religious or academic reasons or because he thought boarding schools were good for boys, their father decided to send them to Hodder Place, the preparatory department of Stonyhurst College, the Jesuit public school in Lancashire. Bessie and Laetitia protested against their going but were overruled. As far as Desmond was concerned, there was at that time no suggestion that his religious convictions were being undermined by association with Protestants. He was a very devout boy. He made his first Communion in April 1910 in the chapel of the convent where he had been a pupil in Nenagh, and his mother wrote in her diary: 'God bless the poor little boy. He was over-excited, very fervent, very serious. May he always remain an honest and sincere man.'[8] As a boy Desmond was always very conscientious about religious observances and never took off his scapular, even for swimming, as other boys did.

Desmond and Kevin were taken to Hodder by their parents. It is beautifully situated on the banks of the Hodder river, and the Jesuits received them kindly with tea and hot buttered toast, but Desmond was very distressed when his parents left him there. He suffered badly from homesickness and every time he left Brookwatson it was a wrench. At the end of each holiday, he would pay ceremonial farewell visits to all his favourite places, including a small garden walk laid out in memory of Fiona. However, once back at Hodder, he came to enjoy the life. As he wrote, 'Hodder was made by the gentleness of Father Cassidy. We roved freely over the valley and by the turbulent river. We played and learned and were good boys. Too good. I bought a missal, flagellated myself with nails, and had a knotted rope tied round my middle until I found, by the anger that was roused in me, that it was rather an occasion of sin than a penance. I started the practice of cutting grottoes to Our Lady in the hillside, and a fraternity of Perpetual Adoration of the Sacred Heart. And all the while I was designing aeroplanes and making plans for universal war.'[9]

Other sources give a picture of a happier time, with fishing, the October bonfire (when the boys burned Lucifer), the activities of the 'Museum boys' of whom Desmond was naturally one, and other pleasures. Next autumn, however, he was transferred to Stonyhurst where he was miserable. 'At Stonyhurst I learned nothing but the joys of prison life. A regularity that extended even to defaecation. Dark corridors, wandering priests, terrible sermons on sins that must not be named lest they be practised. But against that there were forests of flaming candles, golden vestments, and the resonant chanting that tore out the soul for God.'[10]

Desmond's letters home must have revealed what was happening for, after only one term, his mother persuaded his father that he and Kevin should leave Stonyhurst and return to the Diocesan School in Nenagh, where they spent a year before being sent back to England to Bedford School, in Bedford, in January 1914. It seems likely that several factors contributed to the decision to remove the boys from Stonyhurst. First, Desmond was unhappy and, second, he was set on learning science, of which there was none in the early forms at Stonyhurst. 'I never thought that childhood should be happy, to me it seemed that it must be a miserable process of lessening impotence. So that I would have endured Stonyhurst at its worst if it had given me what I had wanted, the knowledge and the power of science.'[11] A third factor was that he seems to have carried his religious devotions to excess, and it has been suggested that his mother was afraid that Desmond would become a priest if he stayed there.

Life at Bedford does not seem to have been happy, but at least the harping on sin was lessened and mathematics and science were competently taught. From the start, Desmond was a good scholar, but his background and inclinations made him different from most of his schoolfellows, and he was uncomfortably conscious of this. He did not make friends easily and was the victim of a considerable amount of ragging. He hated organized games though he was a competent oarsman and enjoyed both rowing and bicycling.

The boys had not been at Bedford long before war broke out. The

food became bad and the boys had to spend a lot of time drilling with
the OTC (Officers' Training Corps). The poor food was in some ways
more acutely felt by the Bernal boys than by others because at home
they continued to eat well, not to say luxuriously, as in Ireland there
was never a shortage of food for those who could afford it. Moreover,
they had the produce of their own farm and garden as well as gifts of
dried fruit and other delicacies sent by their relatives in California. The
boys were not too badly off: their parents sent them food parcels from
Ireland and they always returned to school after the holidays with a
well-stocked tuck box. As time went on, the boys had to go out and
work on a farm near Cardington, which Desmond seems to have hated,
despite coming from a farm himself. He was, however, learning science
and mathematics. But the war was ever-present. All the time the older
boys were leaving school and going into the forces. Sometimes they
were decorated, sometimes they were killed, sometimes they would
come back to the school and talk to the boys. One even crashed his
plane near the school, but he was not hurt.

In retrospect it seems that Desmond considered his life at Bedford
to have been one of almost unmitigated unpleasantness. 'There I could
be tortured, humiliated, waste my time and my interest on dully
repetitive games and military drill, but I could learn everything I
wanted. I lived like a hostage in an enemy land. My companions were
cheerful thieves and liars, and furtive sexual perverts. I merely thought
they were English and kept my hatred of the race closer to myself.'[12]
He found himself suspecting that all was not as victorious as it seemed
for the Entente and secretly rejoiced at the great stand Germany was
making against the world and hoped for the destruction of England's
power. The diaries he kept regularly from 1917 for several years do not
paint quite such a gloomy picture, but they are not cheerful except
when they recount progress in science and mathematics or the proceedings of the science society and the debating society. A contemporary recalls that, during their last years at school, several of the boys
used to go out to a local tea shop where they would discuss the affairs
of the day. So long as those present were those he was used to,
Desmond would come out of his shell and take part, but as soon as an
unfamiliar person joined the group, he would retreat again into silence.

Two friends deserve a mention. Probably Desmond's closest tie was
with Broughton Twamley, with whom he was friendly for his last year
or two at school and who went to Emmanuel College, Cambridge, with
him. Twamley later became a distinguished historian, but died prema-
turely, after a very long illness in 1943. The other was Lovell Hodgkin-

son who wrote home, 'He [Bernal] is the cleverest chap in the school. . . . He is not a bit conceited – he has got a very keen sense of humour and is a simply topping chap.' Hodgkinson was said to be very well-read and was noted for 'extreme originality'. The two boys discussed philosophy and politics and had plans for a serious magazine in the school, but Hodgkinson tragically died of pneumonia before they could implement them.

On the whole, though, these were happy times. There were always the holidays to look forward to. The Bernal boys passed through London travelling to and from Dublin and usually managed to visit the British Museum or some other place of interest, such as Gamages or the Bassett Lowkes toyshop. Their father did not give them much pocket money and insisted on every penny being fully accounted for, but they were sometimes able to buy things that were not available at home. During this time Desmond had his own laboratory at home where he performed experiments in physics and chemistry. Here, too, he kept the microscope which he had bought at the age of fourteen through *Exchange and Mart*. He began to make detailed drawings of diatoms and other microscopic organisms which his younger sister Geraldine helped him to collect in muddy ponds. Desmond also collected flowers, fossils, insects and minerals, and he helped Kevin in the building of a model railway.

The farm prospered though their father's health was not good, and the boys had to give quite a lot of help during the holidays. In 1916 the boys had a few days' extra holiday at Easter because it was dangerous to pass through Dublin on account of the uprising.

When they did get there they saw the 'smoking ashes of the town' and had to obtain permits to cross it from a 'fierce military officer in Kingstown with a revolver in front of him and guarded by boy scouts'.[13] But in spite of everything, life in County Tipperary went on much as before. There were the same bridge evenings, tennis parties and dances, as well as the paper chases and hockey matches at which everyone seemed to be trying to get as wet and muddy as possible. These were followed by delicious teas, the best of which were those provided by Bessie Bernal, who was something of a gourmet and who, although relieved of the routine preparation of household meals by a cook, liked to make her own special dishes for festive occasions. The Bernal family were not molested during the Troubles, though some of their neighbours had their homes ransacked and burned.

Each term, the boys had to go back to England and each term brought nearer the day when Desmond would leave school and be

expected to take a commission in the British army. In the Christmas
holiday of 1917–18, however, the *Leinster* was sunk in the Irish Sea and
the Bernals decided that it was too dangerous to send the boys over to
England. They wanted to send them back to the Diocesan School in
Nenagh. The school was able to accept Kevin but, by this time,
Desmond had specialized in science and they had nobody on the staff
competent to teach him the physics, chemistry and mathematics he
needed to enter Cambridge University, which was now his goal. In the
end he was sent to Dublin where he lived in a hotel and was coached
by teachers in Trinity College. This seems to have been no more
enjoyable than life at school. He wrote in his diary, 'I am bored to
death with this hotel. I have hardly spoken two words all today.'[14] He
spent most of his evenings at the theatre or cinema. He was also
coaching a fellow pupil from Bedford called French, and sometimes
visited his home in Bray where he enjoyed walks with French and his
two sisters.

By Easter the Bernal parents' fears had been somewhat allayed and
Desmond, who had been awarded a scholarship of £60 by Bedford
School, returned there for the summer term. Even now, at the age of
seventeen, he had to endure 'ragging and torturing that I brought on
myself by my pale untidy ugliness and my obstinate insistence on being
different'.[15] He also became a prefect but did not enjoy the experience.
He was not good at keeping discipline. As a refuge from these humili-
ations, he 'read indiscriminately' nearly every book in the school
library. 'I had no favourite authors for the simple reason that it never
occurred to me to look to see what the author's name was. Another
escape was the stars. I would watch night after night, sitting before the
transit instrument with the lamp beside me and my chronometer on
my knees for those spots of light sailing so smoothly, so surely, over the
field past each of the five cross wires. They had the constancy of old
friends, Arcturus, Antares and Lyrae. And I would forget there in the
quiet garden the apple-pie bed that was awaiting me in the dormitory,
the jug of water they would surely try to pour down the telescope.'[16]
But there were compensations, notably the opportunity of working on
his own in the laboratory. He wrote later, 'I was not at ease with my
fellow beings of any class or sex. From men I expected only contempt,
ignorance, and brutality, which I could only escape by keeping my
secret self inviolable. From women I expected nothing [but] trivial
conversation, a humiliating proficiency in dancing, a vague, indefinable
distraction that did not even become sexual.'[17]

It is true that he had outgrown the rather anti-intellectual pursuits of the young people of the Nenagh neighbourhood and that most of the older ones were more interested in bridge playing, which bored him, and in hunting and shooting, which disgusted him, than in intellectual discourse, for which he longed. At school, too, he felt the lack of intellectual companionship: 'I left [school] as amazingly ignorant of the world as I went in. Religion and Conservatism were assumed but of course never discussed, and though being a Catholic Irishman saved me from being dominated by them, I had no inkling of any other alternatives. Only the words of President Wilson deceived me for a while into an interest in wider politics.'[18]

With the end of the war, the threat of military service was finally lifted from Desmond. In 1919 he won several school prizes and a scholarship to Emmanuel College, Cambridge. Meanwhile Samuel Bernal's health, which had been poor for some time, deteriorated, and Desmond spent much time that summer reading to his father and looking after him. He had to go to London to take examinations in September and on the day of his last paper, his father died. Desmond returned at once to Ireland to deal with the funeral and business affairs but soon had to go up to Cambridge to begin his studies there.

Cambridge, at that time, was an especially exciting place to be a student. With the war over, ex-servicemen, older and more experienced than the usual undergraduate, were mixing with the youngsters straight out of school. There was a feeling that the world was changing irreversibly. The Russian Revolution seemed to have shaken the old order to its foundations. The artistic and intellectual world was in ferment. For Desmond, the revelation came halfway through his first term, in the course of an exhilarating all-night discussion with his friend Henry Dickinson on 7 November 1919, the second anniversary of the Russian Revolution. As discussed in subsequent chapters, the conversation with 'Dick' was to have an enormous impact. Later Desmond wrote:

This socialism was a marvellous thing. Why had no one told me about it before? And Dick knew it all, explained it all so simply in a few hours. The theory of Marxism, the great Russian experiment, what we could do here and now, it was all so clear, so compelling, so universal. How narrow my Irish patriotism seemed, how absurdly reactionary my military schemes. All power to the Soviets. It was the people themselves who would sweep away all the things that I hated, smash the arrogance of the English public schoolboy gentleman. It would bring the Scientific World State. But where did I come in? Clearly nowhere. One served the movement as one could in the place where one was called. I saw my whole life of vanity. My universe was broken into bits![19]

At that time, Desmond was becoming susceptible to the charms of women. Shortly after the exposition of socialism by Dick, he was invited to tea by Twamley to meet two young women whom they had both known in Bedford. He describes a conversation with one of them which seems to have been largely a monologue by himself. 'I talked of capital and labour, of control and nationalization and entered into eulogies of the Bolsheviks. She attacked me with all the old fallacies. I decided to switch on to ideas so I started on justice and hope and love, in the spirit of self-sacrifice. I talked of my broken universe and we were out of the magic glade for, in my explanation, I brought in pure science and my discursiveness led me to talk about Einstein and hormones and what not.'[20]

'All Cambridge was a liberation,' he recalled later. 'All the richness of thought was open to me. I was quit of my old companions (except when they raided my rooms and threw Dick into the river) and I could meet for the first time intelligent people and be accepted by them. I saw that I knew nothing but mathematics and science. I had everything else to learn and experience. I read and talked violently, discursively. I never fitted in with my industrious co-scientists, but consorted with socialistically inclined economists or historians. Night after night, sitting over a fire and drinking more and more diluted coffee or strolling around moonlit courts, we talked politics, religion and sex.'[21]

Desmond was also active in many of the university clubs and societies. The one that seems to have interested and influenced him most was the Heretics, the object of which was to 'promote the discussion of Religion, Philosophy and Art'. Membership entailed 'the rejection of all appeal to Authority in the discussion of religious questions'. Speakers at this society were mostly either already well known and respected in their fields or were to become so later. They included Jane Harrison, Roger Fry, Bertrand and Dora Russell, Eugene Goossens, A. J. Eddington and others of similar calibre. Perhaps the society's most spectacular meeting was that addressed by Marie Stopes. She had mistakenly supposed that she was expected to lecture in full evening dress and, at the end of the meeting, insisted on distributing leaflets on contraception to all and sundry from the door of the Kings Parade Café where the meeting had been held.

For many Catholics the loss of faith can be a traumatic experience and it has been suggested that this was true for Desmond. He later wrote, however, 'I lost my faith by a gradual process. First God, then Jesus, then the Virgin Mary, and lastly the rites of the Church disappeared, regretfully, inevitably. It was not from any scientific reason that it had to go, because I had long before reconciled the phenomenal world of science with the transcendental symbolic world of revelation. Now I had a quarrel with the Church because I could not help seeing it as an active agent of political reaction throughout the world. I began to examine its claims historically as to origins and psychologically as to underlying forces. Seen as the one background of the unexplained of life, the Church was impregnable, but seen in its place and time it appears an ideal human construction, as compelling and as fallible.'[22]

Another interest was psychoanalysis. Desmond, together with Henry Dickinson and Allen Hutt, formed part of a group that had gathered round Jonty Hanaghan, whom Hutt described as 'a sort of maverick missionary with a considerable load of what, in those early days of Freudian enthusiasm, passed for Freudianism'. According to Hutt, 'A whole lot of us, including Des, myself, Dick and Eileen gathered for what one could almost call, I suppose, a species of prayer meeting where this strange, rather magnetic man, Hanaghan, addressed us on the basic problems of what, as far as I can recall, were the relationships between the sexes, which we found very encouraging and mostly set us, as it seems to me now, on the right path at an early stage.'[23] Desmond was very much impressed by Jonty and in his diary referred to the group as 'people after my own heart, poets, theologians, mystics, fanatics and rationalists. I can talk (with them) to my heart's content, the most abstruse nonsense, and live up to my universally acknowledged nickname, the sage.'[24] In view of the content of the meetings and the fact that the group consisted of about fifteen men and five women, all of more or less undergraduate age, one is not surprised to learn from Desmond's diary that, when Jonty left Cambridge, 'triangles have given place to the most complexly joined polygons and jealousy is by no means absent'.[25]

One member of these triangles and polygons was Eileen Sprague, then working for a secretarial agency in Cambridge. After a rather stormy interlude Desmond had a long talk with Eileen, at the end of which he discovered that he was in love with her 'but not enough to make me miserable'. They had many friends and interests in common and the relationship flourished. They were married in 1922. Desmond was only twenty-one, and still had another year of undergraduate work ahead. Bessie Bernal disapproved and raised all the usual objections – they were too young, too poor, had no prospects, et cetera – but Desmond replied with a very characteristic letter: 'I do not discount the risk, but I take it willingly, counting the prize and ready to endure the punishment of failure. I must live my life in my own way – my life is my work – Science. This means that I must have a position, food and clothing and be left alone; if I am not happy, if I cannot do my best work without a mate, then I must have one. For these ends, I take what I can, not greedily, my wants are few, but what I take, I feel is my right and I am not ashamed to have it gathered from the pennies of the poor by my own mother's work. What I give to the world is my own work's worth, however poor it may be, I cannot give more. You see obstinacy, folly, selfishness in what seem to me determination, insight, purpose.'[26]

All the turmoil and distractions within Cambridge absorbed much of his energy and, to his disappointment, Desmond got a 2.1 in Part One of the Mathematics Tripos in the summer of 1920 instead of the first he had hoped for. The next year he switched to the Natural Science Tripos, studying Physics, Chemistry and Mineralogy. The continuing distractions were not enough to prevent him from getting a first in Part One of the Natural Sciences Tripos and going on to take Part Two in Physics, under Alexander Wood and Arthur Hutchinson, while carrying out work on his ideas on space lattices. To quote from a lecture given by his former colleague Dorothy Hodgkin in Dublin in 1980, 'The work continued at intervals of "free time". It resulted in a thesis called "On the analytical theory of point systems" and involved, essentially, the derivation by the use of Hamiltonian Quaternions of the 230 space groups of crystallography, the possible symmetry types determining the arrangement of atoms in space in crystals. It was very bulky and, since the space groups had been found in other ways, it was never published.'[27]

Although Desmond only got a second in Part Two of his degree, this did not hinder his career. Eileen typed his thesis for him and it was submitted for a college prize which was awarded to him on 19 May 1923. Alexander Wood sent this thesis to the Nobel laureate Sir William Bragg, who with his son had founded X-ray crystallography, and Hutchinson also recommended Desmond to him. On the strength of these recommendations Sir William offered Desmond a place at the Royal Institution to which he was just moving as its director. There was no money attached to the position but Emmanuel College gave him a short-term grant. Eileen was earning money as a secretary and his own family helped, so he was able to hang on until a grant became available.

All through his time at Cambridge, Desmond continued his habits of introspection, note-taking and formulation of his view of the world and of his own place in it. Many examples of these self-examinations are to be found in notebooks dating from about this time or a little later.

Here is one: 'We might all find out [how to live a good life] experimentally, by thinking and trying for ourselves how to live. In fact few do. They are moulded in their homes and schools and, for the rest, they copy the done thing. A few people break away here and there. They are the oddities and, no doubt, I shall be counted one of them. . . .

'I live, all is implied in those two words. My thoughts and words and actions, all come from my own joy in life. I seem to see before me thousands of possible thoughts, words and actions, all worthwhile doing, all delightful, and my life fills itself thinking, saying and doing them. That is the essential spring of life in me that will stop, I hope, only with my death. Maybe it will be sooner. Poverty or disease or unhappy relations might kill it. But I would have to be hard hit, so closely have I knit my life and my love of life. There is the driving force, leading anywhere, everywhere, but, with me, two things guide it. Science and people.'[28]

Notes

I have, where reasonably possible, omitted material included in Professor Hodgkin's biographical memoir. Apart from a typescript in English, 'Men of Today, Pioneers of Tomorrow', written by Bernal in 1955 and published in German as 'Verantwortung und Verpflichtung der Wissenschaft', in Elga Kern, ed., *Wegweiser in der Zeitwende*, most of the material used is unpublished. It consists of Bernal's own diaries covering his schooldays and undergraduate days and his mother's diaries from 1898 to 1910. These are in the Cambridge University Library, as is a miscellaneous collection of autobiographical notes and reminiscences and the letters. One, in particular, often referred to as 'Microcosm' and written when Bernal was about twenty-five, is the source of many of my quotations.

1. J. D. Bernal, 'Microcosm', typescript, J. D. Bernal Archive, Cambridge University Library, B4.1.
2. Ibid.
3. Ibid.
4. Ibid.
5. Contribution by J. D. Bernal to 'Verantwortung und Verpflichtung der Wissenschaft' (Responsibility and Obligation in Science) in Elga Kern, ed., *Wegweiser in der Zeitwende* (Signposts in a Changing World), Ernst Reinhardt Verlag, Munich Basel 1956, pp. 104–69.
6. Bernal, 'Microcosm'.
7. Ibid.
8. Elizabeth Bernal's diary.
9. Bernal 'Microcosm'.
10. Ibid.
11. Ibid.
12. Ibid.
13. J. D. Bernal, Diaries, J. D. Bernal Archive, Cambridge University Library.
14. Ibid.
15. Bernal, 'Microcosm'.

16. Ibid.
17. Ibid.
18. Ibid.
19. Ibid.
20. Bernal, Diaries.
21. Ibid.
22. Bernal, 'Microcosm'.
23. Reminiscences of Allen Hutt.
24. Bernal, Diaries.
25. Ibid.
26. Personal papers.
27. D. Crowfoot Hodgkin FRS, 'Microcosm: The World as Seen by John Desmond Bernal', *Proc. Roy. Irish Acad.*, vol. 81B, part III, 1981, pp. 11–24.
28. Bernal, 'Microcosm'.

2

Irish Roots

Roy Johnston

The purpose of this chapter is to provide an understanding of those influences on John Desmond Bernal that were specific to the Irish background and that contributed to his subsequent development as a Marxist analyst of science, technology and society.[1] It is based primarily on a study of the Bernal diaries in the Cambridge University Library; for supplementary information I am indebted to his brothers Kevin and Godfrey, and to the Earl of Rosse.

I begin by sketching some of the Irish social and scientific background, with particular reference to the factors influencing the choice of school. I go on to follow the development of Bernal's thinking on the Irish national-democratic revolutionary struggle during the period 1918–1921. Finally, I outline the career options that were open to him at the end of 1921, touching on the brain-drain aspect and suggesting reasons for Bernal's subsequent relegation of Irish questions to low priority, despite his evident concern in this critical formative period.

Social and Scientific Background

Notwithstanding Bernal's own description of his own background as 'small farmer', it is evident (for example, from the status of the Bernal family house and grounds) that the Bernals were verging on landed gentry status, though not to the extent of recognition in *Burke's Landed Gentry*. It is appropriate to fill in some of the implications of this status for the development of a creative mind. The Irish landed gentry are in the main descended from those of Cromwell's army who accepted land in lieu of pay. They would have been of plebeian origin, planted among a population whose culture was as remote from theirs as that of the Red Indians (the Rhodesian supporters of Ian Smith would be a

modern parallel). They developed mostly into reckless colonials, culti-
vating to an extreme the sports of the English aristocracy in the wilder
Irish environment. Some of them, however, took a practical interest in
improving their estates, and thus were better able to survive the
economic stresses of the nineteenth century (having weathered the
political storms of the eighteenth). Such 'improving landlords' (as they
were known colloquially, to distinguish them from the sporting or rack-
renting varieties) were often distinguished amateur scientists. When
they went professional, by the channel provided for them and their
hangers-on by Trinity College, Dublin, they often became internationally distinguished, the prime example being Sir W. R. Hamilton,
Ireland's Astronomer Royal.[2]

In Bernal's neighbourhood (or within an easy bicycle ride) were at
least three such sources of scientific influence. The major influence
pointing Bernal towards microscopy and microstructures was Launcelot
Bayly of Bayly Farm, about five miles south of Nenagh, where Hamilton
had written his 1834 paper on the 'Least Action' principle.[3] Bernal in
his diary referred to visiting Bayly Farm on 6 January 1918 and being
inducted into the use of the polariscope in the microscopic study of
diatoms. He had previously noted being there on 27 April 1917. Later
that summer, on the way back from Roscommon, his father picked him
up in Athlone and they drove home by car via Birr. 'We stopped at
Birr,' Bernal noted in his diary for 16 August 1917, 'and walked in the
Castle grounds. The telescope has been taken off its bearings and the
mirror is at London.' He was evidently aware of the status of the Earl
of Rosse's telescope, which was the largest in the world when it was
built in the 1840s. A few years later, on 2 February 1921, Bernal
referred to cycling over to Castle Lough to discuss geology with a Mr
Parker, in whose car they went on geological trips to local quarries.
Thus Bernal seems to have been familiar with at least three sources of
amateur science in the 'improving landlord' tradition, and to have
absorbed the impression that a scientific interest in natural phenomena
was by no means uncommon, if not the norm.

It is worth digressing to go into the background of the Birr telescope.
By the time of Bernal's recorded visit the observatory had fallen into
disuse; the Fifth Earl had lost interest in astronomy and in any case was
away in the Great War (he was killed in 1918). The previous Earl had
kept up the astronomical work until his death in 1908, and up to about
1912, the observatory still attracted a trickle of visitors. In its heyday,
however, the Birr telescope was a remarkable enterprise. The Third
Earl embarked upon the project in the 1830s with a view to taking the

'witchcraft' out of optical system production. He set up workshops and a foundry at Birr Castle, and succeeded in casting in speculum metal a 72-inch reflector, which he polished to the necessary optical standard in his workshop, with locally made machinery.[4]

His youngest son was Sir Charles Parsons, who served his time in the Birr workshops and then went on to university in Dublin and subsequently to Cambridge, ending up in Newcastle where he began a remarkable career as an engineer and inventor. His best-known invention is the steam turbine which, after the famous Spithead demonstration of 1897, became the preferred motive power for the Royal Navy. Parsons also went into the development of military optical technology, as also in parallel did Howard Grubb, the son of Thomas Grubb FRS, the Kilkenny optician who contributed to the construction of the Birr telescope. The Grubb optical firm was based in Rathmines, Dublin, up to 1920, when it moved to St Albans for strategic reasons because it supplied gunsights to the Royal Navy. It was baled out by Parsons in 1925, becoming the progenitor of the present firm Grubb Parsons Ltd. Thus the technological spin-off from the 72-inch telescope evolved into two major firms feeding the imperial war machine.

Some aspects of the Parsons link may perhaps merit further research. According to Kevin Bernal, Desmond's brother, the 1917 visit to Birr did not involve contact with the Parsons family. Kevin is insistent, however, that Desmond did visit Birr Castle subsequently as a house guest, possibly around 1928, at the invitation of Lady Rosse, the widow of the Fifth Earl and grandmother of the present earl.[5] This visit would have provided an opportunity for a direct (though tenuous) link with the Parsons family tradition.

In *Science in History* Bernal refers to Parsons and the turbine, but does not take up the Birr connection, labelling Parsons as British. We are therefore left with an enigma: if Bernal was aware of the full science/technology background of the Parsons family, why did he not use it as grist to his *Science in History* mill? There is no reference in the latter, in connection with Parsons, to Rollo Appleyard's 1933 biography, on which the above summary is based. It is therefore just possible that Bernal might have been unaware of the full Parsons story, despite his familiarity with the tradition of the telescope itself.

As well as the gentleman-amateur scientific influences from which Bernal picked up much of his initial interest in science, practical aspects of the running of the estate contributed to Bernal's insight into technology, unusual in a pure scientist. He also frequented local industry, being on social terms with the owners and managers. Much of his work on the farm centred round the maintenance of farm machinery, particularly the 'reaper and binder', the hydraulic ram (on which the house water supply was based) and the 3 horsepower oil engine. In his diary there are repeated references to these tasks which would have given Bernal a good practical feel for the basics of mechanical engineering.[6] On a visit to Scott's Mill, managed by Carlo Moynan, on 4 January 1917 Bernal remarked, '... the motive power is a water turbine or an oil engine and the machinery is modern'. By this he meant that the turbine was the prime source and the oil engine was a standby, for when the water supply was low in the dry weather. On 1 August 1917 he visited the old mine works at Shalee and remarked on the Cornish beam pumping engine there; he visited it again the following summer. Bernal developed early a sense of the importance of the primary energy source. This was reinforced by his visit to the Siemens hydroelectric project at Ardnacrusha in 1928. Both in *Science in History* and in *World without War* Bernal develops the global significance of these early insights into primary energy technology.

Thus Bernal's background at the fringe of the landed gentry, in its peculiar Irish form, was a positive stimulus towards creative thinking, both in science itself and in the dynamics of science, technology and society. There were however other factors at work which contributed to his alienation from his Irish background, and channelled him towards a career in exile. These factors were somewhat similar to those that made it necessary for other famous creative Irishmen to make their careers abroad.[7]

Educational Options

Broadly speaking, the upper or 'imperial' landed gentry in Ireland, many of whom had a family military tradition, sent their children to English public schools, followed by Oxbridge or Sandhurst. Typical of this group was the French family of Rossmore, County Roscommon, with which the Bernals were on social terms, the son, 'Tulip', being a schoolfellow of Desmond at Bedford. Bernal remarked in his diary that the Frenchs had been high sheriffs of Roscommon since 1600. There were also Frenchs at Frenchpark, County Roscommon; this was the family seat of Lord French, who served on the Imperial General Staff during World War One. The Anglo-Irish were the East Prussians of the British Empire.

The smaller landed gentry tended to send their sons to Irish public schools modelled on the English system, typical being St Columba's, near Dublin.[8] Science teaching at St Columba's had been initiated in 1893 with the aid of John Joly FRS, a physicist at Dublin University who is perhaps best known as the founder of modern geophysics, and as the first person to give a realistic estimate of the age of the earth. By 1906 the school had a fully equipped laboratory. If the teaching of science alone had been the determining factor, the class position of the Bernals would have indicated St Columba's. Among the smaller landlords, however, the Protestant ascendancy principle was paramount, and this expressed itself acutely in the atmosphere of the Irish-based public schools that catered for their educational needs. The lot of a Roman Catholic in St Columba's early in the twentieth century would have been unenviable, even if Catholic pupils were officially tolerated.

The Bernal family farm was substantial, and supplied the town of Nenagh with milk (in his diary entry for 8 August 1919, during his father's last illness, Bernal recorded a family conference about what to do with the dairy; in the end they decided to employ a manager). They owned a car, and on 10 September 1915 they went on a day trip to Fermoy, Cashel, Cahir and Tipperary, returning to Nenagh through the mountains, an adventurous trip in those days. Bernal repeatedly refers to the chore of mending punctures. Thus while being in the car-owning classes in 1915 put the Bernals in the top 1 per cent or so, they did not employ a specialist mechanic (a 'chauffeur') as the upper landed gentry would have done.

Nevertheless, car ownership gave the Bernals mobility and status, and Samuel Bernal could credibly aspire to mix with the upper landed gentry, especially as his Australian period had given him a sense of identification with the Empire. This upward mobility was reinforced by the decision to send the children to school in England.

Not only had they to 'try harder' for acceptance because they were Catholic, but the Irish educational system appropriate to their class was so Protestant as to exclude them. Thus it had to be England, as the Irish Catholic system left a lot to be desired. According to Kevin, their mother visited Clongowes (the pinnacle of the system, serving the Catholic bourgeoisie and professional classes, and immortalized by James Joyce in A Portrait of the Artist as a Young Man) and dismissed the system as 'too rough'.

Thanks to the Bernals' acceptance of the social mores of the imperial landed gentry, they were among the three Catholic families who were accepted at the Nenagh tennis club, at the boat club at Dromineer on Lough Derg, and generally on the Protestant Ascendancy social network, which around Nenagh at that time was almost as exclusive as that of the white elite in contemporary South Africa. (To this day the Bernals are regarded as social climbers by the more traditional members of the declining Ascendancy set.)

Insofar as there was on Desmond Bernal a shadow of military service during the war years it was due to social pressure from this network, not to conscription; this was never extended to Ireland because of the pressure of separatist politics. The threat to introduce it in 1918 precipitated an anti-conscription movement that was a major factor in the 1918 Sinn Féin election landslide. Bernal was lucky in his date of birth; had he been born a year earlier he would have avoided the military machine only with difficulty.

The Irish National-Democratic Revolution

While Bernal escaped the Great War, he did not escape contact with the developing national consciousness in Ireland. In his later writings Bernal refers to a ten-year period of adolescent romantic nationalism. Which ten years did he mean? On the evidence of his political development in Cambridge it would seem 1912 to 1921. A child of twelve or so living in Ireland during the Redmondite Home Rule period, in the lead-up to August 1914, with the arming of the Volunteers, could not fail to pick up a feel for the importance of military affairs.

That Bernal's early initiation into political thinking was Redmondite is suggested by a note in his diary entry for 6 March 1918: 'Poor John Redmond is dead. Like all Irish leaders he was in the end deserted by the progressive party, but in his case not by his own fault.' This remark suggests sympathy with Redmond and a perceived parallel with Parnell, at a time when the mass of public opinion had rejected Redmond for his pro-war propagandist stance; his party was to be wiped out later in 1918 in favour of Sinn Féin.[9]

It emerges from Bernal's contemporary diaries that his understanding of Irish politics deepened appreciably up to the end of 1921, by which time he was intellectually (but not actively) a supporter of the objectives of the Irish Republican Army (IRA) (namely an independent Irish republic) and an admirer of De Valera. Some of his own writing appears to indicate that he rejected Irish nationalism in 1919 in favour of Marxism, seen as an alternative. 'How narrow my Irish patriotism seemed,' Bernal wrote, describing his conversation with Henry Dickinson in November 1919.[10] If, however, this was written retrospectively it must be seen as a natural consequence of his post-1921 period of withdrawal from concern with Irish affairs, rather than as reflecting his actual views in 1919. The date of his withdrawal from concern for the Irish national struggle is important. If that had occurred in 1919, when the struggle was beginning to build up, it would imply that his Cambridge Marxism was little more than an undergraduate fad, and his grasp of it superficial. If on the other hand he remained intellectually loyal to the Irish struggle right up to the Treaty, despite the imperialist background to the Cambridge intellectual hothouse, his Marxism was made of sterner stuff. I believe the latter to be the case.

On the evidence of his contemporary diaries, Bernal was maturing his Sinn Féin thinking and integrating it with a Marxist understanding of the class struggle in Britain. To suggest that in a flash of insight, under the influence of 'Dick' in November 1919, he cast aside his Irish 'narrow patriotism' and adopted the wider world-outlook of Marxism in the light of 'the great Russian experiment' is to oversimplify his developmental path to the point of falsification. Consider the following entry to his diary, written on 8 December 1919; it records a conversation with 'PW', his tutor: 'on learning that I was spending the vac in Ireland he asked me if I were a Home Ruler. No sir, said I, I am a Sinn Feiner. This is the first time I have dared admit it to a superior.'

Let us go through the contemporary diaries.

13 December 1919: on the way home through Nenagh he noticed the workers on strike, with red badges, for 15 shillings wages.

5 January 1920: at a social event Mary Waller (an upper-class neighbour) remarked 'the Irish ought to be strafed'. Bernal wrote, 'I took it up immediately but ran into the argument that the Irish were ignorant, changeable, dissatisfied and vindictive and this for lack of knowledge I cannot deny convincingly.'

6 January 1920: Bernal got himself up in fancy dress as a Bolshevik with a toy gun, for a joke, to frighten the young people at a party. He found himself hiding from real British soldiers.

12 January 1920: passing through Dublin he picked up Gigi (his sister) at Alexandra School, which is opposite University College Dublin, where Dáil Éireann bonds were on sale publicly, an open revolutionary gesture by the underground republican government. Bernal noted, 'NUI building, covered with invitations to buy Dial Erin [sic] bonds . . .'

His was an instinctive evolution, without detailed knowledge of the minutiae. Thus although a declared Sinn Féin supporter he found himself unable to argue the case, or even to spell correctly the name of the national assembly after seeing it written.[11] His English education had left him short on Irish history. He took steps to remedy this, but without informed guidance it was a slow process, unable to compete with the rapidly developing influence of the Cambridge culture.

Later on that 12 January he saw a film, Zangwill's *Melting Pot*: 'scenes of Czarist pogroms strengthened enormously my wavering faith in the Bolsheviks'. Back in Cambridge on 17 January: 'Dickinson . . . cheered me up no end about Russia, Ireland and everything else'. On 21 February he defended the Bolsheviks in a student argument. On 22 March, on the way home to Ireland, he got into anti-war discussions with soldiers in the train. On 27 March he argued socialism with his aunt Cuddie, who predicted he would be a Tory by the time he was forty.

Let us continue with the 1920 Irish entries.

30 March 1920: 'the conversation as usual runs on the recent murders. He believes that the Government is not trying, and that the sentiment of the country is against Sinn Fein. I am doubtful of both [statements]. But for all that I am rather appalled, these murders seem such a waste, and done in so cruel and cowardly a manner . . . "hang the lot of them" is the cry, with a trace of fear that it may be their turn first.'

1 September 1920: Mr Galway Foley, a neighbour, had been shot by the IRA. Bernal defended the action in family arguments, generating in the company an attitude of 'absolute loathing' to the extent that he 'could not go and meet all the people at the tennis'. The atmosphere persisted on 4 September, when he 'dodged tennis and went to work in the oats field'.

16 December 1920: Bernal noted lorryloads of Black and Tans with prisoners. On the whole, by now it seems he had learned to avoid embittering arguments with the family. According to Kevin 'there was none of this anti-British nonsense about Desmond'. So he must have managed to avoid imposing his views to the extent that they would be memorable to a younger brother.

An incident on 17 December sharpened his appreciation of the contradictions between his class and his philosophy: a local labourer stole firewood; Bernal caught him and remonstrated, but only feebly. On 1 January 1921, by way of New Year resolution, he records: 'about religion I have decided to do nothing, about Virginia nothing, about Ireland nothing, in all things "wait and see"'. Thus socialism was not yet prominent among his priorities. During 1921 his perception of socialism sharpened, but not in such a way as to oust or supersede his developing republican position. Both concepts developed in parallel; indeed Bernal saw the importance of the link between British labour and Irish republicanism with clarity rare for the time.

On 2 January, geologizing with Mr Parker, he noticed a burnt creamery, also a memorial to 'James Lucas, 1st Batt IRA, Ballina Marble Works died of wounds Sept 1919, RIP'. He wondered if it was genuine, though it could hardly be otherwise. The fact that he expressed doubt is a further illustration of the cultural gap between the English-educated aspirant-landed-gentry Bernal family and the red-republican workers in the quarry where he was geologizing.

The label 'red-republican' is appropriate, as in this period workers were taking over creameries and running them by means of workers' committees; one such, at Knocklong near Thurles, went down in history as the Knocklong Soviet. Thus there was no contradiction between Dickinson's Cambridge Marxism and Bernal's Irish grass roots. The location of the contradiction was between the then-developing Irish revolution and the Bernal family class position.

On 8 January 1921 he noted: 'Sinn Fein is again uppermost (in my mind) because of the sight of a gang of Black and Tans roaring drunk on Thursday night. Where does Lloyd George get the likes of them from?' But once back in Cambridge, Ireland seemed increasingly remote. On 17 March, as a footnote to many pages of undergraduate capers, he wrote: 'Patrick's Day without a shamrock for the first time in my life.' Bernal had still not got around to raising his sentiments to the status of convictions with the aid of knowledge. During Easter 1921 he tried to remedy this.

On 2 April 1921 he wrote: 'reading Green's history of Ireland and speculating feebly on the diversity of it'. This history was probably *The Making of Ireland and Her Undoing* by Alice Stopford Green, a formidable and passionate statement of Irish nationhood by an Ulster Protestant, on which generations of Home Rulers and republicans were reared. This could have been the first time Bernal dipped into the Irish historical background consciously seeking political understanding. His note suggests he realized for the first time the extent of his ignorance of Ireland derived from his English education, a sound position of scientific humility.

The April 1921 entries show Bernal at the peak of his understanding of the Irish revolution and its embryonic link with the class struggle in Britain.

1 April 1921: 'The paper with their announcement of a state of

emergency over the coal crisis gave me a shock. Can I dare hope for a revolution and withdrawal of troops from Ireland?'

2 April 1921: 'the first day has passed off quickly, an ominous quiet. The *Times* is not as deliberately partial as usual. I wonder if Northcliff is frightened. De Valera has broken silence and what he says is heartening: "no surrender".' (The Bernals took the London *Times*, not the *Irish Times*, in keeping with the imperial landed-gentry status to which his father aspired.)

8 April 1921: Bernal felt aloof from the entertainment provided by a 'paper-chase', the usually enjoyed cross-country entertainment of the rural youth.

9 April 1921: 'the triple alliance will strike and the reserves are called up. War almost inevitable and the government's deep-laid plans almost make me despair of victory and that means another century of slavery. If it is war, what will I do, another insoluble problem, for the present at least nothing but watch and wait.'

11 April 1921: Bernal went on his bicycle up in the hills behind Ballinaclough: 'I pass a few carts coming from market and notice curious glances. I hope I will not be taken as a spy.' This is more evidence of Bernal's isolation, a lone settler in Indian country. Later, at home, he saw the papers on the coal strike. '... negotiations, compromise. It is all going to fizzle out, bluff on both sides. I am furious.'

Then followed the usual Cambridge interlude, followed by a trip to Paris in July. Bernal remained conscious of Ireland, as the entry for 29 July shows. On that day he called on Alan Hutt, who introduced him to one Liang, a Chinese: '... far eastern politics which I treated in my usual way, profundity based on ignorance' (an admission that in his dawning understanding of the complexity of the Irish revolution he was beginning to appreciate how much he didn't know). He made a lateral leap, showing a grasp of historical parallels: 'Korea led us to talk of Ireland' (Korea had to face a national struggle, on two fronts, with China and Japan). 'It is a pleasure to put a case before a really impartial judge because I can put the whole truth forward' (in his arguments on Ireland with the English he must have felt it necessary to soft-pedal). 'The chief thing that surprised him was the success of Sinn Fein and the lack of economic motive. I explained the character of the people as best I could.'

Lack of economic motive? Bernal's grasp of the dynamics of the Irish revolution did not extend to this depth; his class position was not the best vantage point for gaining insight into the economic thinking of the petty-bourgeois leadership of the Irish national movement. Had Bernal's thinking evolved along the other optional path suggested in the next section, he would have had this insight, but he would not have been arguing the case in Cambridge.

After an erotic period spent in and around Cambridge, he went to Ireland on 19 August 1921. There is a gap until 2 September; Bernal was preoccupied with no fewer than four women, one of them his future wife. Then, in reference to the Treaty talks, which he must have been following intently: 'I have seen de Valera's reply. I admire it for its honesty and logic but I am afraid these count for little in modern propaganda warfare. It looks like bloody war again.' Then, on 8 September 1921: 'The fate of Ireland is in the balance, but it is apparent that Lloyd George will do his best to get de Valera to begin the fighting. It is strangely reminiscent of the coal negotiations.'

This is the last available reference to Ireland in the diary, which for 1920 and 1921 consists of three fat volumes per year. There is no diary for 1922; perhaps the lovelife got too complicated and it has ended up in the confidential box. So we are in the dark about the further development of his Irish thinking. In the line-a-day summary diary there are glimpses: 'January 3: first skirmish with family about E; March 22: home to meet family wrath; March 29: call on Tolers of Beechwood; I discuss Christian Science and politics and look for buried treasure.'

It is possible from the foregoing analysis both to draw some conclusions and to formulate some questions to which, as yet, there are no clear answers. It is possible to speculate on the answers by listing the options open to Bernal at the end of 1921, and by weighing up his probable attitudes, in the light of the evidence to date.

We can conclude that Bernal's Irish republican thinking during the period up to the end of 1921 was developing in close association with his socialist thinking derived from his Cambridge contacts, and that this was accepted and approved of by Henry Dickinson – 'Dick'. Further, he achieved a level of understanding of the interconnection between the struggle of the Irish people for national emancipation and that of the British labour movement for social emancipation that at the time was most unusual, being found in the thinking only of the most far-sighted members of the leadership, such as Liam Mellows.[12] This understanding remains elusive to this day, despite the best attempts of T. A. Jackson, C. D. Greaves and others to propagate it in historical analysis.[13] We can further conclude that Bernal was precluded from

acting on the basis of his analysis within the Irish context by constraints
of family and class.

Industrial Technology

Before dealing with the career options that Bernal faced once he had
completed his degree, it is worth considering further his embryonic
interest in industrial technology, as recorded in his contemporary
diaries.

I have already noted Bernal's links with the gentleman-amateur
scientific network, and I have mentioned the factors which contributed
to his practical feel for technology. There was a further factor of which
the young Bernal was consciously aware: the scientific element in
contemporary productive industry (in some cases local).

I have already referred to the motive power in Scott's mill mentioned
in Bernal's diary entry for 4 January 1917. The manager was Carlo
Moynan, elder brother of Raymond Moynan, who had been to school
with Bernal in Nenagh. Bernal remarked, 'Carlo has taken charge of
his father's works and brought them up to a very modern standard. He
has a good chemical lab where he tests the malt and other products of
the mill.' Later Carlo drove over with a view to a motoring trip, but
'the engine had a leak in the water-gauge, so we played poker'. The
sixteen-year-old Bernal socialized with and looked up to Moynan, who
was three or four years his senior. (The Moynans were by no means in
the nationalist vanguard: Bernal noted on 4 April 1918: '. . . party at
Moynan's. The officers and most of the girls were there, Major Whalley
the CO' [Nenagh was a garrison town]. By 1920, however, Bernal was
writing (April 20): 'I am tempted to give up the Moynans as hopeless;
if they have any brains they conceal them perfectly'. With his Cambridge background, Bernal had 'leapfrogged' the intellectuals of provincial technology; their relative stature was shrinking.

On 23 July 1920 Bernal visited Handley's mill at Ballyartella: 'a
curious old place . . . made a deep impression on me. Turns out very
good tweeds.' However, he remarked on a girl of twelve working there,
with a very curious expression, and a group of handloom weavers bent
by years of toil. He remarked, '. . . if this is the way home industries are
carried on we had far better not have them.' On the way home he
fantasized a concept of a circular electric loom, to take out the
drudgery. Evidently Bernal's contact with local Irish industrial technology had led him to evaluate it as relatively undeveloped.

In the summer of 1920 Bernal went through an inventive phase. On 20 August he referred to devising a machine for making three-dimensional graphs. On the way back to Cambridge on 23 September he went to a machine tool exhibition at Olympia: '. . . it taught me an enormous lot about the workings of metals, more than I could have learned from books in a week. Furnaces in full blast, hygrometers, an integrating water meter.' He commented that the exhibition lacked scientific arrangement (a hint of his future interest in Operations Research).

An inventive vein continued to dominate his thinking during 1921. At home on 29 March, full of his Cambridge undergraduate chemistry, Bernal wrote: '. . . in the cowhouse pacing back and forth. I see at last the whole of the process for recovering the chlorine from the ammonia soda plants. A fortune within my grasp. I give imagination rein. Money, money, money. My eyes glisten, my mouth waters. . . . I can see no flaw. Can it? Might it? No, if it is right it must have been tried.' But this love affair with the entrepreneurial potential of chemical technology had come to an end by 7 May in Cambridge, where he wrote, 'I think it is going to be physics, my mind delights in taking readings. Chemistry is too full of smells and geology of names.'

It follows that as late as 1921 a career in industrial scientific technology remained an option, and if the environment of the Anglo-Irish political settlement had been congenial, he might have looked for career opportunities within it, had the chance of pursuing his crystallographic interests in England not turned up.

Career Options

Within the post-Treaty Irish Free State, as it was called, there were some limited opportunities for practical-minded scientists to turn to technology, thereby helping to build up the economy of the fledgeling nation that had been crippled by partition and civil war. An example of this process was the career of Dr T. A. McLaughlin, who was Bernal's contemporary.[14]

McLaughlin had taken his primary degree in physics in University College Dublin and then had retrained as an engineer while lecturing in physics in Galway. He then went on to Siemens-Schuckert in Germany, where he picked up the technology of the high-voltage transmission and distribution of electricity, based on the then-novel 'national grid' concept. McLaughlin worked on a 'grid' project in

Pomerania that was on a scale comparable to Ireland. By 1923 he was involved in hydroelectric system design; he got hold of hydrological data from the Shannon and produced a proposal that combined a generator near Limerick with a national grid to supply a projected load (based on population and industrial statistics) that was three times the total power then in use. He sold this bold concept to the Irish government of the time, and construction began in 1925. Bernal visited the generator in 1928, when it was in process of commissioning. (This was also probably the occasion of his visit to Birr Castle at the invitation of Lady Rosse.[15]) So why did he not follow some path along these lines?

Despite the family's heavy investment in English-based education, and implied bias towards a career in the English university environment, Bernal at the age of twenty-one had at least three career options open to him, in theory. These were (a) to take over the estate, manage the farm and keep up his science on the gentleman-amateur basis that was the norm among his class; (b) to steer his interest in physics towards industrial uses of electricity, or other high technology appropriate to the industrial needs of the time, particularly in the context of the needs of a newly liberated nation; or (c) to follow his instinct for basic scientific research, keeping alive his interest in technology vicariously. In practice, however, two of these options rapidly closed off during 1922.

The first option was what 'all Nenagh' expected him to do on the death of his father. His failure to take it was the primary cause of his being locally regarded as the black sheep of the family, not his politics. He clearly did not want to take it; he was bored by the local pastimes and excited by the intellectual life he got in Cambridge. Lacking the 1922 diaries we can only conjecture that he chose to manufacture (or at least welcomed) a pretext for a break with the family, and that the engagement with Eileen Sprague and their subsequent civil marriage served this purpose. Eileen herself had an interest in pinning him down; her telegram to Nenagh mentioning the engagement fed the local grapevine from the post office, and helped to generate social pressure.

Their interests coincided and the wedding took place, despite the disapproval of Bessie, Bernal's mother; option (a) was closed off, and the farm passed to the second son, Kevin. It turned out that this was in fact a good solution, Kevin having become an excellent farmer, though at the time he was somewhat young. Had Desmond stayed, neither he nor Kevin would have been fulfilled.

Consider the second option: this was hypothetically touched upon by

T. P. Hardiman, then chairman of the National Board for Science and Technology, when responding to Dorothy Hodgkin's memorial lecture in the Royal Irish Academy in October 1980. Despite the Cambridge bias, given Bernal's developing political understanding of both Ireland and Marxism it was a real option, as is confirmed by the career of T. A. McLaughlin. Had the Bragg opportunity not come he would have had to consider it.

In this context Bernal might have begun, like McLaughlin, with a teaching post in an Irish university. Had he gone to Trinity College Dublin (TCD) he would have joined a physics department with a staff of three, the best-known being John Joly FRS, whose pioneering interest in geophysics Bernal could creditably have developed. In University College Dublin there was J. J. Nolan, who had worked with Rutherford in the 1890s on atmospheric electricity. Compared with Bernal's actual career, however, it is not unfair to state that physics in Dublin was a provincial backwater, although Bernal might have been in a position to transform it, given appropriate political support.

There must have emerged factors in 1922 sufficient to block the second option. The evaluation of these factors, in the absence of the 1922 Bernal diaries, is a matter for conjecture, but it is highly likely that the key ones were the combination of the break with the family and the post-Treaty political reaction which rapidly developed into civil war. The effect of this lethal combination was to remove Ireland from prominence in Bernal's intellectual agenda, though he always retained an interest in progressive causes relating to Ireland.[16]

Bernal's assessment of the post-Treaty environment as hostile and obscurantist was confirmed in a personal context in 1953, when a left-wing student society in TCD (with which prior to that year the present writer had been associated) invited him to speak at an inaugural meeting. This was banned by the college authorities. According to my late father, Joseph Johnston, then a Senior Fellow and privy to the discussions, Bernal was objected to on the grounds that he possessed an unacceptable combination of attributes, to wit, being Irish, a lapsed Catholic and a Communist.[17] This combination was lethal in TCD at the time (though English Marxists were regular visitors) because there were negotiations going on with the state whereby the college would get adequate funding, giving up its old Protestant Ascendancy status dependent on rents and endowments (which had declined disastrously after two world wars). The Board seemed to think that to entertain Bernal would rock the boat.[18] I have seen no evidence that Bernal's name was blacked generally in Ireland for political reasons, although of course ultra-conservatism was not confined to TCD.[19] It is, however, doubtful on the whole whether Bernal would have found an academic position in Ireland congenial.

Having taken the third option, Bernal became a world figure and a subject for various biographical treatments, of which it is to be hoped the present essay will form a part. The world is the richer for his decision, and Ireland not necessarily the poorer, especially as the work of Bernal on science, technology and society is increasingly being appreciated in Ireland today.

Bernal on Ireland

In his political writings, Bernal in later years showed little consciousness of his Irish grounding. Study of the references to Ireland in Bernal's major works show them to be absent where they could usefully be present or, if present, to be dismissive.

As an example, consider *Science in a Developing World*, published in 1959 by the World Federation of Scientific Workers. This is Bernal's account of a symposium on 'Science and the Development of the Economy and Welfare of Mankind' held in Warsaw. This seminal work stated the problems that are now the common currency of the United Nations development agencies, including the 'brain drain' problem. Thus:

> ... there is one particular danger, a very real one, in the smaller countries, at least in the capitalist realm, namely, that as there is not sufficient scope for progress in every field of science, people who are interested in a particular field have to leave the country and there is a continual drain. For instance, from one of the oldest scientific countries, Scotland, it has been going on for centuries.

This is an exact and perceptive statement of the brain-drain problem. It precisely describes his own position. Yet he chooses to substitute Scotland for Ireland, a curious example for an international audience which, if political, might be expected to be aware of the Irish national struggle but hardly the Scottish. Scotland, however, would have presented itself to Bernal's consciousness in the English context, as migrant Scottish intellectuals were highly visible on the English academic scene, more so than Irish intellectuals. Emigré intellectuals from Ireland in Bernal's time would have tended to end up in the Indian civil service, or in the African missions of the Roman Catholic Church, rather than in England, except the occasional outstanding ones (Yeats, Shaw or indeed Bernal himself). Bernal's awareness of the Scottish scene was contemporary, while his Irish consciousness was dated, or perhaps even suppressed as a result of the emotional trauma of 1921–22.

In *Science in History* Bernal's view is global: science is a supra-national phenomenon. Yet the nationality of a scientist can be important, when a scientific principle is developed into a new technology, especially in war. Scientists emerge to explore new frontiers that a new technology has rendered accessible. A scan of *Science in History* for Irish references gives a somewhat meagre yield. The only explicit reference to Ireland relates to the role of Irish monks as a source of continuity of civilization during the Dark Ages. The only names mentioned with explicit Irish connections are those of Boyle[20] and Edgeworth;[21] Hamilton,[22] Johnstone Stoney[23] and Walton[24] each get a passing reference without Irish attribution; Parsons[25] is labelled as British.

One could hang a history of the science/technology interaction in an imperial/colonial system around that selection of names. The fact that Bernal did not take this up in the context of *Science in History* suggests that in taking the global view he had either (consciously or subconsciously) suppressed his Irish background, or (more likely) was simply unaware of the rich Irish veins of raw material to be mined for his analytical mill.

The modern revival of interest in Bernal in Ireland is a healthy indication that science and technology in Ireland are maturing and becoming aware of the socio-economic dimension. His growing acceptance in Ireland constitutes a reason for those of us who aspire to build on his tradition to seek to assert his Irishness, and to sift the records of his early thinking in search of insights. This chapter will achieve its purpose if it stimulates further analysis of the role of science in developing countries, and of the nature of the process whereby science becomes socially necessary and relevant technology in a postimperial environment.

I conclude by stating a problem. Bernal's life and work arouse interest (a) among scientists interested in the philosophy of science and the socio-economic implications of the related technologies; (b) among Marxists and philosophers of science; and (c) among economists and sociologists interested in what has become known as Science Policy Studies. Bernal, however, is scarcely known among technologists and engineers. Indeed, in group (c) above, Bernal is only seldom referred to, where by rights he should be a founding father, a cult figure.

This cultural gap must be attributed to the Cold War and his World Peace Council period, when it could be argued that he misjudged the nature and scope of the market for his ideas. The *Science of Science* Festschrift, though influential, did not restore the balance.[26] Thus there remains important unfinished business (initiated by McLaughlin in the Irish microcosm), namely, the building of an effective bridge between science and socially useful technology, to the extent that technologists and decision makers at all levels are consciously aware of the science-technology-society interaction process in peacetime, for peaceful purposes.

There is a hint to the solution of this problem in the 'classical' operations research of World War Two, and Bernal was at home in the dynamics of the process. To solve the problem in such a way as to avoid World War Three, and to convert modern science to socially useful technology in a global context, remains on the agenda. It can perhaps be solved within the Bernal paradigm.

Notes

I wish to acknowledge the following assistance in the preparation of this chapter: the Irish National Board for Science and Technology, for expenses covering the visit to Cambridge to consult the Bernal archive; Trinity College Dublin for a period of research leave; and Bernard Loughlin, for permission to use the Tyrone Guthrie Centre at Newbliss, County Monaghan (a retreat normally reserved for writers and artists) in the final revision of this chapter.

1. The writer can claim a background having some parallels with that of Bernal: a youth spent on the fringe of the rural Protestant Ascendancy network, during a world war, with local energy sources a continuing technological problem; subsequent exposure to the Cavendish tradition (under Walton and O'Ceallaigh, who had worked under Rutherford in the 1930s); concern for creative social application of scientific discoveries leading to identification with the Marxist revolutionary tradition; interest in operations research as a dynamic bridge between scientific discovery ad technological application. We differ in that Bernal chose to develop his Marxism and his science in the imperial heartland, while the writer chose to attempt a synthesis within the Irish postcolonial context.
2. Sir William Rowan Hamilton (1805–65) was the son of the land steward of a radical landlord who had supported the 1798 insurrection, the aim of which had been to establish an independent Irish republic on the French pattern. He was a gifted mathematician, contributing a unifying intellectual framework for optics and dynamics (the 'least action' principle) in an 1834 paper, and going on subsequently to invent 'quaternions', the first non-commutative algebraic system and the father of all linear associative algebras. He became Astronomer Royal for Ireland, based at Dunsink Observatory, near Dublin; after marrying Helen Maria Bayly on 9 April 1833

he spent time at Bayly Farm, just south of Nenagh. Thomas L. Hankins's biography, *William Rowan Hamilton* (Johns Hopkins Press, 1989) gives an excellent account of the nineteenth-century gentlemanly scientific environment.

3. Launcelot F. S. Bayly was born in 1869 and died in 1952; thus the Hamilton connection would for him have been within the 'living memory' of people he knew and may have contributed to his scientific interests. He is reputed, however, to have been a dilettante.

4. It remains for someone to analyse why, despite the well-developed infrastructure, this remarkable local enterprise failed to 'spin off' into local industrialization with Birr emerging as a world centre of optics, like Jena. (Sir W. R. Hamilton's involvement with the Birr telescope and the Earl of Rosse is discussed in Hankins's biography cited in note 2.)

5. According to Desmond's brother Godfrey Bernal, Desmond and Eileen visited Ireland in or about 1925. Eileen does not recollect the Birr Castle visit, but Godfrey does.

6. See for example 29 August 1917, 21 December 1917, 30 August 1918.

7. Bernal recorded his Ardnacrusha visit; Godfrey Bernal recollects his conversing in German with the Siemens engineers, some of whom lived in Nenagh while the construction was going on.

   There is perhaps a thesis for someone in the comparative study of the alienation factors at work in the case of the scientists (e.g. Tyndall, Kelvin, Bernal), the technologists (Dunlop, Holland, Ferguson etc.) and the writers (Shaw, Wilde, Joyce etc.). The technologists listed were respectively responsible for the pneumatic tyre, the submarine and the tractor with hydraulic linkage to control the depth of ploughing. Few would know them as Irish, and few Irish would know to rank them, and the scientists, as being of significance comparable to that of the writers.

8. G. K. White, *A History of St Columba's College*, Old Columban Society, 1980..

9. This is not to suggest that the family was Redmondite – they would undoubtedly have been Unionist. Desmond's evolution away from the family politics towards Marxist republicanism would have been initiated by a youthful admiration for Redmond in the Home Rule period.

10. See this volume, p. 11.

11. Bernal was careless with spelling, but not usually with keywords.

12. See C. D. Greaves, *Liam Mellows and the Irish Revolution*, London, Lawrence and Wishart, 1971.

13. See T. A. Jackson, *Ireland Her Own*, London, Lawrence and Wishart, 1971; C. D. Greaves, *The Life and Times of James Connolly*, London, Lawrence and Wishart, 1961.

14. I am indebted to the Irish Electricity Supply Board, Dublin, for providing a transcript of an autobiographical broadcast by Dr McLaughlin, reprinted in the *Irish Times* of 11 January 1938.

15. See note 5.

16. According to C. D. Greaves, Bernal was a consistent supporter of the Connolly Association (a left-wing political organization of Irish emigrants based in London).

17. I am indebted to Paul O'Higgins of Cambridge for this recollection of a conversation with my father.

18. Mr G. Giltrap, Secretary of the college, has obliged by searching the board minutes for that period, but could find no reference. It would have been practice for such a decision to be a Provost's ruling and not necessarily minuted.

19. C. D. Greaves, then working as an industrial chemist in energy technology, recollects being asked in or about 1950 by H. M. S. 'Dusty' Miller, an engineer working for the Irish state Peat Development Board, to get information from Bernal. The implication of C. D. G.'s communication was that direct contact with Bernal would have caused embarrassment. I took this up with 'Dusty' Miller (he is now manager of a consultancy firm) on 18 June 1982: he was adamant that there were no such considerations. The Peat Development Board had then, and still has now, an open progressive Research and Development policy, having strong links with peat work throughout the world,

especially the USSR. The enquiry to Bernal was in connection with operations research, in which then-novel field Bernal had a reputation. It went via Greaves because Miller knew Greaves on the fuel technology network, and Miller did not know where Bernal was. Operations research was subsequently initiated in the Peat Development Board.

'Dusty' Miller and Professor Patrick Lynch of University College Dublin were the joint authors of a seminal report, *Science and Irish Economic Development*, published in 1966 by the Organization for Economic Cooperation and Development (OECD). This laid the basis for direct Irish state support for science. Both acknowledge the influence of Bernal. So it could be argued that the tribute by Hardiman referred to on page 31 has historic roots in the development of Irish science policy over the past two or three decades.

20. Robert Boyle, seventh son of Richard Boyle, First Earl of Cork, born in Lismore in 1627, was Irish in the sense that Edmund Spenser or the Duke of Wellington were Irish.

21. R. L. Edgeworth was in communication with the Birmingham Lunar Society. His daughter Maria achieved fame as a writer. An 'improving landlord' in the best tradition, he sympathized with the republican objectives of 1798. Among his inventions was a track-laying vehicle for use on soft ground. His estate was at Edgeworthstown, County Longford.

22. See note 2.

23. George Johnstone Stoney (1826–1911) was physics professor at Queens College, Galway. He identified and named the electron in 1891 prior to its discovery.

24. In 1932, along with Cockcroft, E. T. S. Walton converted lithium to helium by bombarding it with protons accelerated artificially by a high electrostatic potential. (The imprecise journalese 'split the atom' has entered the lore, but I dislike using it.) Walton subsequently returned to Dublin where he ran a 'string and sealing wax' physics department in Trinity College, in the early Cavendish tradition, until the Irish government woke up to the need to finance science in the late 1960s (thanks to the OECD report of 1966).

25. Parsons 'served his time' in the Birr Castle workshops during his vacations from the Trinity College engineering school. Subsequently he claimed, with pride, to have invented the 'sandwich' principle of engineering training. Though his inventive career was British-based, he was as Irish as Bernal himself.

26. M. Goldsmith and A. Mackay, eds., *The Science of Science*, London, Penguin, 1964, was a '25-year after' tribute to *The Social Function of Science* by a group of leading scientists who had been influenced by it.

3

Political Formation

Fred Steward

J. D. Bernal was a prominent intellectual figure in the politics of the left from the 1930s to the 1960s. His roles in the anti-fascist movement of the 1930s and the peace movement of the 1950s and 1960s were very influential. The critical analysis of the role of science in society was established by Bernal in a series of highly original studies. Underpinning this range of activism and scholarship was a commitment to a Marxist and Communist outlook. Although his political ideas have been the subject of biographical and critical comment, this has tended to focus on their expression after Bernal's prominence had been established: their origin and evolution during his earlier years have received less attention. With the availability of his personal diaries and unpublished papers going back to his initial involvement in politics as a student, it is now possible to uncover this formative process more fully. The story reveals a richer and far more contradictory pattern of political development than many observers have recognized. In particular his intellectual and practical experiences of the emergence of British socialism and Communism during the 1920s played a critical role in the shaping of his political outlook which was to endure throughout his life.

'Socialism . . . a marvellous thing'

Bernal's engagement with socialist ideas began at the age of eighteen during his first term as a student at Cambridge University. Both his diary of the time and subsequent accounts identify a discussion with his fellow student Henry Dickinson on the night of 7 November 1919 as the event that opened the door to the left: Bernal wrote, 'This socialism was a marvellous thing. Why had no one told me about it before?'[1] The encounter, which lasted through the night until 5 a.m., had an intense impact. The following morning Bernal wrote, 'Got up . . . perfectly happy. My old life was broken to bits and the new lay in front of me.'[2]

The dramatic nature of this conversation is evident; yet in order to assess its significance consideration is needed of the nature of his nonsocialist outlook prior to this moment and the character of the socialist commitment into which he had entered. Bernal subsequently portrayed his outlook before the 'new light and life' of socialism as one of Irish nationalism combined with a militaristic view of social change.[3] Certainly his direct observations of conditions in Ireland made a vivid impression on him as a teenager travelling between his Irish home and his English school: 'Once again on Irish soil, the contrast with England is striking. The dirty streets, the idlers and barelegged ragamuffins.'[4] So too did the ferocious power of modern war: 'sheltered from an air raid in Euston tube station for three hours listening to the bombs and the guns'.[5] Bernal contrasted the 'universal . . . theory of Marxism' to his earlier 'narrow . . . Irish patriotism'. He also contrasted a commitment to popularly based social change – 'It was the people themselves who would sweep away all the things I hated' – to his previous 'absurdly reactionary military schemes'.[6]

This is an oversimplification of his earlier views. A disposition toward general social reform and democratic change are clearly present in Bernal's politics before he went to university. In his sixth-form days at Bedford School he was a regular participant in the school's debating society and consistently espoused liberal and progressive views on contemporary social issues. He considered the 'decay of civilization' to be due to 'bad social conditions and if these were remedied it might right itself'. He defended 'the extension of higher education to the masses', equality between men and women, and the benefits of historical progress and economic reform.[7]

These views were beginning to set him apart from the conservatism and complacency of the upper-class English boys who were his fellow pupils. Although clearly an able speaker, during his last term at school he lost every debate he spoke in.[8] The unsympathetic context encouraged political self-reliance in the teenage Bernal. He confessed that his theory of life was to be 'completely oblivious of public opinion'.[9] At school he remained rather insulated from the impact of political events such as the Russian Revolution of 1917 and the emergence in 1918 of the Labour Party as an independent force in British politics.

His interpretation of society was primarily shaped by his perceptions of nature and of the individual, and these relied on the science and religion he had learned in his youth. In science, the combination of Newtonian physics and Darwinian biology encouraged a positivist commitment to materialism and progress; in religion, his Catholic faith legitimized both morality as a principle and the importance of a worldview. These two elements – of progress, from science, and of a worldview, from religion – facilitated a dynamic and universalistic outlook. The antiquarianism of the clerics' presentation of history was counterbalanced by the empiricism of his science teachers' interpretation of the world. The tensions between the material and the spiritual were partly bypassed by a separation of three different domains: science dealt with matter and change in the natural world, religion with the moral and spiritual life of the individual, and politics with the problems of society.[10] The third domain remained the least formed part of his outlook, and coexisted and overlapped with his faith and his science. The revelation of socialism to him in the autumn of 1919 was ultimately to unsettle this compartmentalization.

What was the nature of the socialism to which he now turned, in this period of ideological ferment and political change on the left? During his first year at university, Bernal's socialist outlook was that of the trade unions and Labour Party rather than that of the Marxist sects. From January 1920 he attended the weekly meetings of the Cambridge University Socialist Society (CUSS),[11] and in March 1920 the inaugural meeting of the Labour Club.[12] His friend and mentor Henry Dickinson ran the socialist pacifist Union of Democratic Control (UDC) and at this time wrote to the student journal, *New Cambridge*, expressly to point out that the socialist movement of which the UDC was part was 'not associated with the Communist (Bolshevik) movement'.[13] During the Easter vacation Dickinson attended the national conference of the University Socialist Federation. The following term he reported 'the split between reformists and revolutionaries'. Bernal wrote, 'He has joined the first and I am with him on every point'.[14] Reform through parliamentary representation and trade union organization was seen as the path forward.

Sympathy with the Russian Revolution was common to many socialists of the time, and so at this stage Bernal's 'eulogies of the Bolsheviks'[15] were not synonymous with being a Communist or Marxist. Conversely, as far as right-wing students were concerned, anything remotely to do with socialism or pacifism was labelled 'bolshevism' and was subject to virulent hostility. This hostility often degenerated into violence, and Bernal directly experienced this violence during his second term, when the inaugural meeting of the UDC – addressed by the well-known pacifist and critic of World War One Norman Angell – was so severely disrupted that police cleared the hall.[16] A few days later Bernal and fellow members of the CUSS organized a 'protective force' for the prominent socialist Brailsford after discovering a plot to kidnap him.[17]

It is unlikely that Desmond Bernal read Marx, Lenin or other Communist writers during his first year at university. Political books that impressed him included Keynes's *Economic Consequences of the Peace*, Pickering's *The Claims of Labour* of 1844, the Webbs' *History of Trade Unionism* and the Labour manifesto on the bacon trade.[18] He read his first exposition of Marxist theory in March 1920, and that was Enrico Ferri's *Socialism and Positive Science*.[19] Written in the early 1890s, this argued that social Darwinist conceptions of biological inequality, competition and survival of the fittest should not be rejected by the left but could be located within a socialist framework of equality and collectivity.[20] This application of concepts from natural science to social problems appealed to Bernal and reinforced his enthusiasm for the evolutionary interpretation of history that he found in H. G. Wells's *Outline of History*.[21] The other deep impression made by the book was its assertion that Marx's 'genius' was his theory of 'economic determinism'. The primacy attributed to economics as 'the determinative basis of all moral, judicial and political manifestations of human individual and social life' complemented the ILP leader Brailsford's exposé of the role of capitalist conglomerates in World War One in *War of Steel and Gold*, which Bernal had read earlier.[22] Ferri's support for electoral success by an independent socialist party as the 'principal means of social transformation' and his attack on 'communist anarchists' who favour 'revolt' would certainly have fitted in with Labour's perspective.

The Labour Party made significant advances in the local elections of autumn 1919 and March 1920, and this is where Bernal placed his hopes. He saw Marxism at this point as a general theory of historical change rather than a different political strategy. At a Cambridge University Union debate on nationalization of the coal industry, Bernal admired the speech of a fellow student who 'gave it to the capitalists hot and strong'.[23] In August 1920, Bernal was 'immensely cheered' by the formation of the TUC–Labour Party Council of Action to resist British military intervention against the Bolshevik regime in Russia. He added, 'I only hope they live up to their name.'[24]

The socialism embraced by Bernal at this point posed no challenge to his science; indeed, in the form expressed by Ferri and Wells, it positively embraced nineteenth-century mechanical materialism. Ber-

nal's own scientific beliefs were also in this mould (including the view
that Einstein's relativity theory was 'rather a barren subject').[25] Initially
he did not perceive a contradiction with his religion, and at High Mass
shortly after his conversion by Dickinson he 'gradually worked up into
a fine state of religious and socialist enthusiasm and joy of renuncia-
tion'.[26] However, he began to become acutely aware of tensions regarding the Church's social teaching. He listened to Father Bede Jarrett
argue that on social progress 'the Church is eternal and above human
stirrings and ambition',[27] yet a fortnight later noted that the Advent
pastoral read out at Mass 'was permeated throughout with a fear of
Bolshevism'. He talked to his local priest, Father Marshall, about
socialism. Although Marshall was progressive in outlook, his view that
'the only way to remedy social wrong is by moral persuasion' seemed
to Bernal to be 'rather slow'.[28] However, in spite of these doubts he
continued to maintain his faith and participated actively in the Church.

His science put his faith under greatest stress. An argument with the
landlady on his journey back to Ireland at Christmas 1919 which
centred on religion and science left him 'very agitated internally'.[29] In
January he read the major Catholic writings on science: Walsh's *The
Popes and Science* and Windle's *The Church and Science*.[30] These did not
quieten his concerns, and in February a sermon that 'dwelt on the
value but the insufficiency of the method of scientific research' left
him 'rather depressed all day';[31] yet his attachment to the Church
remained solid during his first year at university. On returning to the
church in his old school town of Bedford he wrote that 'it was just the
same as ever' and he experienced 'a higher pitch of emotion' than he
had for a long while. In the spring of 1920 he was still expounding his
conception of God and the Trinity.[32]

Bernal perceived no fundamental conflict between his new-found
socialism and his more established nationalism. If anything the reverse
was the case. At the end of his first term he declared to his tutor that
he was a 'Sinn Feiner' and wrote: 'This is the first time I have admitted
it to a superior.'[33] The general strike in Ireland in April 1920 reinforced
his optimism about a symbiosis between nationalism and socialism,[34]
and he saw 'an outlook full of hope'.[35] The pursuit of Irish politics was
far more serious, in a personal sense, than the espousal of Labour's
cause. Bernal was intensely aware of this and went through many
agonies in making his decision. He was in touch with Sinn Féin activists
and was confronted directly over the level of his commitment. In March
1920 he entertained some Irish friends in his college rooms: 'I asked
them about Sinn Féin. They said they would swear me in any day.'[36] He

never felt able to make this commitment and experienced guilt for not facing the risks that it would involve. In June 1920 he was seized 'by an intense depression by thinking of the S[inn] F[éin] war and the possibility of death'.[37] His aunt Cuddie chided him during an argument on Ireland: 'If you have got views like that, I can't see why you don't join Sinn Féin'. This upset him greatly: 'The shaft stuck, what had I done for Ireland? Nothing at all. Sometime the call for action will come and I will be too late. Thoughts like these, coupled with the dreadful prospects of what will happen if I do act, made me ... obsessed ...'[38] He considered himself 'too much of a sensible coward' to join the Volunteers.[39]

But increasingly he felt that the nationalist cause was insufficient in its own terms. He was unhappy with the 'limitation' that 'everything is regarded in the light of its relation with Ireland'.[40] National independence without economic transformation was not enough. After observing the 'life-breaking useless toil' in an Irish woollen mill he declared, 'if this is the way home industries are carried on, we had far better not have them'.[41] His social position cut him off from the Irish community, so that when his aunt called the Irish 'ignorant, changeable, dissatisfied and vindictive', he said, 'This, for lack of knowledge', he could not 'convincingly' deny.[42] As discussed in Chapter 1, he also felt ambivalent about the form of struggle pursued by the IRA: 'I am rather appalled, these murders seem such a waste, and done in so cruel and cowardly a fashion.'[43] Nevertheless, nationalism remained central to his political outlook after his adoption of socialism. During August 1920 the Council of Action called for no intervention by Britain in support of Poland against Russia. Bernal was aware that this put class before country: 'if the red armies are beaten, the soviet will fall and ... reaction will have its turn. And yet I admire the Poles' plucky stand. I don't like to see a country conquered – even for its own good.'[44]

The relationship of intellectuals to socialist politics was a central concern. Bernal and his Cambridge friends were far removed from working-class life. A tea party with Maurice Dobb was described: 'a lot of éclairs ... we talked socialism'. Getting ready for the college dinner, Bernal wrote, 'I was in despair about tying my tie. . . . Dobb performed the office'.[45] The social context made Bernal self-conscious about his position. He was reassured one week after his conversion to socialism to hear G. D. H. Cole point out 'the need for educated people in the labour movement'.[46] At a 'Hands Off Russia' meeting, he was also impressed by Professor Goodge, who gave sketches of the Bolshevik leaders he had met when visiting Moscow. Bernal could not remember their names but 'all were men of high intellectual standing, working day and night'.[47] The message was that there was a positive role for intellectuals in the labour movement. His contact with working-class socialists was limited but he was greatly impressed by Ernest Bevin when he spoke at the Guildhall in May 1920 on the aims of Labour: 'An enormous man with a fine voice, immensely inspiring.'[48] Bernal's political activity did sometimes extend beyond university circles; in a debate at Chesterton Workmen's Institute he advocated that 'the problem of poverty can only be solved by the socialization of land and capital'. The realities of popular opinion were revealed when 'the motion was lost'.[49]

'A Communist . . . in theory'

Bernal's second year at university was in sharp political contrast to his first. The tensions with his religion and his nationalism became acute and had to be resolved. His socialist commitment changed from identification with the broad cause of labour to a choice between sharply contested political strategies. It was a year of emotional and mental turmoil, which in 1921 was to lead to Bernal's first explicit adherence to Communism. The first Unity Convention of Marxist Organizations in Britain was held during the summer of 1920 with a view to the establishment of a communist party. Leninism began to be expressed more vigorously as an alternative to Labourism. During the autumn term, Bernal began to consider this option. Trotsky's *History of the Russian Revolution* provided him with his first systematic account of Bolshevism in practice. Eden and Cedar Paul's *Creative Revolution* was the first work he read from the new generation of English Communist intellectuals.[50] By December 1920, some contact had been made with the infant Communist Party, and his friends Brooke and Dickinson attended a meeting with working-class militants from the town. Bernal was closely in touch with these activities and was outraged by the student newspaper's sensationalist treatment and its call for Maurice Dobb to be expelled from the university.[51] He became friendly with Allen Hutt, a fellow student who had become committed to the Communist cause. Hutt became secretary of the CUSS in January 1921 and Bernal heard him speak on the Communist view of 'the transition to socialism' in February.[52]

In parallel with the emergence of Communism was a sharpening contest between Fabianism and guild socialism. Bernal heard Sidney

Webb expound the Fabian approach in January 1921 and heard
G. D. H. Cole's countercase for guild socialism the following month.[53]
Bernal was an active participant in contestation over the future direction of socialist politics. His choice was influenced by the practical
political experiences of 1920 and 1921. The municipal election results
in early November 1920 were very disappointing for the Labour Party:
'It has gone badly with us throughout the country', he recorded. In a
'long pessimistic discussion' with his friends he concluded, 'Labour is
on the downgrade. Politically there will be a split and a LibLab party
will be formed.'[54] Such doubts as to 'Labour's answer' were magnified
decisively in April 1921. The betrayal of the miners by their partners in
the Triple Alliance on Black Friday shattered Bernal's hopes: 'Negotiations, compromise. It is all going to fizzle out. Bluff on both sides. I
am furious.'[55] A fortnight later, at the Cambridge May Day Festival, his
bitterness at the defeat is evident: 'Packed in a crowd of well-dressed
proletarians with the sight of green trees and blue sky beyond, to listen
to insipidities and party lies for two hours.'[56] A week later he made his
political choice. The occasion was his reading of Raymond Postgate's
The Bolshevik Theory. 'After this,' he declared, 'I am a Communist',
although he added cautiously, 'in theory.'[57] Postgate was a member of
the newly formed Communist Party of Great Britain and his book
espoused the new Leninism; it was sharply critical of both the 'bourgeois democracy' relied on by Fabianism and the emphasis placed on
structural reform of the state by guild socialism. Soviet democracy was
presented as the alternative to both these options. By May, Bernal was
part of a student group receiving the party newspaper, the *Communist*,
edited by Postgate. This represented some sort of link between revolutionary working-class politics and the rarefied life of Cambridge
students. At the end of July, in the East Anglian village of Fenstanton,
Bernal and his friends 'formed a subsection of the pavement club
outside the garage to the great scandal of the villagers. We ate choc-
olates. I read the *Communist*, and Barnes ingeniously removed some
portion of her underwear. We got off in under half an hour leaving
the *Communist* behind for propaganda.'[58] The split between Communists and reformists, 'reds' and 'pinks', resulted in conflicts within the
university labour group. At the end of the summer term, 'Hutt and I
rush off to the Labour Club election to find the place packed with
pinks and our candidates . . . defeated.'[59]

Bernal's adoption of Communism in 1921 was accompanied by
disengagement from his religion. During the trade-union struggles of
April 1921 his exasperation with the Church's distance from social

affairs came to a head: 'As I sat in church crushed by the weight of an appallingly dull and empty sermon . . . I wondered what force of inertia was preventing the congregation from rising and slaughtering the preacher on the spot.'[60] A few days later he made his first explicit declaration of the abandonment of his faith and described himself as 'an atheist'.[61] This admission shocked him into an intense period of reflection. His reluctance to make a clear break was evident: 'I am faced with two primrose paths, religion and irreligion, each of which I know I should not take because I want to follow both.'[62] Eventually in August he did make the break, telling his priest of his 'descent from religion'. His emotions for the Church were still present but were no longer in tune with his reason. 'As I walked along the dark passage I wanted badly go to into the familiar chapel to pray to his God and mine, but I passed the door and stepped down the stairs. Outside there were people and streets glistening with rain. The God of my childhood is long departed from me.'[63]

In the same year Bernal finally rejected active involvement in the cause of Irish nationalism. By October 1920 the centrality of nationalism was receding: 'I argued without the usual sense of insecurity on the strategic unimportance of an independent Ireland.'[64] During his Christmas vacation at the end of 1920, he witnessed at first hand the brutalities of the Black and Tans[65] and the destruction of the creamery and other buildings in his home town. By January 1921 he had decided firmly against any involvement: 'I intended to do nothing.'[66] From then on, Irish politics occupied a declining portion of Bernal's interests, especially as his family ties weakened and his life focused more and more on England. The concerns expressed in his vacations in Nenagh in April and September 1921 were followed by disengagement.

Psychology, the New Religion

During his first two years at university Bernal's views on sexuality and personal life also underwent a profound change. The results were to have implications for his political outlook as well as for his attitude to personal relations. Sexual frustration precipitated a severe bout of depression at New Year 1920 which led to an interest in psychology. His friend Lucas 'psychoanalysed' him and suggested that 'love was the cause of his despair'.[67] The scrutiny of hidden emotions appealed to him and shortly afterwards he wrote, 'D stayed late in my bedroom discussing various aspects of personal ethics and egoism. We both feel better after dissecting our inner selves and exposing the most diseased organs to each other's critical examination.'[68] His interest in psychology at first remained superficial. In a conversation with a female acquaintance, Jose Kingsley, he noted, 'I dropped into psychology, palming off my garbled second-hand knowledge culled from MacDougall or Freud about various instincts which I called emotions.'[69] At home, psychology began to challenge politics as the topic for family argument: 'Freud had quite a vogue this evening. Snicoanalysis [sic] proves even more fascinating than consequences but Mammy and Cuddie persist in calling it dangerous and childish in the same breath.'[70]

Over the same period he read the feminist Olive Schreiner's *Women and Labour*. Although he felt 'enlightened and influenced' by Maude Royden's *Women and the Sovereign State* and found Cicely Hamilton's *Marriage as a Trade* 'illuminating on feminine psychology',[71] this period is not marked by a growing interest in feminism. Instead Freudian psychology comes more and more to the fore, with sexual liberation a strong motivating interest. Bernal's first love affair – with his second cousin, Virginia Crawford – ended at the beginning of March 1921;[72] there followed a number of relationships with women whom he met through CUSS – 'Grey', Sylvia Barnes, and Eileen Sprague.[73]

The primacy of Freudianism over feminism had important consequences for Bernal's political as well as personal outlook. By the beginning of his second year at university his psychological interests were beginning to intrude into his political discussion. 'I find myself more of a Freudian than any of the others, though I never read a word he wrote.'[74] In fact it was June 1921 before he read any Freud. The only psychological work that he had read was Bonsfield's *Psychoanalysis – Theory and Practice*.[75] One month after his commitment to Communism, Bernal read Freud for the first time.[76] *The Interpretation of Dreams* had a profound impact on him and within two days he writes: 'I expound the new religion.'[77]

By his third year at Cambridge Freudianism evoked more enthusiasm in Bernal than Marxism. His adoption of Communism in May 1921 was not immediately followed by a deeper study of Marx; his interest in psychology, however, did lead him to read more Freud, including *The Psychopathology of Everyday Life*.[78] He found mainstream Marxism wanting. In January 1922 he was not impressed when he heard Raymond Postgate 'rather unconvincingly describe Communism at the CUSS'.[79] In April he criticized the dominance of economics in Marxist social theory: 'The pretended saints, the economists, are only witchdoctors, dupes of the magic they pretend to explain. Marx did begin to understand. The Marxists do not. . . . Freud will give us enough to start on: money = excrement.'[80] Later that year, in a paper given to the Political Science Society using analogies from physics, he also stresses the importance of individual psychology. Humanity is considered as 'a system of particles' with 'external forces (resources) acting on all the particles and internal forces (the new psychology) acting between every pair of particles'.[81]

In 1923, in an essay, 'Psychology and Communism', he predicts that the place of economics as the 'chief scientific basis of Communism' will in the future 'be taken by psychology'.[82] Psychology is considered to provide a framework for the analysis of the 'mechanism of man's reaction to the environment' while the reactions themselves are 'the scope of economics and history'. Bernal seeks to 'formulate a mass psychology' by applying theories of individual behaviour to the activities of social groups. The result is an elitist view of social change. The 'mass of proletarians' are held back from revolution by the 'deference and submissiveness' caused by Oedipal repression, and 'most of the psychological motive power for revolution is contained in the hate aspect of the Oedipus complex although it requires special circumstances to set it free'. The special circumstances are created by a social group that although 'always very small' is 'of considerable importance'. This is 'the rebels' who have diverted their Oedipal hatred against 'teachers, employers, clergymen, governments'. This group is the source of political power for social change: 'However small this revolutionary nucleus is, it may succeed in liberating the whole of the repressed energy of the great bulk of the revolutionary proletariat if it is sufficiently bold and thorough. In this lies the power of extremist or Bolshevist activity.' Political change is not seen as a rational process since 'half of men's activities were dictated by the unconscious'. Communists 'implicitly recognized' this to be the case in 'abandoning the parliamentary labour attempt to persuade their enemies that they are in the wrong'. Instead of persuasion the objective of politics should be emotional identification: 'One of the chief aims of the Communist Party . . . is to promote in the bulk of the proletariat this identification of themselves with each other and with a suppressed class struggling to be free.' Bernal felt the traditional Marxist term 'class consciousness' to be 'rather unfortunate' since the binding strength of a group is 'emotional, not consciously rational'. The Communist Party itself had a special status, in Freudian terms: that of a 'free group': 'It is sufficient for the small nucleus of Communists to be class conscious in the full sense of the word, in order to guide the great proletarian group along the paths of constructive revolution.' The repugnant consequences of such a role are shown in Bernal's espousal of techniques for mass manipulation: 'The present aim of all Communists in all bourgeois countries should be to form the proletarians into a group by the use of suggestive propaganda, the simpler and less intellectual the better, to assist in the identification of (a) the capitalist class with the adversary, the tyrannical father, and (b) the proletariat themselves with the leaders of a revolutionary party.'

Russia is seen as the 'invaluable [and] successful example' of this approach in its 'promotion of . . . a cult of the personality'. The distasteful by-product of the 'love' of 'the leader such as Lenin' is that it not only 'holds the group together' but also makes its members 'capable of regressing to a primitive form of behaviour' and likely to turn their hate feelings 'more particularly against people belonging to closely allied groups'. Such features are justified by the need either to hold onto power ('the fruits of the revolution could never have been retained') or to overcome disunity: 'a united front is only possible when the hatred of the external enemy is sufficient to cause the internal hatreds to be repressed'. Political manipulation of the masses is justified if the objective is liberation: 'such exploitation is only excusable if the party in power makes use of it, not to perpetuate its own dominion, but to free the proletariat from their mental dependence'.

These views of a 23-year-old student reveal an enthusiasm for the power of mass psychology with little awareness of its dangers. Even at the time, his friend Dickinson expressed scepticism at Bernal's optimism about the benevolence of the Communist Party.[83] Nevertheless, Bernal did recognize that members of the party might be afflicted by some of the 'neuroses' of the society in which they live: 'the Communist cannot help to cure society while he himself is ill. The Communist must first face the facts of his inner life and it is here that the new psychology comes to his aid.' Bernal was evidently not comfortable with the characteristics of some communist activists and felt that psychology could help them change. On balance, however, psychology began to be seen as instrumental for the party's influence over the individual.

'I am a . . . relativist'

The fundamental shifts in Bernal's outlook on religion, politics and psychology in 1921 were accompanied by a new relativism in morality and science. He embraced a new moral relativism. In a heated discussion at the CUSS, Bernal opposed the existence of an 'objective good'. He was in the minority. His hostility to moral absolutes reflected his turn from Catholicism: 'Being purely rational it can have but little effect on conduct and the result of its application is the warping of souls that you see all round you, especially among religious people.'[84] 'Morality', he concluded, is 'not absolute' but 'determined' by the 'environment'.[85]

Like his political and religious outlook, during 1921 his view of science also began perceptibly to shift. The certainties of the positivism and materialism of nineteenth-century science became less convincing to him at a philosophical level in spite of his growing engagement with the practice of science. Early in 1920 Bernal had been struck by a lecture in which Eddington pointed out: 'that most of the laws of nature were laws of human mind projected on to realities'.[86] The relativism inherent in this view implied a far more human and fallible notion of science than that which Bernal had held before. It began to occupy a more central place in his own outlook. In the spring of 1921 he found himself arguing with Dickinson on the philosophy of reality. 'I find my views peculiarly clear. I am a monistic relativist. Everything is the same because we know nothing about it.'[87] Bernal's adherence to Communism was not, therefore, accompanied by the view of 'positive science' that he had read in Enrico Ferri's book the previous year. Instead his new philosophical outlook implied a more problematic role for science in the analysis of society.

'Scientists . . . limited views and conventional lies'

The intellectual and emotional turbulence of Bernal's days as a university student had contradictory consequences for his involvement in science. At the end of his first year he was devastated that he had only achieved a second: 'the hardest blow my self-esteem has ever had to bear. I was disgraced.'[88] By the end of his second year he had regained his confidence after shifting from mathematics to physics: 'In this glorious new world of science there may be some paths left for me to walk.'[89] By his third year he was busy trying to relate the role of science to his new political commitments. In 1923 he explored these views in an essay, 'The Failure of the Scientist'.[90] Although Bernal believed that science could usefully be applied to social issues – 'the intellectual powers of first-class scientific workers could . . . lead to valuable contributions to social theory' – he saw little evidence that this had been achieved with any success. 'Scientific men either never think clearly or independently on social subjects or when they do, are afraid to speak out.' Most scientists who think about politics will 'despair of doing anything in a field where there are two sides to every question and be baffled by the lack of data and exact laws and theories'. The result was initially utopianism: 'It is very easy to become impatient of humanity and build little modern utopias containing every possible improvement,' followed subsequently by managerialism and industrialism: 'when this rather childish stage ends as it generally does in disappointment, it is impossible for the scientist not to admire the organization and efficiency of great industrial concerns'.

Scepticism of the contribution of science to social understanding was reinforced by criticisms of the social position of scientists. Their origin from outside the working class was perceived as a profound limitation since they '[were] nearly all drawn from the . . . middle class [and] absorb all the limited views and conventional lies characteristic of that class'. Although scientists were growing closer to the proletariat in economic terms ('the present course of capitalist development [results in] a gradual sink in their status to that of a skilled worker'), their built-in ideological tendency to support 'the present system' blocked 'effective criticism or united action to raise the status of the scientist'. This conservatism was linked by Bernal to complicity in the social wrongs of capitalism: 'As they have made this age what it is, they must be prepared to accept a large part of the responsibility. War, famine, unemployment, sweating, workhouses, slums and prisons are all consequences of the present economic system; a system only made possible by scientific discovery.'

A consequence of this negative analysis was a rejection of science as an appropriate arena for political work. This was an explicit and conscious decision by Bernal since he was fully aware of two areas in which political intervention was possible: trade union organization and science policy. During his first year at university he had attended the inaugural Cambridge meeting of the National Union of Scientific Workers (NUSW) on 26 May 1920 where 'Professor Soddy, an energetic sort, speaks on misappropriation of research funds and preaches rank socialism very convincingly', and joined up as a member.[91] During the same year he heard a talk by Lord Haldane (uncle of the communist student J. B. S. Haldane), architect of science policy reform and in particular the creation of the Department of Scientific and Industrial Research.[92] He also became friendly with Henry Dickinson's father, who worked in the Science Museum, with whom he discussed 'problems of scientific administration and research organization'.[93] Bernal found it ironic that in a 'scientific age' the control of science lay 'in the hands of a relatively small group of persons ignorant of science'. He was pessimistic: 'the position of scientists is not likely to change materially' in the absence of 'catastrophic revolution . . . war [or] economic ruin'.

This perspective downgraded political activity directed specifically at scientists. The NUSW was written off as lacking both resoluteness and power: 'The union of scientific workers [is] a contradiction in terms . . . and . . . is doomed to relative inactivity. . . . A strike of scientific workers would be almost certain to fail.' Bernal also felt that there was 'very little to be expected' of ideological/cultural intervention 'through educational bodies'. Instead of such current activities, the focus shifted to future possibilities of socialist revolution. 'Then and only then will it be possible to apply scientific activity in wider human fields.' The issue for scientists would be a choice with a threat of coercion: 'The lot of the scientist will be hard. In Russia they backed the bourgeoisie and suffered accordingly.' Revolutionary change would need 'a new kind of scientist, one who has learned to think as clearly and fearlessly in human questions as in his own subject'. The conclusion of the analysis was that the socialist scientist should not try to organize fellow scientists but should politicize the workers: 'The current political role of the scientist, then, is to assist in the general political struggle and prepare to contribute to postrevolutionary conviction.'[94]

'People who dare and do things'

Evidence suggests that Bernal joined the Communist Party of Great Britain in 1923.[95] His relationship with Eileen Sprague, whom he married in June 1922, brought him into closer contact with local Labour politics. 'She . . . spoke of what she had seen in the Central Ward. Being a coward and an idler I have no arguments with people who dare and do things. I can only admire them and agree with them.'[96] In October 1921 he argued with friends about the Poplar councillors, and at the beginning of November he participated in the municipal elections at the Chesterton committee rooms.[97] This was probably Bernal's first experience of practical politics outside the university. During his fourth year at university (1922–23) his horizons widened beyond Cambridge to focus on national left-wing politics. His friendship with the Sargent Florences[98] probably assisted this process but he increasingly also spent time in London and mixed with the metropolitan left. These contacts centred on the 1917 Club, where he lunched with friends such as Henry Dickinson and Maurice Dobb.[99]

Bernal's move to London to work at the Royal Institution in October 1923 located him firmly in the capital's left intellectual scene. There were links with the Labour Research Department and a host of new personal contacts ranging from friendship with the Kapps to dinner parties at the Russells'.[100] Membership of the Communist Party, in the absence of any specific role for intellectuals in its ranks, meant general party work through his local organization. His location in Soho during 1923 and 1924 did not apparently involve many demands in terms of activism, but his workload changed markedly in 1925 on his move to Bloomsbury. He became a member of the Holborn Labour Party which was a very active left-wing constituency branch. (At this time it was within the rules of the Labour Party for individual members of the Communist Party to join its ranks.) In the autumn of 1925, however, two events changed things. The Communist Party leaders were arrested and jailed for periods of six to twelve months; and the Labour Party annual conference voted to expel Communists from party membership. Holborn Labour Party along with other like-minded Labour groups resisted this policy through the formation of a national Left Wing Movement. Bernal became actively involved in the local Holborn Left Wing group and on 18 January 1926 was elected to its committee. A few days later he was proposed for membership of the Holborn Labour Party branch committee but failed to get elected.[101]

1926 was the year of the General Strike and Bernal, as an active Communist Party member, became embroiled in the struggle. The miners' strike started on May Day, and Bernal marched in 'The Great Procession' to Hyde Park to hear the General Strike declared; speakers included 'the maenad Saklatvala', Communist MP.[102] A local Council of Action was established by the Holborn Labour Party and the Holborn Trades Council. Bernal took on the role of street-corner orator, giving speeches each day from 6 to 11 May at Leather Lane and New North Street. Although he saw a 'few buses' on the streets on 5 May, by the following day, after an eight-mile walk through the East End to visit Eileen at a nursing home, he observed with satisfaction,

'All is quiet. Complete stoppage.' By 10 May, however, he was dispirited by events: 'A gloomy day. Government propaganda telling. I go to work and do not speak.' A brief flurry of renewed optimism the day after was followed by distress at the collapse of the strike on 12 May.[103]

Bernal became drawn into the internecine arguments and manoeuvres of the strike's immediate aftermath. After disagreements with his friend Clifford Allen he spent a day in solitary wanderings on Leith Hill at the end of May.[104] He continued to be deeply involved with the Left Wing group and the Holborn Labour Party, but increasingly he used terms such as 'plot', 'conspiracy' and 'cabal' to describe events within the organizations.[105] In July the Left Wing group's power within the Labour Party branch suffered a setback, and Bernal subsequently attended a meeting at which 'imminent dissolution' appeared likely.[106] However, at the next meeting in September he wrote: 'The Left Wing group does not break up but I am appointed leader.'[107] His new responsibility coincided with growing demoralization and weakness of the left and was not a happy experience. After the disaffiliation of Holborn Labour Party from the national organization at the end of 1926, he complained, 'The Labour Party saddle me with many unpleasant jobs.' These included 'addressing letters for the belated Miners Appeal'[108] but a lot of 'high feeling' was also expended internally on organizational struggles for positions within the Labour Party branch. His impulse on occasion was to 'attempt conciliation', but the conflicts ran too deep.[109] In February 1927 Bernal carried the Holborn banner to Trafalgar Square but was forced to resign his position in the Left Wing group shortly after as splits developed. After a meeting in March 1927, addressed by Ivor Montagu, his active involvement waned.[110] His spell in the hothouse of the London left ended with his return to a post in Cambridge in the autumn of 1927.

World of Thought and Feeling

The period in Bloomsbury left Bernal disillusioned with what he saw as 'dull years' of involvement in party politics.[111] It led to a 'disaffiliation out of politics except academically'. He had serious reservations about 'the Party attitude', and in particular: 'slurring of difficulties, bad working with non-communists, the Party line'. The sectarian politics of the time held little appeal for him and he admitted as a 'personal bias' his dislike for '(a) Party work, (b) Party members, (c) Discomfort, (d) Fighting, (e) Obscurantism'. These negative feelings were reinforced by the generally discouraging context for radical politics in Britain in the late 1920s: 'In England the resultant impression can hardly be other than depression and foreboding.'

Since to Bernal the importance of political influences was whether they pointed to 'hopeful and constructive things', he looked with renewed interest to the world outside and in particular to 'the countries that are being reborn'. In spite of his chastened view of the Communist Party, the Russian Revolution was top of his list as 'the greatest political phenomenon that is still with us'. He distinguished his attitude towards it from that of either the 'bourgeois' who saw it as 'a horrible tragedy . . . a bad dream' or the 'worker' who viewed it as 'a merciful release . . . a symbol not to be examined too closely'. He was realistic about the situation there: 'Russia is still poor, relatively inefficient and ruled by an oligarchy that relies on force.' However, the overriding positive feature was that it was 'the first attempt at an organized state . . . the revolution in itself would have been merely negative if there had not been . . . men who took it as their opportunity . . . to set up the socialist organized state'. This 'huge . . . unprecedented . . . experiment' justified the shortcomings, since 'that oligarchy having given to the whole people a hope, an incentive to a corporate life has enabled them out of utter disorganization and misery and in the teeth of a hostile world, to maintain themselves and to develop . . . their organized state.'

Bernal's views on organization and planning compared with democracy and liberty were contradictory. He lumped 'individuality and liberty' together with 'property and profit' as bourgeois 'ideals', and disparaged the 'pretence of democracy' in Britain. On the other hand the 'brutality' of the rise of fascism in Italy impressed him with 'the fragility of the safeguards with which democracy protects liberty . . . even in the capitalist countries', which suggests a more positive view of individual freedoms.

His self-consciousness as 'an intellectual bourgeois young man' led him to value 'a life of comfort, security and liberty' but also to accept a notion of sacrifice in the 'disquieting' choice to be made between the 'aristocratic and socialistic attitudes'. He was resigned to the fact that 'no society is perfect' but saw a 'critical division . . . between those societies which we reject and those to whose principles we can be loyal though we may object to many of its practices'. Capitalist civilization had 'sunk too low' to command such loyalty and it was 'the hope of Communism' to capture it. Bernal's continuing support for the new Soviet society was based, then, on loyalty to its principles rather than on endorsement of all its practices.

Bernal's outlook on world politics was far more optimistic than on domestic politics: 'it furnishes a spectacle of interest unequalled for centuries'. While conscious of the significance of fascism as a form of 'violent reaction' using 'an organized party as the basis of its power' with claims 'to control the economic machinery', he saw it as 'unimportant . . . except as an irritant'. This view was based on its 'national' character and that fascisms in different countries were 'bound to be in opposition'. His greatest hope was for China and 'the sure beauty and sanity of its civilization. . . . Once China could break from the web of Western financial and political control, and make Western science and technology her own she would emerge as one of the greatest and most hopeful nations in the world and ready, if need be, to take on the torch of culture from Europe.' The form of liberation of developing countries would depend on the attitude of the Western powers. Opposition and oppression would arouse popular hostility, 'and so long as Russia is there to assist them, a modified Bolshevist type of state is likely to emerge'. Given that this was the usual approach this would 'in the long run . . . destroy the dominance of the Western capitalist powers'. Socialist change, while it had slipped from the British agenda, was still seen as consolingly active on the world stage. His commitment to Communism remains evident, though it is tempered by an awareness of fragmentation: 'the new religion of Communism . . . has passed rapidly its Protestant and internal political stages; it is having its initial heresies and schisms'. The division of the world into Communism and capitalism is also seen as a feature persisting uncertainly and perhaps detrimentally into the future: 'there is nothing in history to show which will win, or whether they will continue side by side in a conflict that intensifies their enmity and resemblance'.

His view of Marxist theory remained very close to the one he had developed in the early 1920s. 'Economic determinism' was seen as its core and viewed as a potent theory: 'one of the most powerful mental weapons the nineteenth century has given us . . . every social and political movement, every development in art or religion or science, can be viewed from the Marxian standpoint, be made to reveal their origin in the economic environment'. Yet, as earlier, he rejected an exclusive reliance on economic determination and was 'sceptical of the attempt to draw all knowledge from it with the more extreme Moscow school' and critical of the 'more ignorant Marxists . . . claiming the supremacy of the economic factor'. His criticisms were directed at simplistic ideas of economic determinism: 'it forms an integral law to which all human activity must statistically conform but which need not appear anywhere explicitly as a causative factor'. Insufficient in itself, 'it is necessary to supplement it with a psychological social theory which will explain individual relations in society' and, as before, 'Freudian psychology probably holds the key'. As in his earlier writing, categories such as 'class' and 'state' were seen as economically derived 'inclusive terms' and rather than seeking to elaborate these more effectively as 'political' concepts, Bernal focused instead on the individual as a psychological concept.

The practical application of science was seen as the key to material progress and machines were welcomed as showing that 'work . . . is essentially unnecessary'. In spite of his criticisms of the 'cultural sterility and political conservatism' of America, Bernal enthused about its positive side: 'the most complete use of natural resources and machines to the increase of comfort and the elimination of drudgery . . . human control of the material environment [of which] the rest of the world, including Russia, stands in admiration and envy'. Technological progress was seen as a great virtue in giving 'a sense of human power and self-confidence as against one of helpless dependence on an unknown and generally hostile universe'. He had little time for the 'host of Anti's (-vivisectionists, -vaccinationists etc.)' and saw the 'vague uneasiness' about technological change, or even 'open hostility as foreshadowed by [Samuel] Butler', as due to 'individual attitudes . . . and secondary effects'. Behind these was 'a fundamental contradiction of the twentieth century: the contrast between the human control of nature and the human control of society'.

Bernal shared his enthusiasm for technology with that for scientific inquiry itself: 'the cold, abstract, analytical mind who will not scruple to violate nature so that she be understandable'. Yet he saw scientific inquiry as only one mode of achieving understanding and argued that there are two kinds of truth: 'the inner truth of mystical intuition, something so purely felt that it is lost in words, and the outer truth of externals formulable through the abstractions of scientific language'. It was important to avoid the 'common confusion' of these two modes: 'the absurdities of the projection of mysticism into the world of experience, or the equally ridiculous scientific nihilism which proclaims the nonexistence of transcendental essences and qualities which were implicitly removed in forming its abstractions'.

As well as making this distinction Bernal elaborated his earlier realization that science is a human construct and not an absolute or objective account of nature: 'causality is an anthropomorphic fiction . . . the most dangerous mental occupation is the manipulation of symbols mistaken for phenomena ... the most scientific method is in reality least so for it is not in our power to exorcise our own demons of unconscious prejudice'. Theories in physics derive their importance 'not by their correspondence with an objective world which they represent badly but by their correspondence with other associations of ideas'. However, the practice of physics is different: 'it is completely objective, giving clear unambiguous answers'. This distinction between theory and practice was important to Bernal; but the main thrust of his epistemology was to reinforce a tendency to relativism and a 'scepticism of human mental activity': 'No human proposition can be either true or false. It must ... contain a mixture of both.' As a result: 'the hollowness of rationalism itself is exposed. Unsuspected motive becomes apparent beneath the most objective statements.' He con- cluded: 'We must abandon finally [the] idea of the rational soul.' This was not seen as a legitimation of the irrational; instead, that 'we are in our humility much better armed against our own ignorance. We know that we are never absolutely right but at the same time we can see in which way we are likely to be wrong.' Nevertheless on the basic conflict between materialist and idealist theories of the universe, Bernal believed, 'at present no one can have any reason apart from tempera- mental for choosing either alternative'.

'The aristocracy of scientific intelligence'

During the late 1920s Bernal's political attitude towards intellectuals in general and scientists in particular changed significantly from his dismissal of them in 1924 as a conservative and fragmented group. He now saw them as likely to share his desire for a powerful, organized state: 'students of science ... see most clearly the need for a unified and controlled world economic machine'. In addition the 'corporate feeling' that he had found 'most completely in the Communist Party' was also perceived as present among intellectuals: 'it exists in a vague way among all those who work by the scientific method ... a company vast, indefinite, confusedly organized but, in an age of fading loyalties, more and more attached to truth and to each other.'

His growing detachment from active socialist politics was accom- panied by an increasingly favourable disposition towards scientists acting as a political force. This is shown in an essay, 'The Growth of Organized Research',[112] in which the technocracy movement of the time is viewed with favour: 'The effect of allowing even a small body of

scientists to draw up schemes ... under official auspices has already
been shown by the case of the technocrats in America who have been
led to condemn the whole of the present economic system as incompatible with any adequate utilization of scientific discovery.'

There were similar elements in his first book, *The World, the Flesh and
the Devil*, published in 1929.[113] This book was Bernal's first public
expression of his views on social development. It is an enigmatic work
since, although it draws upon his intellectual trinity of science ('the
physical sciences'), Communism ('history'), and Freudianism ('the
knowledge of our desires'), these are applied ('for brevity') to only
three fields: 'nature' (the World), 'human biology' (the Flesh) and
'psychology' (the Devil). In contrast to much of his unpublished
writings of the time, the social world of economics and politics enters
side- rather than centre-stage.

Nevertheless his political observations are consistent with a less
exclusive attachment to socialism. The satisfaction of basic human
needs can be resolved through either 'rationalised capitalism or Soviet
state planning'. The 'mass of the people' along with the 'ruling class'
exercise 'a secure hold on tradition' to the detriment of 'genius'. The
future lies with the elevation of the 'scientific expert' to the point
where 'the ruling powers are the scientists themselves'. The path is
recognized by a variety of 'new nations: America, China & Russia',
whatever their social system, though he still reserves enthusiasm for
'proletarian dictatorship' and the 'Soviet state' as probably necessary
for 'the real independence of science' and for 'scientific institutions
[to] become the government'. Along with these more restricted claims
for the benefits of Communism goes a strengthened commitment to
the value of technical progress.

Central to Bernal's concept of 'progress' is 'the replacement of an
indifferent chance environment by a deliberately created one' and an
'ever-increasing acceleration of change'. In contrast to a 'static' and
'stale' future he espouses one of 'daring' and 'danger'. 'The dangers
to the whole structure of humanity and its successors will not decrease
as their wisdom increases, because knowing more and wanting more
they will dare more, and in daring will risk their own destruction.' This
is an approach to life that he shares as positive and necessary: 'This
daring, this experimentation, is really the essential quality of life.'

He does not deny the 'quite real distaste and hatred' aroused by
technological change and admits to such feelings himself in response
to his own forecasts of space globes and 'disembodied brains'. However,
on the crucial choice between 'humanisers and mechanisers', he

declares allegiance to the latter. He obviously took such 1920s critics of technology as Aldous Huxley seriously but, on balance, chose to reject their fears: 'They may be prophets, predicting truly the doom of the new Babylon, or merely lamenting over a past that is lost forever. . . . With these uncertainties before us each must follow his own desires, accepting that his opponent may be as right as himself. The event will show which but only after his own time.'

As to the structure of society, Bernal's view was that 'originality, organising power and industriousness' were 'very evenly distributed' and that social progress would continue to depend on 'the aristocracy of scientific intelligence'. Although this would need to be recruited from 'wider and wider circles' there would be a fundamental division of labour: 'A happy prosperous humanity enjoying their bodies, exercising the arts, patronising the religions, may be well content to leave the machine by which their desires are satisfied in other and more efficient hands. . . . discoveries . . . will give . . . the means of directing the masses in harmless occupations and of maintaining a perfect docility under the appearance of perfect freedom.'

Despite the mischievous tone of speculation in the book, the overall argument is a serious one. On balance, Bernal chose to believe in the beneficence of technological change and the benevolence of the scientific expert. Yet to reduce his outlook to this would be a caricature. In his other writings of the late 1920s he rejected positivism and espoused a social outlook in which historical analysis ('more and more essential in discovery') and integrative synthesis ('the comprehensive ordering of the world of experience is . . . the most important for the individual. It gives the universe . . . meaning.') were central. This set him apart from the mainstream technocratic tradition in spite of his moves to share in it. Even at the heart of *The World, the Flesh and the Devil* lay an explicit rejection of a simple 'rationalism' and of the 'naturalism' that sought to displace it. Instead, Bernal's own desire was for social change, to enable people to 'find the capacity to live at the same time more fully human and fully intellectual lives'. At core, his social outlook strove to integrate social, intellectual and emotional themes.

By the end of the 1920s Bernal had a decade of political experience behind him and had devoted considerable thought to where he stood in terms of social and political analysis. He continued to hold together outlooks derived from Communism, science and psychology yet experienced many tensions between them. At the time he defined his 'present problems' thus:

Freudian v. Communist
Scientific v. Communist
Scientific v. Freudian[114]

It was the impact of political events over the next decade that shaped the resolution of his dilemmas.

'Subordination . . . to a revolutionary class movement'

Bernal's drift from Communism towards technocracy in the late 1920s was rudely halted by the onset of the Great Crash and the subsequent political crisis. This had a series of effects on his outlook. It led to his renewed interest in Marxism's emphasis on the economic factors in history and a reinvolvement with the Communist Party. He rejected the independent organization of intellectuals and redefined the role of the individual.

Bernal now began to define the period 1921–29 as simply a 'temporary stabilization of capitalism'[115] and questioned the direction his ideas had begun to follow during this period. In particular he gave renewed significance to the role of economics. 'It was possible [during the 1920s] . . . to dispute the importance of the economic factor, but no-one . . . can do so today.'[116] The visit of the delegation led by Bukharin to the London Congress of the History of Science in 1931 reinforced this shift and he found the Soviet contributors 'impressive' in their argument that 'political and intellectual history are two branches of economic history'.[117] He emphasized the analysis of 'economic necessities' as the 'great generalising principle of Marxism' and began to study with greater enthusiasm the more detailed features of Marx's economic theory of historical change.[118] Bernal's renewal of interest in Marxist theory was less evident in relation to philosophy, however, than in relation to economics and history. Dialectical materialism did not initially appeal to him. He described the Russian visitors as 'a phalanx uniformly armed with Marxian dialectic'[119] and found that its 'abstract' and 'bald' presentation aroused 'justifiable suspicions of empty pedantry'.[120] And while the economic crisis stimulated a new interest in Marxism, the political crisis rekindled his commitment to the Communist Party. As we have seen, during the late 1920s he had admitted his dislike for many features of Communist politics that he had experienced through active involvement in the Communist Party. In 1931 he summarized the arguments presented by those who 'sympathise with the ends of communism but object to the means':

> Communism is dogmatic and doctrinaire . . . it reproduces the worst features of revealed religion. To accept it is to abandon the right to free thought. . . . The emphasis is always on the mass, not on the individual. . . . The Communist Party is antidemocratic, dictatorial and tyrannical in its government. Is political liberty, won after so many years of struggle, immediately to be abandoned? The membership of the party does not inspire confidence. The Russian members are admittedly heroic but ruthless in politics and muddled and inefficient in economics, while those in bourgeois countries are plainly both fanatical and neurotic. The policy of the party is unnecessarily violent, the emphasis throughout is on hate not love. . . . The propaganda is virulent and overstated . . . the communists invoke revolution, the very prospect of which with its attendant civil wars and famine is too horrible to be borne.[121]

The fascinating thing about Bernal's riposte to these concerns is that he seeks not to deny them but to justify them. In the first place, he argues, the values that the intellectual seeks to uphold no longer exist: 'the intellectual freedom, the peace and harmony with which he is loath to part are . . . even at present, empty illusions'. Second, the 'most criticised' characteristics are those of 'any party of action rather than inaction'. As a result the Communist Party is bound to express characteristics with which the intellectual will not be comfortable:

> With a party striving for fundamental change . . . the effective human driving force must be composed of those who have nothing to lose and know it. Hence the insistence on class fighting and hatred. . . . From its very nature such a party will contain unstable and fanatical spirits, so that the danger to the movement is much more from compromise and schism than direct outside opposition. Hence the necessity for rigid doctrine and close organisation.

The experience of the intellectual within such an organization will not be a pleasant one, since it will involve 'suspicion and dislike of the workers, contact with them which may be awkward and troublesome, and . . . much tedious and apparently unprofitable work in propaganda and organisation'. One should not expect a warm welcome, since 'Idealists from other classes, however disinterested, are suspect because of their traditional and personal cross loyalties.'[122]

Overall, the image of the Communist Party presented by Bernal in the early 1930s is hardly an attractive one and does not conceal some distaste for his own experiences within it. However, whereas this had led him in the late 1920s to emphasize the importance of the individual or the independent role of intellectuals, the new circumstances now led him in the opposite direction. He now felt that a choice had to be made between political influence and personal reservations:

> Is it better to be able to think of schemes of one's own for the betterment of humanity without the slightest power to put them into practice, or to join in a scheme of which one does not altogether approve, but which is being attempted and for which one can work wholeheartedly?

This choice was perceived by him as one of sacrifice and subordination. Pleasant things such as 'comfort, work, sexual relations' were to be replaced by unpleasant ones: 'persecution, loss of work, imprisonment'. What was required was 'subordination . . . of individual opinion . . . subordination to [a] revolutionary class movement'. The compensation for this would be 'a kind of extension of personality':

> The most fundamental effect on the individual of an understanding of Marxism is to make him see and feel his place not only in contemporary society but as an active part of a definite historical process . . . there is a complementary gain in the appreciation of the individual's function in social change.[123]

His dilemma of a few years before of 'Communism v. Freudianism' was being explicitly resolved in favour of the former: 'The whole complex of individualism, self-development and self-expression loses its overwhelming importance . . . these are no longer advanced but reactionary aspirations.' At the same time he repudiates some of his earlier views on the importance of individual 'rebels' in promoting social change: 'the influence of individual protest and exhortation has probably been greatly exaggerated'.

He also now rejected the collective technocratic path, on two grounds. The first was a recognition that knowledge *per se* is not power in a straightforward way: 'if truth is not compelling' then intellectuals are 'powerless'. The second was that the social diversity of intellectuals was more important than their similarity:

> Intellectuals form a small close-knit international society. They have always had more in common with each other than with the people among whom they live. . . . But this unity in contemplation and appreciation covers an even greater diversity in action. Individually every intellectual is tied to his particular environment, to his nation, party and class. . . . Groups of intellectuals . . . nearly always find it impossible to discover amongst themselves a common basis for effective action.

The positions that Bernal was adopting at this time were a response to the gravity of the post-1929 economic and political crisis and, at one level, acknowledge the limits of the individual intellectual approach and the importance of collective and political action. Unfortunately, this process of rethinking occurred in the context of a very sectarian and dogmatic phase of Communist politics, the so-called Third Period of 'class against class'. As a consequence, Bernal's reasonable contention that 'an intellectual can only be effective if he ceases to be exclusively concerned with his work or fellow intellectuals, and mixes as a citizen in the political and economic life of the time' is disfigured by a simultaneous assault on the values of individual freedom: '[The] ideals [of] peace, democracy and freedom of thought [are] delusive fantasies.'

His analysis of the contradictory social position of the intellectual is a sophisticated one: 'an intellectual belongs to a stream of tradition and is tied past, present and future to other intellectuals. But he is tied not to them alone but to political society and material civilisation.' Yet the subtlety of this immediately gives way to an overriding economic determinism: 'The intellectual . . . is usually . . . a parasite of the really functioning capitalist . . . in practice this loyalty [to class] is alone effective.' One reason for this was that the Communist Party's credentials had been enormously enhanced by both the economic vulnerability of world capitalism in the wake of the Crash and the political turmoil surrounding the Labour Party following Ramsay MacDonald's defection.

A second influence was the positive example of the Soviet Union, which Bernal visited in 1931. This was not uncritically glorified – 'Fourteen years of Russian experience shows how complicated and far-reaching the process is' – yet its practical problems of reconstruction were seen to show that the negative features of communist revolution would ultimately recede: 'they will be worn away, as they have in Russia, by the necessity of dealing responsibly with concrete reality'. This process of construction had a powerful appeal in contrast with the position in Britain: 'If there were any attempt at the ordered reconstruction of society, if all existing knowledge and technique were being applied to cope with manifest evils and if all were sharing equally in the labour and the reward, the path of the intellectual would be straight and easy. It is so in the Soviet Union. . . . Here our minds as much as our machines lie idle.'

A third factor was a more general view of the declining importance of the individual intellectual in society. This view was based both on the growing complexity of research ('larger and larger scale, the individual less and less important') and on the idea that knowledge was not purely rational: 'political and economic interests distort men's whole attitude to knowledge. There is no pure knowledge. There is no free private judgement.' Liberal ideals were seen to command less and less allegiance: 'for a comparatively disinterested intellectual, the abolition of exchange economy and the necessary substitution of a party dictatorship for parliamentary democracy, and of a world state for nationalist anarchy, are ideas relatively easy to accept.'

These three factors reinforced the Communist Party's theoretical arguments and kept them at the centre of Bernal's rethinking. In many ways, the analysis of the status of the individual and the position of the intellectual offered by the party was no different to that of the early 1920s, when Bernal first joined its ranks. At that time, as we have seen, the acceptance of the analysis was accompanied by a rejection of political work among scientists in favour of general activism in the party. A similar view still prevailed in the party, but the context was dramatically different.

The economic crisis had a radicalizing effect on many intellectuals and produced what Bernal described as 'shoals' of groups interested in political and economic action. This created many new opportunities for political work by Marxists among students and professionals, and Communist academics were quick to get involved. During 1932, Bernal played an instrumental role at Cambridge in reviving the Association of Scientific Workers and setting up the Cambridge Scientists' Anti-War Group. The result was a stern lecture by R. Palme Dutt in *Labour Monthly* in the autumn of 1932 that there was 'no special work and role of Communists from the bourgeois intellectual strata' only 'the general all-round role of the Communist' in 'the winning of the workers'. The intellectual who had joined the Communist Party was instructed to 'forget that he is an intellectual (except in moments of necessary self-criticism) and remember only that he is a Communist' in order to concentrate on the 'struggles of the workers'.

It is not clear what exactly happened to Bernal's relationship to the Communist Party during this period. He may have lapsed in the late 1920s, and Magda Phillips recalled recruiting him in 1933. Bernal himself wrote of losing his card in 1933. In practice, he maintained close links with the party leadership[124] but was not a public member of the party from the early 1930s on.

'No truce in ideas'

1933 saw the rise of fascism, with Hitler coming to power in Germany. This confirmed Bernal in his commitment to political work among scientists and intellectuals, a process in which his contacts with French intellectuals such as Langevin played an important part. In an article, 'The Scientist and the World Today', published in the Winter 1933–34 issue of *Cambridge Left*, he described the 'uneasiness' of many scientists on the questions of war, unemployment and fascism and urged that it be 'crystallised in discussion and organisation'. After his earlier vacillations on the social position of scientists, he now recognized this as contradictory but with enormous political scope: 'Scientists are in an anomalous position. Socially they belong to the capitalists, culturally to the workers.' He argued for an effective bond 'between worker and scientist'. The 'anti-intellectualism' and 'barbarism' of fascism made him more positive about some of the achievements of 'centuries of middle-class culture'.

Bernal's decision in 1932 to concentrate his political energies on science and scientists was a crucial one. It created the opportunity for the creation of a vital social movement and the development of a novel analysis of the social function of science. The Communist Party only began fully to recognize the importance of this several years later with its recognition of the need for a wider social alliance of popular forces in 1937–38.

Interestingly, the five years between 1933 and 1937, in spite of being a period when Bernal vigorously pursued this independent political path, saw him embrace orthodox Communist theory more wholeheartedly than at any point before. This included strong hostility to the policies of all other political parties as 'shamefaced versions of fascism'. Two unpublished documents from 1934, 'A Criticism of the Manifesto of the Federation of Progressive Societies and Individuals' and 'The Scientist and the Contemporary World', sustain a rather intemperate and sectarian tone. All proposals for national planning are rejected and proposals for democratic planning bodies by groups such as the guild socialists are contemptuously dismissed: 'Control is no substitute for ownership.' The distinction between evolutionary and revolutionary socialism remains critical: 'There can be no truce in ideas' between them.

Bernal's dismissal of the 'forms of democracy' as distinct from the 'power of democracy', his reduction of the state to 'effective political control by producers for profit' and his appeals for a 'world economy', a 'world state' and a 'classless community' remained unchanged from the early 1920s. In practical terms what did this mean? First, an active support for the Soviet Union, which was seen as overcoming 'the major characteristics of capitalism' in spite of 'a grimmer side to the picture' arising from 'inadequate means'. Second, a principal commitment to education and analysis: 'Revolutionary activity is not a matter of conspiracy or insurrection. It is primarily an educational process, an evoking of a conscious appreciation of economic and political forces through the experience of everyday struggles, enlightened and integrated by revolutionary theory.' This remained his mission on the terrain of the politics of science.

The 1930s saw a reversal of the situation that Bernal had sustained during his earlier period of political activities. In the mid-1920s he had loyally played the role of party activist focused on the narrow terrain of labour movement politics, yet had retained an unorthodox and individual stance on matters of political theory and Marxism. In the 1930s he insisted on orienting his activism towards scientists and intellectuals, against the strictures of the official party line; yet, ironically, this was combined with the adoption of the new sectarian orthodoxy of Soviet Communist doctrine. The initial expression of this was the adoption and espousal of dialectical materialism. Bernal committed himself to this in a lecture organized by the Society for Cultural Relations with the USSR in 1934. He argued that 'dialectical materialism is the most powerful factor in the thought and action of the present day'. He was careful, though, to avoid suggesting that it could inform the specific content of scientific knowledge: '[it] is not a formula to be applied blindly either in the natural or human world', or 'a substitute either for experimental method or for the logical proof of laws or theories'. On the contrary, he argued that it could be 'derived from . . . science'.[125] A year later he began to write on Engels's philosophy in relation to science.[126]

In many ways it is possible to see why dialectical materialism held a deep appeal for Bernal. Its concern with the 'origin of the new' and the 'unity of thought and action' touched key elements of his thought which had been long evident. More problematic was the integration of the traditions of dialectical materialism regarding contradiction and the role of the individual. The philosophy seemed to offer a resolution of the conflict of opposites that had constantly preoccupied Bernal in his social outlook – their 'union' and 'transformation' were now promised; but the danger was that less recognition would be given to the insecurity and uncertainty associated with knowledge. This was illustrated by the central contradiction that Bernal was concerned with, that between the individual and the collective. The implicit subordination of the individual that he had adopted in 1931 was legitimized in the new framework. Although he had no doubts about the existence of 'the man of remarkable genius', such a person could only exert influence by 'necessarily being part of contemporaneous movements'.

This new trend in Bernal's thought went against his earlier concerns with the individual and psychology which had been expressed in his interest in Freud. The adoption of dialectical materialism was followed by an explicit and public repudiation of Freudianism. In a remarkable article, 'Psychoanalysis and Marxism', written in 1937, he repudiated all the ideas to which he had subscribed in his mid-twenties.[127] Rather than the 'psychological' and the 'material' aspects of humanity being 'reconciled in a dialectical way as two opposites', Freudianism was just one more form of subjective philosophy and 'must be understood and rejected as such'. It sought to 'set up the individual ... as the centre and measure of all things', while 'the essence of Marxism' was its provision of 'an objective and scientific picture of the processes of social change'. A recent book attempting to reconcile Marx and Freud 'wanders from the path of Marxism' and Freudianism 'is a profoundly dangerous influence, paralysing action and tending to fascism'.

The passage of his own outlook was subsumed beneath a general observation:

> In recent years the Freudian wave has begun to recede. The effects of the world economic crisis of capitalism, and of the close menace of fascism and war, startled the intellectual strata into awareness of the objective world, and aroused a new wide interest in Marxism.

This of course expressed the influence of these world events on Bernal's own trajectory. The expulsion of individual psychology from his social thought partly resolved his earlier agonies of choice between Communism, science and Freudianism. There was a price to be paid for this, but it also served as a release, allowing him to focus on his twin interests of politics and science.

In the mid-1930s Bernal began to elaborate a more systematic critical analysis of science within capitalism. In 'The Frustration of Science' and his chapter of *Britain without Capitalists* he explored the dynamics of the relationship between science and industry in a capitalist economy and began to articulate an alternative vision of science harnessed to welfare rather than profit.[128] His analysis of science was beginning to appear analogous to Marx's view of the proletariat, that is, brought into existence by capitalism but ultimately the source of an alternative society. His political attitude towards scientists had become more positive: 'Only a relatively small number of influential scientists are definitely reactionary'; in the new society, the 'intellectual worker' would be given special status. Science was restricted within the present system: 'Under capitalism . . . invention is very haphazard and is carried on in a frightfully wasteful way . . . improvement is deliberately slowed down.' After 'the revolution' a 'rational' system would lead to 'an increased tempo of advance', and the 'central planning authority' would maintain 'the most harmonious and rapid possible develop-ment'. Science would be an 'unmixed blessing' in the new society.

This new society was defined as 'a Soviet state', initially a 'Soviet Britain' but ultimately 'world-wide' and reflected the Communist Party's stated aims represented in the 1935 programme 'For a Soviet Britain'. Although Bernal refers to 'the political stages through which the revolution in England is brought', the form of these is not discussed and there remain echoes of the abstract revolutionary rhetoric of the earlier period. Such a stance was becoming 'increasingly untenable' for two reasons. One was the growing role Bernal was playing through the institutions and organizations of science and the increasingly detailed programmes for reform that he was elaborating. The second was the looming recognition that the defence of a democratic system against the international menace of fascism was starting to overshadow other objectives of social change. The adoption of Popular Front politics by the Communist Party in the latter half of the thirties, the Spanish Civil War, and the growing involvement of scientists in issues of UK national defence presaged a fundamental shift in political perspective.

This was first expressed in Bernal's original and influential second book *The Social Function of Science*. 'Collaboration between the forces of Science and Democracy throughout the world' is now given much more prominence than the relationship of science with communist revolution. Capitalism is still identified as the key principle of social organisation but seen less as overridingly determinant of every aspect of human activity: 'autonomous traditions of religion, literature and science have grown up, but they are in the last instance dependent on their fitting in with the general scheme'. The politicization of scientists is no longer tied to the adoption of socialism, and Bernal becomes much more critical of sectarian party politics.

The scientist . . . must become a politician, but . . . never . . . a party politician. Only when the parties can get together on a broad programme of social justice, civil liberty and peace, can the full help of the scientist be expected. The scientist . . . can best help by making no exclusive commitments and assisting all progressive parties without favour.

Bernal's own commitment to Marxism remains explicit and strong, though in his interpretation science has come to occupy centre stage. 'It is the chief agent of change in society . . . we have in the practice of science the prototype for all human common action . . . in its endeavour, science is communism'.

A New Age

The onset of World War Two drew Bernal much more closely into an organized national endeavour and led to a reassessment of the role of the state in a capitalist society.

Now, in wartime, we are coming naturally to think and act in terms of directed economic and social organisation. . . . In every industrial region of the world today . . . there exists a form of planned economy. . . . More and more the capitalist states are showing an external similarity in their means of control of production and distribution to the planned socialist economy of the Soviet Union.

This signalled a 'new age' with 'a new awareness of the unity of all human societies and of the practical possibility of . . . cooperation'. This was due to the increase of global 'communications' and 'economic interdependence' and heralded the supersession of 'liberal capitalism' by 'a socialist world order'.

The emergence of this more positive globalistic outlook was still tied to a critique of the democratic structures and values of the capitalist societies. The espousal of 'individual rights' as a political objective in itself 'has no meaning in the modern world'. They were not to be abandoned but could 'only be secured by first attending to the change of society as a whole'. Their achievement was 'not merely a political act, but an economic and technical one as well'. This viewpoint led to a critical attitude to democracy in Britain. 'The ruling class of this country did not go into the war to save democracy; they do not know what democracy is and if they did they would not like it.' Fascism and democratic capitalism are still seen as two sides of the same coin: 'monopoly capitalism, whether fascist or plutocratic'. 'The rise of

fascism was simply the counterpart in less prosperous countries of the
. . . pluto-democracy of the more prosperous ones.'

An alternative form of democracy was required which drew upon the
Soviet model. It would be 'a different kind of liberty' in two respects.
First it would be a direct form of democracy: 'working together in small
groups . . . is, rather than the ballot box or the party system, the essence
of democracy'. Second, that individual choice was subject to a broader
social purpose: 'it . . . works only by virtue of belonging to a larger
scheme . . . coordinated in a planned way for a common purpose. . . .
It will be an ordered and not anarchic freedom.' The implication of
this analysis was that concepts of 'rights' such as liberty, equality and
fraternity should be considered as 'relative' rather than 'absolute'.
Philosophical values were 'relative and determined by social forms . . .
classical values of truth, beauty and goodness depend for their meaning
entirely on the existence of society and change their character with the
change of that society'. 'Scientific truths' as well as social values were
'also relative in a different way . . . [not] static'. 'Relativity' is . . . the
only "absolute" there is.'

The conclusion from this analysis was twofold. First, the values and
ideas of the old system were unsatisfactory and should be changed. 'In
a really new order all the old ideas will be transformed.' Second, this
would involve a 'radical break' and 'violent change', not a process of
reform: 'for years the organised working classes, the Trade Unions and
Labour Party [thought of] social change as a gradual process. . . . Trade
Unions and Labour leaders abandoned any hope of power or even of
radical improvements' and concentrated instead on preventing any
movement that would endanger the position of restrictive monopoly
capitalism. Anything, they felt, would be better than violent 'change in
which working class privileges or their own positions might be lost'. By
1942, then, Bernal's perceptions of a new globalism and common
purpose was still combined with an outlook on politics that embodied
long-established features of Communist politics. These included a
rejection of any positive continuity for the democratic forms established
within capitalism and a marked hostility to the non-Communist left and
labour movement.

'Democracy . . . an absolute value'

By 1945 Bernal's outlook had dramatically changed, under the influence of cooperation between different political forces during the war and the 'fearsome apprehension' resulting from the atom bomb. His new views are expressed in his writings in *The Freedom of Necessity*. In contrast with 1942, he now placed much greater value on the positive features of democratic society ('liberty and democracy must be secured. . . . Our British democracy, from long practice, does enable us to secure the people's will without coercion or bloodshed') and the need for cooperation of different political trends ('A large measure of collaboration between people of different political and economic opinions will be necessary'). Certain values are given a more secure status than merely a relativist one: 'Over and over again, and never more than in these last years, the common feelings of justice, fellowship and liberty have reasserted themselves . . . democracy has an absolute value.' Bernal now writes of beliefs which are 'solidly materalist but . . . none the less humanist'. This humanism entails 'a respect for human individuality' which he acknowledges 'can also be reached emotionally and is embedded in the framework of all great religions'. 'The greatest crime in the world is . . . depriving man of his inheritances of thought and the possibility of full and constructive expression of it.'

This new emphasis on democracy and the individual represented a change in the balance of his ideas rather than in the fundamentals of his overall outlook. He still stressed the importance of the 'economic' as well as the 'political' side of democracy, along with the need for 'a balance' between the individual and the collective, avoiding the extremes of 'anarchy' and 'tyranny'. There is a renewal of some of the ideas that he had begun to express in the late 1920s and in particular the view that there was more scope for significant economic reform within capitalism than he had conceded at any time since 1930: 'Even the capitalist system, however, can be made to organise production rationally . . . in time of war. With the same controls it can be made to do so in peace.' The realization of this 'under a capitalist economy . . . is a difficult but not impossible task', though it was still seen as only a step 'pending its reorganisation on more rational lines'. The ideas expressed in 1945 amounted to a new perspective on democracy, reform and political alliance.

Yet Bernal's basic political framework was not explicitly recast to accommodate them, instead they were incorporated incrementally within it. This was evident in two particular respects. The first was his continued adherence to the Soviet model as the exemplar for social change. This limited the perceived applicability of the Western democratic model to other societies. 'In countries with a long history of tyranny and feud such democracy is unrealisable.' The second was his espousal of dialectical materialism as an individual, alternative world-view. This was combined with continued polemic against the 'unreal and useless tenets' of liberal philosophy which sustained the 'continuation of capitalism' by the maintenance of 'false beliefs'. The education system is seen as 'warped' and 'distorted' with no provision for the teaching of professed critics of the capitalist system. This remains an instrumental view of the development of ideas in society and springs from his view of a conflict between two systems of thought. Such contradictory elements remained present in Bernal's outlook and were open to the influence of changes in the political environment.

Two Camps

The early fragmentation of the postwar unity of 'the great alliance of the United Nations' was to have such an effect. In a radio broadcast in 1946 is evidence that Bernal still held a positive view of political developments in Britain. 'Already in this country we have begun to make a decent, ordered human world. With social security, with a national health service, with new education, with new planned towns and factories and mines, we have something to look forward to.' But the divisions of the Cold War were beginning, and the danger was apparent: 'Unless we can stop them splitting the world into two camps in men's minds, the division will grow.' In fact, the development of the Cold War led Bernal to retreat from the new thinking he had expressed at the end of the war. By 1948 he was warning that 'Marxists always looked with suspicion on any reformulation of the original classics for fear that they might form the starting point of further revisions and softenings', and argued that 'outside the Soviet Union there is nowhere any intellectually reputable system of thought that has gained widespread adherence and is serving as an inspiration for action'.

If the world was splitting into two camps then Bernal was going to champion the 'socialist' side. Bernal's published writings and the positions he took on a variety of issues in subsequent years bore the imprint of his chosen partisanship during the period of international Cold War cleavage. The public positions he adopted on Lysenko and

Stalin were culpable endorsements of the official party line. He was also the recipient of the excesses of the period in his peremptory removal from the British Association's governing body following injudicious remarks about Britain during an overseas trip.

The stance that Bernal adopted in the late 1940s and 1950s had a direct lineage to the political outlook he had formed in the early years of the 1930s. This position embodied a loyalty to the Soviet model and adherence to an alternative world-view expressed in dialectical materialism. The framework remained premised on two conflicting types of social systems and thought. His books *Science in History* and *World without War*, both written in the fifties, represent a disconcerting combination of a vision that retained astonishing breadth and social relevance with a style of partisanship that resonated more and more narrowly in the new political landscape.

His *Marx and Science*, published in 1952, also bore the political marks of the time but interestingly involved a return to primary study of Marx's writings and politics.[129] He found new sources for the view that both nature and knowledge should be regarded as processes, and he sought to reconcile relativism with progress: 'a sequence of relative truths each representing a greater and greater understanding'. He rigorously explored Marx's writings on technological change in the Industrial Revolution to elucidate the relationship between science, industry and society.

Bernal's deep involvement in the peace movement during these traumatic years rendered him particularly vulnerable to the emerging schisms within the socialist camp itself, particularly in relation to Hungary and China. In both cases his sympathies began to tug him from his previous loyalty to the Soviet Union. His heavy international schedule in such a context was increasingly to take its toll on his health.

White Heat and Uncertainty

The 1960s were Bernal's last decade of active political influence. Ironically, but in keeping with many of the earlier periods of his life, it was rich in contradiction. One thinks of his independent social thought as a Communist during the 1920s which conflicted with party orthodoxy and had both psychological and technocratic aspects, and of the conflicts in his political practice within both independent movements of scientists and intellectuals in the 1930s and the government military establishment during the 1940s. In all cases he veered between attempts to modify theory to account for these experiences more convincingly and a continued reliance on Communist orthodoxy.

The 1960s saw new contradictions which were quite different in nature. The unexpected rise of Harold Wilson to the leadership of the Labour Party in the early 1960s opened a new opportunity for the promotion of Bernal's views on the planning of science. Like Bernal, Wilson had been left with a positive view of the potential of rational planning following his experience as a civil servant during World War Two. The ideas expressed in Wilson's famous 1963 Scarborough speech on the 'white heat of the scientific and technological revolution' bore the clear imprint of Bernal's ideas and his continuing practical political influence. His views now had the ear of the highest levels of government.

Yet the promise of new influence expressed in this development was accompanied by world events reflecting the decline of cohesion of Bernal's own chosen political formation. The international Communist movement continued to fragment and Bernal found himself less and less in sympathy with the Soviet Union. In an unpublished manuscript of 1965 he supported the new tendencies within the Communist movement 'to break away from a minority irreconcilable type of Communist Party and make a broader type of united front'.[130] He criticized Communist orthodoxy: 'thinking and expression thus tend to be extremely cautious and scholastic, any attempts at variation will be suppressed . . . the doctrines so carefully preserved cease to have any compulsive effects, still less any inspirational effects'. Yet his thoughts on these issues remained private: 'this essay could not be published anywhere . . . it would cause universal offence'. In 1968 what he called 'the stupid and illegal movement of Soviet military units into Czechoslovakia' led him to criticize 'the reactionary policies . . . of the Soviet rulers'.[131]

More fundamental to this re-evaluation of his long-standing loyalty to the Soviet Union was a questioning of whether the newly emerging global problems that really concerned him could actually be resolved by socialism as he had often promised. He began to find both capitalism and socialism lacking on the problem of feeding the world's hungry: 'to the capitalist powers it seems a shattering criticism of the Free World economics . . . the socialists believe it is an unreal [problem] which will vanish with the destruction of capitalism. Both ignore it in practice.'[132]

The raising of these basic questions in the personal writings of his later years expresses a continued willingness to engage with the nature of political movements and the social role of science which had been of fundamental interest to him from his student days. Bernal travelled a longer and more varied road in his political thinking from the 1920s to the 1960s than has often been recognized, by either his supporters or his critics.

Notes

1. J. D. Bernal, 'Microcosm', typescript, J. D. Bernal Archive, Cambridge University Library, B4.1.
2. Diaries, 8 November 1919.
3. 'Microcosm'.
4. Diaries, 30 July 1918.
5. Diaries, 18 November 1917.
6. 'Microcosm'.
7. Diaries, 31 May, 11 March, 1919; 29 November, 5 October, 1918.
8. Diaries, 29 March 1919.
9. Diaries, 2 March 1919.
10. Diaries, 29 March, 31 May, 7 June, 1919.
11. Diaries, 22 January, 29 January, 4 February, 1920.
12. Diaries, March 1920.
13. *New Cambridge*, 6 March 1920.
14. Diaries, 24 April 1920.
15. Diaries, 21 February 1920.
16. Diaries, 1 March 1920.
17. Diaries 3, 4 March 1920.
18. Diaries, List of Books Read, 19 April, 15 August 1920.
19. Diaries, List of Books Read, 1920.
20. E. Ferri, *Socialism and Positive Science*, ILP, London, 1906.
21. Diaries, 11 September 1919; 20 September 1920.
22. Diaries, List of Books Read, 1920.
23. Diaries, 3 February 1920.
24. Diaries, 14 August 1920.
25. Diaries, 27 June 1919.
26. Diaries, 9 November 1919.
27. Diaries, 16 November 1919.
28. Diaries, 30 November 1919.
29. Diaries, 11 December 1919.
30. Diaries, List of Books Read, 1920.
31. Diaries, 8 February 1920.
32. Diaries, 21, 22 March 1920.
33. Diaries, 8 December 1919.
34. Diaries, 13, 21 April 1920.
35. Diaries, 24 April 1920.
36. Diaries, 10 March 1920.
37. Diaries, 10 June 1920.
38. Diaries, 23 July 1920.
39. Diaries, 1 September 1920.
40. Diaries, 10 March 1920.
41. Diaries, 23 July 1920.
42. Diaries, 5 January 1920.

43. Diaries, 30 March 1920.
44. Diaries, 21 August 1920.
45. Diaries, 6 December 1919.
46. Diaries, 13 November 1919.
47. Diaries, 5 December 1919.
48. Diaries, 11 May 1920.
49. Diaries, 28 January 1920.
50. Diaries, List of Books Read, 20 October, 17 November 1920.
51. *New Cambridge*, 4 December 1920, 'A Page of Treason', p. 139, Letters p. 133. Diaries, 4 December 1920.
52. Diaries, 25 February 1921.
53. Diaries, 20 January, 24 February 1921.
54. Diaries, 2 November 1920.
55. Diaries, 11 April 1921.
56. Diaries, 1 May 1921.
57. Diaries, List of Books Read, 7 May 1921.
58. Diaries, 31 July 1921.
59. Diaries, 9 June 1921.
60. Diaries, 10 April 1921.
61. Diaries, 14 April 1921.
62. Diaries, 26 April 1921.
63. Diaries, 13 August 1921.
64. Diaries, 27 October 1920.
65. Diaries, 16, 18, 21 December 1920.
66. Diaries, 4 January 1921.
67. Diaries, 18 January 1920.
68. Diaries, 19 February 1920.
69. Diaries, 22 March 1920.
70. Diaries, 8 January 1921.
71. Diaries, List of Books Read, 1920.
72. Diaries, 2 March 1921.
73. Diaries, 6 June, 2 June, 11 June 1921.
74. Diaries, 27 October 1920.
75. Diaries, List of Books Read, 29 September 1920.
76. Diaries, List of Books Read, 3 June 1921.
77. Diaries, 5 June 1921.
78. Diaries, List of Books Read, 7 December 1921.
79. Diaries, 20 January 1922.
80. 'The Great Society', 1922.
81. The Dynamics of Human Society', 1920, J. D. Bernal Archive, B4.2.
82. 'Psychology and Communism', 1923, J. D. Bernal Archive, B4.10.
83. H. D. Dickinson, Annotation to J. D. Bernal, 'Psychology and Communism'.
84. Diaries, 18 August 1921.
85. Ibid.
86. Diaries, 19 February 1920.
87. Diaries, 6 May 1921.
88. Diaries, 21 January 1920.
89. Diaries, 3 August 1921.
90. 'The Failure of the Scientist', unpublished typescript, 1923. J. D. Bernal Archive, B4.8.
91. Diaries, 10 February 1920.
92. Diaries, 26 May 1920.
93. Diaries, 8 June 1921.
94. 'The Failure of the Scientist', 1923.
95. Letter to Gary Werskey, 1968.
96. Diaries, 18 August 1921.

97. Diaries, 11 October, 2 November, 1921. The Poplar councillors were refusing to levy an extra rate in the East End, to pay relief to the unemployed, unless the richer London boroughs levied it too.

98. Diaries, 29 March 1923.

99. Diaries, 7 April 1923.

100. Diaries, 1925.

101. Diaries, 18, 22 January 1926.

102. Diaries, 1 May 1926.

103. Diaries, 6–12 May 1926.

104. Diaries, 31 May 1926.

105. Diaries, 6, 7 July, 9, 13, 20 February 1926.

106. Diaries, 3 August 1926.

107. Diaries, 15 September 1926.

108. Diaries, 10, 17 November 1926.

109. Diaries, 9 February 1927.

110. Diaries, 12, 13, 20 February, 3, 6, 10 March 1927.

111. Bernal's views in 1927 are contained in the 'Microcosm' manuscripts.

112. 'The Growth of Organized Research', typescript, c.1929/30. B4.7.

113. *The World, the Flesh and the Devil*, Kegan Paul, 1929.

114. 'Microcosm'.

115. UH.

116. 'Scientist in the World Today', Chapter 2, 1934.

117. 'Science and Society' *Spectator*, July 1931.

118. 'British Scientists and the World Crisis'. *Labour Monthly*, 14, 702–9, 1932.

119. 'Science and Society', 1931.

120. 'What is an Intellectual?', 1931. B4.13.

121. Ibid.

122. Ibid.

123. Ibid.

124. Interview with Brian Pollitt, 1984.

125. 'Dialectical Materialism', in *Aspects of Dialectical Materialism*, London: Watts 1934 89–122.

126. J. D. Bernal, *Engels and Science*, Labour Monthly pamphlet, 1935.

127. 'Psychoanalysis and Marxism', *Labour Monthly*, No. 19, July 1937.

128. J. D. Bernal, 'The Frustration of Science', in *Science and Industry*, London, Allen and Unwin, 1935. 'Science in Education', in 'A group of economists, scientists and technicians', *Britain without Capitalists*, London, Lawrence and Wishart, 1936.

129. *Marx and Science*, London, Lawrence and Wishart, 1952.

130. 'For New Prospects of Peace and War', ms, 1965. B4.93.

131. The Doctrine of Peaceful Counter-revolution and Its Consequences, 1968. B4.108.

132. Enormity or Logic and Hypocrisy in the Ultimate Solution, October 1967. B4.101.

4

The Scientist

Peter Trent

To assess a man's contribution to human knowledge by his life's work, an interval of time is often needed. This is particularly so in science, which advances not in a steady, uniform way over the whole range of studies, but in small steps in one area, larger ones elsewhere. There are sudden rushes in one subject, where changes in approach rapidly bring new results. Then, such detailed advances may change our overall picture and bring a much fuller understanding of the inter-relation between what had previously seemed unconnected fields of work. Hence, to gain a perspective of scientific advance, and an individual scientist's contribution to it, we may need to allow some time to elapse.

But appreciation of the work of J. D. Bernal, who lived from 1901 until 1971, never needed any such interval. Even by the 1930s it was clear that his was a major contribution to twentieth-century science. Formal recognition of this came in 1937, when he was elected a Fellow of the Royal Society, though, as we shall see, the signs of his genius were clear and plentiful well before then.

His own account of the choice of science for his life's work goes back to primary school days:

> There was no science at school, no hint of it except in one reading book that had extracts from Faraday's lectures. I heard from my cousin at Oxford of liquid air. They talked at breakfast one day of X-rays. I thought of the bright sun – through my hands showed black bones against dark red flesh, and that night, piling white books around the lamp that I was given to read by, I tried to get the mysterious rays. The books fell down, the lamp fell off the table and broke. Daddy came up and thrashed me because he was so frightened I might have burnt myself. But inventions were even more exciting, engines and boats, cannons and electricity. You could see how they worked, how they were thought of. I could even think of new ones myself.[1]

It was at Bedford School, in 1911, that the young Bernal's knowledge of science really began. He was always an avid reader. Forty years later, he was heard to complain faintly that he could not read as much as he used to – no longer an average of a book a day. To those who knew and worked with him, what was even more remarkable was his power of recall. He could and did quote page references to scientific papers or articles even though he might have read them only once, and that several months or even years before. At Bedford School he read nearly every book in the school library. It is clear that his interests at school were much wider than the formal teaching in mathematics, physics and chemistry. Nights were spent stargazing with a telescope lent to him by his housemaster.

At home in the holidays there were expeditions to collect fossils, and much microbiology, carried out with a microscope that he had bought for himself when aged twelve. There were also chemical experiments of one sort or another, such as demonstrating to his sister Geraldine the properties of sodium by dropping it into hot water: 'after a bit it caught fire. Kate and Brigid (who worked in the kitchen) screamed and he took the saucepan off the stove – actually nothing really happened.'

Bernal spent his first year at Cambridge reading Mathematics, and took Part One of the Tripos with a 2.1. This was mainly because he was attracted by other lectures – those of the mathematician J. H. Grace FRS, a Fellow of Peterhouse – and did not attend the lectures he was supposed to attend. Grace's lectures raised problems of symmetry and space that were useful to Bernal later. For the time being, Emmanuel College, disappointed, switched him over to Natural Sciences and he took the first part of the Natural Sciences Tripos in Physics, Chemistry, Geology and Mineralogy in 1922 with first-class honours.

It was during the summer vacation of 1921 that Bernal first began to think about the question of atoms in space. His interest in the study of crystals came not only from the practical side, in mineralogy, where he was encouraged by the lecturer in mineralogy, Dr A. H. Hutchinson, but especially from the more abstract geometrical aspects. The latter interest was to continue through his life, underlying work such as the investigation of liquid structures that he pursued at Birkbeck College towards the end of his career, but he was already working on the geometrical aspects in 1922, while still an undergraduate. There is part of an early paper by Bernal in the Cambridge University Library called 'The Vectorial Geometry of Space Lattices'. This work – on the number of symmetrical ways of arranging atoms in space in three dimensions

– continued during his last undergraduate year, and in May 1923 he submitted a thesis, 'On the Analytic Theory of Point Systems'. This was essentially a derivation of the 230 space groups of crystallography by quaternion mathematics, and was an alternative derivation of these space groups to those produced by Federov, Schoenflies and Barlow at the end of the nineteenth century.

This thesis marked the start of Bernal's research career. It greatly impressed Dr Hutchinson who advised him to work with Sir William Bragg, then at University College London. Bragg saw the thesis and was sufficiently impressed to invite Bernal to come to London. Emmanuel College awarded him a research grant which supported him for a while in London and then Sir William, who had moved meanwhile to the Royal Institution, managed to obtain a salary for Bernal as a research assistant.

Later, Bernal would tell two stories about this time. The first told of Sir William Bragg's response when Bernal asked him what he thought of the space group thesis. 'Good God, man, you don't think I read it,' was said to have been Bragg's response. The first page was clearly sufficient to show that the young Bernal was worth encouraging in research. The other story relates to the only other man, apart from Bernal, known to have read the thesis. This was Carl Hermann of Stuttgart, to whom a copy had been sent. Five years later, Bernal met Hermann, and in the course of conversation he asked whether he remembered the thesis. 'Certainly,' said Hermann, 'there is a mistake on page ——'. Whatever the virtues of the thesis, it was a clear first expression of Bernal's lasting interest in the analytical aspects of the geometrical structures that were closely connected with so much of his later work.

Before describing some of the research that was led and inspired by Bernal over nearly four decades, it may be useful to outline briefly the bare bones of crystallography.

Crystals, in one form or another, had been known from primitive times. The shape of gemstones, their constancy of form when cleaved, the growth of crystals of identical shapes from chemical solutions, and many other aspects of the crystalline state had long been known. But an understanding of the underlying structures had to wait until the late eighteenth century, when the development of modern chemistry was made possible by the atomic theory of John Dalton.

This theory put forward the idea of atoms of different elements combining together in various proportions to make compound atoms, or molecules as we now call them. The theory shed light on the earlier observation, by Steno in the seventeenth century, that the angles between the faces of crystals of particular substances were always the same. The Dutch scientist Christian Huyghens saw the key implication from this: that a crystal could be built up with *regular* arrays of identical molecules.

In 1800, a French abbé, René-Just Haüy, showed how such molecules could be associated in different kinds of crystal. Later in the nineteenth century Mitscherlich showed that chemically similar compounds could be found in nearly identical crystal forms. The science of crystallography had truly begun.

At the end of the nineteenth century, crystallographic observations used the techniques of the chemistry, physics and geology of that time. Chemical techniques were well established. The foundations of chemical reactions were known, techniques for the analysis of substances were routine, the breaking down of compounds into atomic constituents and the synthesis of compounds were both accepted. Also, the growth of crystals from solutions and from melts were established phenomena, as were the ways in which the same substance could appear in different forms. The most often quoted example of this is probably carbon, since carbon atoms can be arranged so as to form substances appearing as totally dissimilar as diamond and graphite.

The various optical properties of crystals were of considerable interest. Most observations were made with visible light, and the ways in which rays of light of various colours could be absorbed, reflected and refracted in their passage through crystals gave valuable information on crystal structures. Of course, the knowledge acquired was not solely about the crystals themselves. Much was learned about the nature of light itself, with properties such as polarization being observed.

An obvious field for investigation was the passage of electricity through gases at various pressures. The production of light by the passage of an electric current through a tube containing gas at low pressure was observed and then, in 1895, when Wilhelm Konrad von Roentgen in Würzburg was conducting such experiments he discovered the production of what he called X-rays. His first paper was published in 1895, and was followed by another in 1896. Roentgen's papers detailed all the properties of X-rays that were to be known for the next decade. In particular, X-rays possessed some of the properties of visible light in that they travelled in straight lines, casting sharp shadows. However, they could also pass through materials that were opaque to light. What was more important was that whereas visible light had been known since Newton's time to be deflected when it passed through boundaries between different transparent substances, X-rays showed no such effect. Nor, in the early experiments, did they show any evidence of the diffraction and interference effects that were displayed by visible light.

For example, if light of one colour, that is, of one particular wavelength, falls on a set of close parallel lines evenly ruled on a surface, the light scattered at different angles will differ in strength depending on the angle of scattering and on the closeness of the ruled lines. When the separation between the rulings is about the same as the wavelength of the light, the light waves scattered from the successive strips of the surface – and not obscured by the ruled lines – can, at certain angles, be in step and so reinforce one another to give a maximum intensity. At other angles the waves can be out of step and can cancel one another, and the intensity of the scattered light will be less. This phenomenon, termed interference, enables us to calculate the wavelength of the light from a knowledge of both the separation between the ruled strips and the angles at which the scattering (or interference) maxima are seen.

With the precision of the machines available at the end of the nineteenth century, it was possible to rule reflections gratings with a separation between strips down to approximately the same length as the wavelength of visible light. Consequently, the wavelength of the light could be calculated from the interference effects observed. Now, since Roentgen's X-rays were produced by the passage of an electric current through a discharge tube, it was a reasonable supposition that they were some sort of electromagnetic radiation. To establish their wave nature it would be sufficient to show that they could exhibit interference effects in a similar manner to visible light. Roentgen himself looked for such an effect but could not find it. Several other workers in the years immediately following pursued this line of work, and the conclusion was reached that if the X-rays were an electromagnetic radiation, they were of extremely short wavelength.

So the situation rested for several years. Max von Laue had been working on the theory of one-dimensional and two-dimensional diffraction gratings. In 1912, he was discussing with Ewald the latter's work on the optical properties of regular three-dimensional arrangements of small particles – such as were assumed to be present in crystals. Von Laue asked what sort of distances there were between the particles. From Ewald's reply he realized that these distances were approximately the same as the probable wavelengths of X-rays, as recently calculated from diffraction by wedge-shaped slits. By extending the theory of optical two-dimensional diffraction gratings to the three-dimensional arrangement in crystals, von Laue became convinced that diffraction of the much-shorter-wavelengths X-rays must occur.

This theory provided the inspiration for the experiments of Friedrich and Knipping. They passed a beam of X-rays through a crystal of copper sulphate on to a photographic plate. When the plate was developed, in addition to the dark spot produced by the undeflected beam they also found an irregular pattern of black spots, less intense but clearly visible, arranged about the axis of the primary beam. X-ray interference had been observed, and a new chapter in crystallography had begun. In order to be able to calculate the arrangements of atoms in crystals several further developments were necessary.

W. L. Bragg showed that the equations required were equivalent to those describing reflection from planes, and his treatment provided a simpler way of thinking about diffraction effects. He solved simpler structures, such as those of sodium chloride and zinc blende, essentially from first principles. Immediately after World War One, P. P. Ewald called a meeting of the small group interested in solving crystal structures – Bragg, C. G. Darwin, R. W. James, and himself – and they worked out the factors affecting the diffraction effects, the intensities contributed by the scattering by the electrons in the individual atoms according to their geometrical positions in the crystal.

By the early 1920s W. H. Bragg had started to build up a team of research workers investigating different aspects of X-ray crystallography. Bragg suggested to Bernal that he work on the structure of graphite, on which, hardly surprisingly there was conflicting evidence. This was because more than one stable form of binding of a particular kind of atom could clearly occur (graphite and diamonds being forms of carbon, for example) and distinguishing the different structures underlying different forms was far from simple.

It was not always possible to produce, for X-ray analysis, specimens of crystals that were large enough to be mounted in an X-ray camera in a precise position relative to the direction of the X-ray beam and the film position. Consequently a technique developed by Debye and Scherrer and by Hull had been used for the graphite measurement. This consisted of using the substance in the form of a powder mounted on a noncrystalline fibre and relying on the random positions of the very large number of small crystals that constituted the powder to ensure that enough were in the correct attitude to produce a visible X-ray diffraction pattern. The contradictory results that had led Bragg to suggest the problem of graphite to Bernal had both been obtained by powder photographs. Bernal decided that it would be essential to use single crystals, and he managed to dissect out some small crystals from the graphite specimens available to him. He also decided to use a new method for categorizing individual reflections, which was fully developed and described later in a Royal Society paper.

From the beginning, Bernal moved on to some more general problems in X-ray crystallography. First, he transformed the method of interpreting diffraction patterns. In 1913, Ewald had introduced a new concept in this key problem – the reciprocal lattice. In the reciprocal lattice each plane in the crystal is represented by a point and so can be correlated directly geometrically with a reflection of a plane, that is, one X-ray diffraction spot on an X-ray photograph. Bernal derived charts that permitted reading from the photograph of reciprocal lattice coordinators for each reflection – and then correlating the intensities of the reflections with the atomic positions in the crystal. The formal mathematical relations between the two had been established at the crystallographic meeting near Munich just after World War One and could easily be applied in simple cases to derive the atomic positions by trial calculations. A little later it was shown how to use them to derive directly the electron density distribution in crystals by calculations involving the substitution of heavy atoms in isomorphous structures.

In the case of graphite Bernal first derived the correct unit cell and then the correct structure. In a paper published in 1924 in the *Proceedings of the Royal Society* Bernal pointed out how this structure corresponded to the structure of the carbon atom that had been developed a little earlier by Niels Bohr. He went on to discuss how the structure he had found fitted in with the mechanical, electrical, thermal and chemical properties that distinguish graphite.

Bernal was not only an innovator in the interpretation of X-ray diffraction patterns. The equipment on which he had carried out his earlier work (originally using an alarm clock to rotate the crystal) was developed by the expenditure of much time and effort. To improve it further, he designed an X-ray photogoniometer, which was built and marketed by the Cambridge firm of W. G. Pye and Co. This machine was to remain the main equipment of X-ray investigation in many laboratories for years to come. Bernal published four papers in the *Journal of Scientific Instruments* over the period from 1927 to 1929, giving details of its construction and use.

This work did not occupy all his energies, for during this time he also worked on the structure of alloys of copper and tin, publishing a paper in *Nature* in 1928. It was then that the accident occurred that led to one of the most-often-repeated Bernal stories: that of the lost crystal. He was concluding his work on a single crystal of delta bronze by determining its density. This involved weighing the crystal in a cage, in air, and then immersed in a liquid and, after that, removing it and finally weighing the crystal on its own. In his own words, 'I lifted the wire and the cage from the liquid carefully, delicately, and then without knowing, my hand shook. It had disappeared, gone. The crystal was gone. The crystal that I had worked on for months. The crystal about which I knew everything but that one final weighing, and now it was gone and those months of work with it. "Stupid," I thought and as I gathered all the dust of that room, looking through it, grain by grain, for my crystal, "if you had thought more this could not have happened". It was stupid to arrange things so that any weakness or wavering might lose everything.'[2] Bernal's power of vivid description, which was such a characteristic throughout his life, stands out clearly from that passage.

He left London and the Royal Institution in 1927, returning to Cambridge and a lectureship in structural crystallography. This lectureship had been created when his earlier adviser, Dr Hutchinson, became Professor of Mineralogy in 1926 and proposed the creation of a new lectureship in structural crystallography. C. P. Snow gives an account of Bernal's appointment which presents a realistic and vivid picture of the young Bernal early in his career:

> I remember Arthur Hutchinson, then Master of Pembroke and Professor of Mineralogy, describing the interview. Yes, Bernal had come into the room and sat down. His head had sunk into his chest (he had always been superficially shy, detesting introductions, or any of the small change of social intercourse). He sat there, looking sullen, so far as they could see his face at all. Yes, he admitted his name was Bernal. Yes, he admitted the routine facts about his career. They couldn't get anything else out of him. Finally, in despair, Hutchinson, who was in the chair, asked him what he would do with the sub-department if he were given the job. At which Bernal threw his head back, hair streaming like an oriflamme, began with the word NO (as he has usually begun his best speeches), and gave an address, eloquent, passionate, masterly, prophetic, which lasted forty-five minutes. 'There was nothing for it but to elect him,' said Hutchinson.'[3]

There were difficulties at first, of course. The buildings were not easy to adapt for the work, and the equipment was at first made up from whatever could be obtained from other laboratories, the whole giving a very Heath Robinson impression. But as Bernal's reputation grew he got grants for apparatus with which the laboratory was slowly equipped.

In the summer of 1928, Bernal made his first major tour abroad, visiting laboratories in Germany, Holland and Switzerland. Even at this early stage in his career, Bernal was much more than a gifted research worker in his chosen field. He was a prophet of science with the power to inspire all those with whom he worked, and with the vision and ability to bring together people and ideas from all over the world. He came back from his tour with ideas and inspiration that he transmitted to Sir William Bragg with such effect that next spring Sir William called a conference of X-ray crystallographers at the Royal Institution, which was held in March 1929. This set up three committees, one on publications, one on nomenclature, and one on abstracts. These met several times in the following years and organized the production and publishing of the International Tables for the Determination of Crystal Structures. Bernal was secretary of all three committees. In addition, he continued to work on the structures of metal alloys – especially on bronze, the work he had started in London. In 1929 he gave the opening paper to a Faraday Society meeting on 'The Problem of the Metallic State'. He could have continued with this work as a main line; indeed he started and encouraged several research workers in this and related fields, on inorganic compounds of various types. But the focus of his main interest was shifting towards the more complex compounds to be found in biochemistry, for at Cambridge he had made new friends with biological interests, notably J. B. S. Haldane, Joseph and Dorothy Needham, N. W. Pirie and R. L. M. Synge, who was then an undergraduate.

Bernal was never short of ideas, interests and problems. One example will suffice, relating to the structural changes that must occur when a liquid solidifies on cooling. In this connection, water was a particularly interesting substance and he directed the work of a research student, Helen Megaw, on the structure of ice. This work led to several related problems. The hydrogen atom, simplest of all atoms, is to be found everywhere, either on its own or together with a single oxygen atom, when it forms what is called a hydroxyl group. The way in which hydrogen atoms and hydroxyl groups join with other atoms is of much importance when considering the structures of whole ranges of substances. This was fertile ground for work on many structures, both simple and complex. Bernal's work on water in the early thirties led to a masterly paper published by himself and R. H. Fowler in the *Journal of Chemical Physics* in 1933. The short summary with which this paper commences is worth quoting:

On the basis of the model of the water molecule derived from spectral and X-ray data and a proposed internal structure for water, the following properties of water and ionic solutions have been deduced quantitatively in good agreement with experiment:

(1) The crystal structure of ice.
(2) The X-ray diffraction curve for water.
(3) The total energy of water and ice.
(4) The degree of hydration of positive and negative ions in water.
(5) The heats of solutions of ions.
(6) The mobility of hydrogen and hydroxyl ions in water.

And the following inferred in a qualitative way:

(7) The density and density changes of water.
(8) The explanation of the unique position of water among molecular liquids.
(9) The dielectric properties of water and ice.
(10) The viscosities of dilute ionic solutions.
(11) The viscosities of concentrated acids.

This quotation is a good example of the way in which Bernal could always present a comprehensive view, not just of one aspect of a problem, but of all aspects, combined together in a full picture. He could always see the wood *and* the trees. His own story of the origin of that paper lays it at the door of the weather. In 1932 he was invited to take part in a meeting to be held at the School of Physical Chemistry in Moscow. It was here that he made his first full contact with leading Soviet scientists, and the conference was clearly valuable in this respect. But his work on the paper on water quoted above started when the British group at the conference left to come home by air. Bernal and Fowler arrived together at the airport early in the morning, only to find travel impossible because of thick fog. So they walked up and down and talked. The fog was an obvious topic of conversation and when Bernal extemporized a theory to account for the behaviour of water in connection with its behaviour as fog, and ice also, the collaboration started. Bernal wrote later, 'It was, in fact, the only time during our visit when we had time to think.' The occasion 'would have led to nothing but talk, if it had not been that the person I was talking with was Fowler. He said, "You must write it up." '[4] Before moving on to the work on biochemical substances that occupied much of Bernal's efforts in the thirties, it is worth pointing out this was the origin of his work, after World War Two, on the structure of liquids, a subject that always fascinated him.

The work that occupied the rest of his time at Cambridge was largely with substances that were being isolated by the various biochemical

groups with whom he had connections. His friends sent him crystals of
the substances they were isolating and he made many observations,
with varying success, on these. A considerable effort went into his work
on the sterols and on sex hormones. Crystals of the sterol calciferol
(vitamin D_2) were supplied by J. B. S. Haldane and of the sex hormone,
oestrin, by S. Zuckerman.

Bernal's first major success in analysing structures of biological
interest came with his measurement of six sterols, including calciferol.
His methods provided him with precise measurements of the size and
shape of the unit cells of their structure. But he carried his observations
further and formed the impression that the molecules were thin and
up to 20 angstrom units (Å) – or, 2 thousandths of a micron long, with
overall dimensions of about 5Å × 7.2Å × 17–20Å. He looked up the
chemical literature, which had normally been used as a source for the
structural formulae of molecules, and realized that the accepted formulae represented much thicker, shorter molecules and could not be
correct. Bernal's work led to the rearrangement of the rings in choles-
terol – to that accepted today by Rosenheim and King, Wieland and
Dane. Crystallographically his work was checked much later by his
students Dorothy Crowfoot and Harry Carlisle. The change in the ring
system made it easy to see the connection between the sex hormones
and the sterols.

His interest in biological molecules had started with amino acid
crystals, seen as building blocks of proteins, and he shared this interest
with Astbury, with whom he had worked in the Davy Faraday Laboratory
at the Royal Institution. From 1934 Bernal went on to open up the
whole field of X-ray analysis of proteins, viruses, nucleic and other still
more complicated biological systems by showing that it was necessary
to keep such materials *wet* in whatever mother liquor these were found,
usually water; otherwise order would be lost and diffraction effects with
it.

The first crystallographic work had been carried out on fairly simple
crystals, where the X-ray diffraction patterns were produced by the
regular repetition of rather few atoms. It had been found quite early,
however, that other substances also produced X-ray diffraction patterns.
Work on these aspects of the subject started in the twenties. In the
Kaiser Wilhelm Institute in Berlin, Meyer and Mark were studying the
X-ray patterns produced by textile materials such as cotton and silk. In
1926 Bragg had the idea of going beyond well-formed crystals, and
gave a lecture on 'The Imperfect Crystallisation of Common Things',
which included mention of fibres, amongst other things. It was Bragg's

request to Astbury for help in getting X-ray photographs of wool that led Astbury to pursue further work on biological substances, in the field that he himself named 'molecular biology'. Astbury and Bernal both worked in this field of common interest during the thirties.

The difference between X-ray diffraction pictures of simple crystals – such as those of common salt or diamond, or even relatively complex organic compounds like tartaric acid – and of biological molecules arises from the complexity of the arrays in biological substances. The simplest units of biological molecules can contain not just a few but many thousands of atoms. Thus the molecular structure of biological substances can be very complex. The problem for the X-ray crystallographer in this field becomes one of determining the various subunits that make up the biological molecule and the way in which they are attached one to another. The work on the structure of the nucleic acids DNA and RNA, now well known in connection with the genetic code, is an example of this, but there were many similar projects.

Bernal's perception of the range of applicability of X-ray technique was expressed in a paper he published in 1930 in *Radiology*. In it he wrote:

> Ultimately the diffraction of X-rays is due to the interaction of the radiation with the electrons in the atoms of matter and so the deepest information it can give is on the position of the electrons in the individual atoms, or, more correctly, information as to the average distribution of density of negative electricity. Here it is most closely related to pure physics and leads to the possibility of the complete description of the state of any piece of matter. But without going nearly so deeply X-rays can show the position of the maxima of electrical density, that is, the position of each kind of atom, in a regular crystalline structure. It is in this field that the chief work of X-ray crystallography has been done. . . . Besides these two types of information, the diffraction patterns of crystals may be used as in ordinary crystallography to fix the nature and orientation of the minute crystals which go to build up common substances, with the advantage that there is no limitation to the substances that can be examined in this way, and a means is provided for analysing all the textures of natural and artificial products. This is the type of information which links X-ray crystallography to biology on the one hand and technology on the other.[5]

It was this type of information that was sought by both Astbury and Bernal in the thirties. In 1933 Astbury obtained an X-ray diffraction picture of pepsin, mounted in the dry state. The picture showed two broad rings and, from the spacings of these, Astbury concluded that peptide chains were present in the crystals, that they were straight, and possibly broken to one length, to correspond with the molecular weight, estimated by an osmotic method to be 34,500. Then, in 1934, a biochemist from Oxford, John Philpot, working in Uppsala, managed almost by accident to produce some large crystals of pepsin. A tube of these, still in the mother liquor from which they had crystallized, was brought to Bernal in Cambridge for X-ray analysis. He first attempted to obtain an X-ray picture from a dried crystal, but this gave no clear diffraction pattern. He examined the crystal under a microscope and decided that the drying had impaired the crystal structure. He then made an X-ray picture from a crystal in a fine glass tube, sealed into the tube with some of its mother liquor. This gave much better X-ray reflections, and Bernal deduced the shorter constant of the crystal lattice to be 67Å. The longer lattice constant was much harder to estimate, because of overlapping reflections which confused the picture. His estimate was 154Å (or a multiple); and, in fact, the multiple was later found to be nearly two.

The seminal importance of Bernal's work on pepsin was not just for the structure of the protein itself. This protein is of high molecular weight, so the pattern is complicated and to this day it has not been solved completely. The observation of a discrete X-ray pattern from the hydrated crystal, however, opened up the enormous field of X-ray analysis of the atomic structure of protein crystals.

The work on proteins was then fairly launched and carried on through the following years in Oxford, Leeds and Cambridge. The Cambridge work was carried out by Bernal, I. Fankuchen and M. F. Perutz and was on chymotrypsin, ribonuclease, and on haemoglobin. Dorothy Crowfoot moved from Cambridge to Oxford in 1934 and the American Fankuchen came to Cambridge from Manchester to continue with some work on sterols that Crowfoot had left behind, and later to work on viruses. Perutz came, trained as a chemist, from Vienna, whence he was sent by Mark with the instruction to learn crystallography and work with Bernal.

The rationale of this work was clearly stated in a lecture by Bernal to the Hungarian Chemical Society in 1953:

> The structure of protein molecules remains the key problem of crystal chemistry, as of biochemistry. Protein molecules containing from a thousand to a million atoms are the largest quasi identical units produced by organisms; they are also the most complicated. At the same time they play a central role in the metabolism of organisms. They are the basis of all, or nearly all, enzyme systems. They provide the characteristic features which provide the phenomena of immunology, the antigen anti-body reaction. As structures

they are responsible for active muscle and nerve, and passive connective fibres and hair. Their importance biologically and industrially justifies the attack on the problem of their structure by all methods, chemical, physical and crystallographic. One method alone will not suffice, all need to be employed together and coordinated.[6]

The tobacco mosaic virus supplied by N. W. Pirie was worked on by Bernal and Fankuchen in the late 1930s, first at Cambridge and then at Birkbeck. It was studied at various dilutions by marvellously precise and ingenious apparatus, and the work was published during World War Two in the *Journal of General Physiology* – where its complexities have been largely buried. Very fine monochromatic X-ray beams were passed through the various preparations held in Lindemann glass tubes at various concentrations. At low concentrations the virus sols would flow through the tubes, and the rod-shaped units would orient in the direction of flow parallel with it and arranged in a hexagonal close-packed lattice to fill the space. The sols could be dried to wet and dry gels in which structures characteristic of particles – tactoids, positive and negative – formed. At small angles, sharp reflections characteristic of rod close packing were observed, the distance between rods varying down to 150Å with the virus concentrations. At wide angles, the same wide-angle pattern was observed, indicating the arrangement of sub-units within the individual particles.

Bernal and Fankuchen suggested a spiral arrangement of small units roughly 11Å × 11Å – distances characteristic of packing in proteins. Present-day research has shown this description to be only roughly correct, the small units being short stretches of alpha-helical protein chains. The overall organization is not in a regular crystallographic spiral but in a wide helix repeating non-crystallographically at seventeen repeats around a central tube-like cavity round the edge of which the nucleic acid winds rather more irregularly.

Bernal and Fankuchen did not fully recognize the character of the helix, and sought to explain the complexities in terms of a structure that shows too few repeats in two dimensions to give a sharp diffraction effect. The correct interpretation was only realized after the diffraction patterns of the alpha helix had been derived by Pauling, Branson and Corey.

In the same papers Bernal and Fankuchen examined the long-range forces between virus particles in concentrated gels. This was the beginning of much subsequent work on polyelectrolyte gels such as muscle and cornea. Although the basic principles were established early on, Bernal was continually developing ideas for the refinement of X-ray

measurements and for the interpretation of the diffraction pictures. To
give a simple example, the X-ray diffraction patterns give information
about the intensities of the X-rays scattered from various places in the
crystals under investigation. The scattering depends strongly on the
number of electrons in the atoms on which the incident X-rays are
directed. So, in order to find out more about the detailed structure of
some crystals, a technique was developed for substituting heavier atoms
with more electrons at a particular place in the crystal. Consequent
changes in the diffraction pattern, together with a knowledge of the
chemistry of the substitution, made it possible to get information about
the normal structure of the crystal. Bernal used this technique with
considerable success in several of the structures being investigated at
this time.

The years from 1934 to the outbreak of war in 1939 were a productive
period for Bernal and his collaborators. At Cambridge, Bernal, Fankuchen and Perutz worked on chymotrypsin and haemoglobin, and later
on the tobacco mosaic virus. At Oxford, Dorothy Crowfoot Hodgkin
worked on insulin, whilst Astbury continued protein work at Leeds.
The scene seemed set for continued development and discoveries in
the field of crystallography. In 1937, P. M. S. Blackett, another of the
leading figures of physical science in the thirties, moved from London
to Manchester and Bernal was appointed to follow him in the Chair of
Physics at Birkbeck College. He had only one year – hardly time to play
himself in – before war broke out. Although the move to London
necessarily slowed down the experimental work, the flow of
publications continued and even increased. Bernal was rapidly drawn
into wartime activities. His scientific contribution to the war effort is
described by Ritchie Calder in Chapter 7 of this volume.

Immediately after the end of the war there were many problems of
reconstruction to be faced. Birkbeck College, to which Bernal was
returning, had suffered considerably in the blitz. It had lost several
laboratories, its library and its theatre, all by fire. This was on top of
the fact that the accommodation, then in the City between Chancery
Lane and Fetter Lane, was old, built in 1883–85 and even without the
war damage was hardly suitable for a twentieth-century science department. Birkbeck's greatest asset was its uniqueness and its spirit. Its role
in London University was to provide face-to-face education for persons
engaged in full-time work earning their living – the lineal descendants
of the mechanics and artisans for whose self-improvement George
Birkbeck had founded his London Mechanics' Institution in 1823, later
to become Birkbeck College in the University of London.

Because of the nature of its students, Birkbeck could not move out of London for the duration of the war as other colleges did (although Bernal did move two research students and X-ray equipment to Oxford). There was no major problem in moving the Oxford apparatus back to Birkbeck at the end of the war, and acquiring more new apparatus. The war had prevented the completion of the college's new building, which was left as a steel skeleton just to the north of London University's Senate House. At this time there were considerable shortages, and building materials especially were in short supply, so the college's work was carried on in improvised accommodation during the wait for its new building to be completed. For offices and lecture rooms, it was possible to use commercial office buildings near the main building. To replace the science laboratories, however, was not easy. To provide new laboratories for the development of crystallography was even more difficult.

Fortunately, the university owned a number of buildings in Bloomsbury, and made over two houses in the war-damaged remains of the east side of Torrington Square, numbers 21 and 22. While they were being repaired, Bernal was able to start his research for one year at the Davy Faraday Laboratory, and then at the Medical Research laboratories in Hendon.

More important than buildings were the people working in them. The pre-war groups of crystallographers had in any case broken up. Fankuchen was back in America, Dorothy Crowfoot Hodgkin was at Oxford, Perutz at Cambridge. So, starting almost from scratch, a new group was brought together, who were to develop crystallography at Birkbeck over the late forties, fifties and sixties. There were of course many students and visitors who were not permanent members of the department, but amongst those working in Bernal's laboratories over this period were Ehrenberg, Carlisle, Franklin, Jeffery, Klug, Levine and Mackay.

In the strategy of science in Britain in the first half of the twentieth century and in the acquisition of knowledge and understanding, a major role was played by the professorial heads of departments. They met in various learned societies, such as the Royal Society, which was perhaps the most important, but also in the more specialized bodies such as the Royal Society of Chemistry, the Royal Statistical Society, and so on. Their exchange of ideas through meetings and the learned journals provided the background to research and the development of knowledge. In the deployment of university resources and the building of research teams and their direction on to specific problems, the major role was played by professorial direction. Bernal thought out the mixture of research he wanted and chose the people. His team at Birkbeck was directed over a complete, interconnected spectrum of work.

There were problems, of course. To quote Aaron Klug, now a Nobel laureate, who was then working in Torrington Square with Rosalind Franklin on virus structures:

> I have already mentioned the attic rooms in which we provided a rather peaceful place for Rosalind Franklin to exercise her great ingenuity and skill at preparing better and better specimens and for myself to contemplate some of the theoretical problems involved in the helical diffraction analysis. The X-ray tubes were down in the basement and sometimes the condensation on the pipes resulted in drops of water falling from the ceiling and I have a vivid picture of Rosalind sitting under an umbrella carrying out the virus preparation in a camera.[7]

From the practical viewpoint, the requirement was clearly to develop equipment and techniques for structure investigations. A principal part was played here by J. W. Jeffery, who was responsible for the building and equipping of a mechanical workshop of adequate refinement to produce X-ray cameras, spectrometers and the other necessary hardware. But to reproduce existing cameras and so on would not have been enough. Bernal supported and encouraged the work of Werner Ehrenberg, later to become Head of the Physics Department at Birkbeck. Ehrenberg's work was, amongst other activities, directed to producing better X-ray tubes and cameras for the diffraction studies.

It had been shown earlier that X-rays, although electromagnetic waves, were hardly deflected at all when passing through matter, unlike electromagnetic waves such as visible light. The changes in direction of light passing through materials such as glass can be employed in order to make lenses that can focus light; but no such possibility then existed for focusing X-rays. To produce sharp X-ray diffraction pictures, the only practicable means used in the pre-war period had been to use small X-ray tube anodes (this is the target in an X-ray tube which, when bombarded by electrons, emits the X-radiation), and to narrow the X-ray beam by passing it through lead sheets with small holes in them. Fairly fine beams could be produced in this way, but they were not very intense.

Ehrenberg's work was directed at this problem, initially by using the techniques of electron optics to shape the electron beams used for bombarding the anodes in the X-ray tubes, and so reducing the area of the anodes that emitted the X-rays. This fine-focus tube produced finer and more intense X-ray beams. Ehrenberg also made use of the fact that X-rays could be reflected when they fell at glancing angles on to some surfaces. By using metal and glass reflectors shaped in large-radius curves, he was able to produce focused X-rays.

The fine-focus tubes were to prove very valuable. To quote Klug again:

> The first photographs were taken on a Philips microcamera modified by Franklin for low-angle work using one of the Ehrenberg-Spear X-ray tubes which had been developed at Birkbeck some years before. . . . It was the use of these tubes and focused radiation that produced the beautiful X-ray pictures of Rosalind Franklin which have become familiar through very wide reproduction. . . . In fact, a good deal of the success of the investigations can be attributed to the introduction by Franklin of one of these non-conventional X-ray tubes and cameras, which provided high intensity and fine resolution from very weakly scattering specimens.[8]

Bernal's role in the inspiration and encouragement of this work is best summarized by a final quote from Aaron Klug:

> During these exciting years which may be said to have concluded the initial stages of the programme on rod-shaped and icosahedral viruses initiated by Bernal in the 1930s, it was always a great encouragement to have the Professor come bounding up the three flights of stairs day after day with a characteristic 'What's new?' Of course, one could not, except perhaps in the bumper years 1955 to 1956, always provide him with something new, but his constant interest and questions were a source of great inspiration to us. Moreover, he laboured on our behalf finding money and space and equipment, problems which are now more easily solved now that Molecular Biology, and the structural analysis of biological molecules, is taken for granted; it was not so in those days and a great debt is owed to Bernal, not only for starting the work, but for seeing to it that conditions were created for it to be carried on.[9]

In 1953, at a discussion meeting on the structure of proteins held at the Royal Society, Bernal outlined a technique for obtaining more information from X-ray diffraction pictures. Although this depended essentially on the use of computational methods that were at that time impractibly cumbersome, the preparedness to modify existing techniques and develop new approaches were utterly characteristic of Bernal's work.

The foregoing may give a slightly distorted picture of Bernal's work in the fifties. In fact, although perhaps the major achievement of the Birkbeck laboratories at this time was the work on virus structures of Franklin, Klug and those who followed them, this was only part of the whole spectrum of Bernal's activities. At that time his department had much work going on in other fields of classical crystallography. For example, the inorganic group started by Helen Megaw and then led by Jim Jeffery worked on the structure of various silicates. These compounds are involved in substances in day-to-day use such as bricks and cements, so this work had much value in giving a better understanding of the best use of materials common in the building and construction industries. Close connections were developed with the Building Research Station. Or, another example, work was carried out on the structure of the very fine ash produced by the furnaces of coal-fired power stations. Disposal of this material, called descriptively 'fly ash', had presented a tiresome problem to the power industry for a long time. With the help of the work of Jim Jeffery's group, uses were found for this material, for example as a filling material for lightweight thermal insulating building blocks.

Bernal's lifelong interest in the generalized aspects of the process of diffraction by individual atomic and molecular arrays which underlies crystallographic science continued throughout this period. This interest was also applied to the structure of liquids. He had first published on this topic in 1933 in a paper referred to earlier, by Bernal and Fowler, and then again in 1937. He now initiated a considerable programme of work, with much emphasis on developing simple practical models from which measurements could be made to compare with data obtained from scattering measurements on real liquids. He reviewed this work in 1962 when he gave the Royal Society's Bakerian lecture. His title was simple: 'The Structure of Liquids'. This work was always one of his favourite fields, in such time as he had over from his many other commitments.

When Bernal set up various research groups at Birkbeck after the war, he himself supervised the work on the structure of liquids, with two or three research students and technicians to help him. He considered that the kind of calculations then being carried out by others on the liquid state gave little insight into the actual molecular arrangements likely to be present. In the absence of the theory of statistical geometry that he saw as necessary for the formal solution of the problem, he set out to find the answer by model building. He built one of his major models himself, using the evidence from X-ray diffraction experiments on liquid metals of the distribution of distances between nearest neighbours in the liquids. He had a set of rods cut with the same distribution of lengths and joined balls together with the rods in an irregular way ensured by making one join every time he was interrupted in his office, on an average every five minutes, by someone on the telephone or coming for consultation. The model so constructed had approximately the right density and energy. Other very ingenious models were the one composed of Plasticine spheres, covered in chalk and compressed together in a football bladder, or the similarly compressed heap of ball bearings, dipped into black paint, which outlined their contacts with one another. From these models he made several discoveries about the characteristics of homogeneous, random, close-packed assemblies. Bernal saw his results as demonstrating how the random state differed from the regular state in an abrupt way requiring change of state, and how the arrangements that appeared in the random state had elements promoting fluidity.

Whilst crystallography was the prime focus of his scientific work throughout his life, his breadth of vision and tremendous energy meant that he was to be found actively participating in many, many fields of science. He was passionately concerned with the application of scientific work for the benefit of mankind, continually demanding the adequate funding of science so that such work could be carried out. Whilst carrying a vast scientific workload, he travelled widely in Europe, India, China and the Americas. He was a founder member of the World Federation of Scientific Workers, and a tireless advocate of the use of science in the tasks of peace. In 1945 he was invited to give one of the Friday Evening Discourses at the Royal Institution. He chose as his theme 'Lessons of the War for Science'. In this lecture he indicated directions in which valuable advances could be made by properly directed scientific effort, but he also sketched out the organizational essentials and the way in which the effort in the natural sciences could be best directed.

He wrote and spoke much also on the place of science in a nation's culture. From his inter-faculty lecture 'Science and the Humanities', delivered in Birkbeck in 1946, through correspondence on the document produced on the 'Finance of Fundamental Research in Britain', to broadcast discussions in the early sixties on the future of science in Britain, or in his correspondence on the formulation of the Labour Party's policy on science, there was a steady pattern of activity. In the field of direct scientific work he was active on the problems of classification, storage, and retrieval of the growing mass of scientific information that resulted from the great post-1945 expansion of scientific work.

With characteristic foresight, he had encouraged the work of A. D. Booth, carried out in one of Birkbeck's Torrington Square buildings.

Donald and Kathleen Booth were pioneers in the development of digital computers. The vast use now made by present-day crystallographers of computers and computing techniques was foreseen thirty years ago by Bernal, as were the databases that are now an essential part of the research scientist's equipment for handling and extracting information from an ever-increasing volume of research work.

He tried to use the early computers himself (with the assistance of Mike Bernal) in the liquid work, but he was defeated by their inadequate power. But the basic ideas he used at that time (the early 1960s) in programming the machines are now used widely in studies not only of liquids but also of noncrystalline arrays in general, which are currently of both industrial and biological interest in addition to their fundamental scientific attraction.

Today, the work he started on liquids in general, and water in particular, has in a way come full circle. He always said that his interest in liquids arose from his desire to understand the operation of biological molecules, and that required an understanding of water, the fluid that dominates the biological environment. Yet water was too difficult – in those days the understanding of simple liquids was insufficient, and so that is where he had to start. Today, we understand simple liquids fairly well and can begin to model water – both pure and in a biological context – with the aid of the current generation of computers whose advent and importance he foresaw.

Two aspects of his work on liquids seem to stand out today as being of prime, and lasting significance. The first concerns his identification of the central importance of the repulsion between molecules – what he called, after Lewis Carroll, 'impenetrability'. No atom could approach its neighbour closer than a minimum distance (the 'molecular diameter'), and this, combined with a lack of crystallinity, essentially determined the structure of the liquid. It is now widely recognized that in this beautifully simple conceptual model is the basic physics of simple liquids, and the underlying ideas have been adopted by others in perturbation theories of liquids. Concerning water, the availability of powerful computers has over the past decade and a half led to an enormous volume of modelling calculations on liquid water and aqueous solutions, all assuming a basic model for the water molecule itself. Most of these models are similar, and resembled that proposed in Bernal's classic 1933 paper with Fowler. One is tempted to conclude that had he had the computing power now available, he would have solved the problem of water decades ago. Even 'impenetrability' now appears to have strong relevance to water structure. Alice's Wonderland has interesting reality at the fundamental molecular level.

By the end of the fifties, Birkbeck College was considering the building of a new extension to the main building. In this building its crystallographers could, for the first time, enjoy purpose-built laboratories. The building was completed and occupied in the 1966–67 session. The work of the college's crystallographers transferred and continued to flourish there.

But in July 1963 – after an all-night flight from New York, followed by an Academic Board meeting in Birkbeck – Bernal suffered a stroke. He made some recovery over the rest of the summer, but inevitably his tremendous level of activity was drastically reduced. Nevertheless, he continued working with his collaborators in the new Department of Crystallography, now finally separated from the Department of Physics. He suffered another stroke in 1965, with more serious effects, and from then until his death in 1971 he became progressively less able to communicate by speech. Despite this increasing disability, he continued his work, producing books, papers, articles, reviews and other written work.

Much of his published scientific work at this time was still directed to problems of the structure of liquids. Bernal worked much on this, using many varied models to investigate the packing of particles in liquids. The conclusions from these various model experiments could then be tested against experimental observations and the comparisons could lead to correction or refinement of the theory on which the initial models were based.

He also continued his interest in the biological field. During the late fifties and early sixties there was much speculation on the origin of life. The development of satellites and orbiting spacecraft and detailed investigation of the nature of meteorites all stimulated interest in the topic. Bernal was involved and active in this field, contributing to conferences, writing articles and reviews, right up to his retirement from Birkbeck in 1968. During this time, as told elsewhere in this volume, his involvement in the history and politics of science continued, too.

Perhaps the hardest thing to bear, for his colleagues and friends, but above all for Bernal himself, a true intellectual giant, must have been frustration at a progressive impairment of communication from a still-active mind to the world outside. The only consolation for us must lie in the mass of his published work – so much to read and reread time and again. This work is his best memorial.

In 1964, in a volume to commemorate the twenty-fifth anniversary of Bernal's *The Social Function of Science*, C. P. Snow wrote:

People have sometimes asked, just how will he rank in scientific history in the narrow sense. I think the answer is that in natural gifts he stands very high; he is the most learned scientist of his time, perhaps the last of whom it will be said, with meaning, that he knew science; he has enormous imaginative sweep and deep insight; he has a major scientific purpose. And yet his achievement, though massive, will not dominate the record as it might have done. This is partly owing to a peculiarity of his nature. He likes to start something, drop an idea, get the first foot in – and then leave it for someone else to produce the final finished work. The number of scientific papers, all over the world, published under other names, which owe their origin to Bernal is very large. But he has suffered from a certain lack of the obsessiveness which most scientists possess and which makes them want to carry out a piece of creative work to the end. If Bernal had possessed such obsessiveness he would have polished off a great deal of modern molecular biology and won Nobel Prizes several times over.[10]

Notes

1. 'Notes for Boris Polevoi', typescript, J. D. Bernal Archive, Cambridge University Library, B4.68.1.
2. J. D. Bernal, 'Microcosm', J. D. Bernal Archive, B4.1.
3. C. P. Snow, 'J. D. Bernal: A Personal Portrait', in M. Goldsmith and A. Mackay, eds., *The Science of Science*, London: Penguin, 1964.
4. See J. D. Bernal and R. H. Fowler, 'A Theory of Water and Ionic Solution, with Particular Reference to Hydrogen and Hydroxyl Ions', *Journal of Chemical Physics*, vol. 1, no. 8, 1933.
5. J. D. Bernal, 'The Place of X-ray Crystallography in the Development of Modern Science', *Radiology* (St Paul, Minnesota), vol. XV, pp. 1–12, July 1930.
6. Lecture to the Hungarian Chemical Society, 1953. J. D. Bernal Archive, A3.117.
7. Aaron Klug, 'Virus Research at Birkbeck', unpublished paper, 1965. J. D. Bernal Archive, C.9.
8. Ibid.
9. Ibid.
10. Snow.

5

The Social Function of Science

Chris Freeman

This chapter discusses Bernal's seminal book *The Social Function of Science*. It was seminal in the full meaning of the word: it gave rise not only to many new ideas in the social sciences and to a new subject – the 'social studies of science' – but also to new policies and indirectly to many new institutions both inside and outside government. To go through the contents list at the beginning of the book is to draw up the agenda of most of the debates on policies for science and technology since World War Two.

Much of the detailed discussion is, of course, now dated and of interest mainly to historians, but the issues which the book raises are still the fundamental issues half a century later. The main reason for this is that Bernal was the first to see 'science' clearly as a social subsystem, to attempt to define and measure its boundaries, to assess the problems of managing and planning the subsystem as a whole, and to relate all this to the wider social system in its historical development and possible future.

This was an ambitious objective and it is not surprising that the book is open to criticism on many points of detail, as well as on major judgements of contemporary and future developments. Nevertheless, it remains an outstanding original intellectual achievement and perhaps Bernal's greatest single contribution to our understanding of the contemporary world. Its quality was widely recognized at the time of publication (1939) and that recognition was never confined to Marxist circles, although the response was most enthusiastic there.

This chapter represents a personal evaluation of Bernal's book (and other closely related work) rather than the account of an historian or a definitive biographical study. For this reason it begins and ends with personal comments on *The Social Function of Science*. In the introductory remarks I try to convey the central message of the book and its significance for the social sciences as it appeared at the time and in retrospect. I then follow Bernal's subdivision of this book into two parts: 'What Science Does' and 'What Science Could Do'. These sections attempt to demonstrate Bernal's achievement in rather more detail in the light of subsequent developments in the theory and practice of science policy.

The discussion of 'What Science Could Do' leads inevitably to the fundamental political problems of the planning of science, which aroused intense controversy at the time of the book's first appearance and have continued to do so ever since. In the fourth section of the chapter I attempt to relate this controversy to subsequent debates on the freedom of science and civil liberty more generally.

This entails finally some attempt to put *The Social Function of Science* in the context of his life and work as a whole and in particular its brilliant forerunner *The World, the Flesh and the Devil* (1929). Even if there were a fundamental consistency in the pattern of Bernal's life, he constantly responded to new developments in science and in society. The direction in which he modified his views is of importance as the learning process of an outstanding scientist striving to implement his youthful ideal in a complex and changing world.

As a student at the London School of Economics I was fortunate enough to hear Bernal lecture several times before, during and after World War Two on the social function of science and related topics. These lectures were not part of the regular curriculum; they were events organized by various student societies. But I learned more from them than from any lectures in the regular curriculum with the possible exception of some by Harold Laski.

One lecture in particular stands out in my mind. It was in 1947 and as an ex-service student I was dissatisfied with much of the regular economics curriculum which seemed to me and most of my fellow students to be even more remote from the real world than when I started economics in the 1930s. Keynesian economics was still only grudgingly accepted and even young lecturers such as Kaldor, who taught the full range of economics theory brilliantly, scarcely mentioned 'research and development'. It was therefore extraordinarily refreshing to hear Bernal talk about the role of scientific research in the economy and in war from first-hand knowledge of government, the armed forces and industry.

But even more stimulating was his discussion of the future role of scientific research in the civil economy. It was from Bernal and not from any economist that I first learned about the concept of 'research intensity' – a measure of the relative scale of resources committed to R&D by different industries and different firms. In his lecture Bernal described the differences between the 'research-intensive' industries such as electronics and chemicals for whom new product development was an accepted part of firm behaviour and other industries which had not yet internalized the functions of research, invention and development. It was an elaboration of Chapter 6 of *The Social Function of Science* based on the new developments during the war.

His picture of the growth of industries and the competitive behaviour of firms made far more sense to me than any of the orthodox models of perfect competition that were the staple diet of undergraduate economics students. Only in our courses on economic history and the history of economic thought did we get some notion of a more dynamic and realistic picture of the rise and decline of industries and firms related to the progress of science and invention. These ideas were, of course, not original to Bernal: they were the central theme of Schumpeter's model of economic development, and of the Marxist model from which this was derived.

But Bernal went beyond Schumpeter and Marx in his perception of the extent to which the R&D function had become professionalized and internalized both within industry and in government. It was this theme that was the organizing principle of *The Social Function of Science* and enabled him to describe, measure and criticize the social subsystem of science and technology. In Marx's time this specialization and professionalization of the science subsystem had hardly begun, and even for Schumpeter before the First World War, research and invention were still largely exogenous to the firm. But by the 1920s the 'invention of a method of invention', as Whitehead called the R&D laboratory, was generating a 'research revolution'.

Bernal's principal contribution to economics and to the other social sciences was his clear perception that the allocation of resources to the various branches of organized R&D and related scientific and technical services, and their efficient management, had become crucial for the development and performance of nations and enterprises in war and peace alike. Around this central idea he was able to build up a critical analysis of the use and misuse of science and technology in Britain and in other countries and thereby to establish 'science policy' and 'technology policy' as an important issue of public debate and government intervention.

What Science Does

A quarter of a century after the publication of *The Social Function of Science* a group of his admirers published a set of essays, under the title *The Science of Science*, to commemorate Bernal's achievement. They also set up the Science of Science Foundation to attempt to continue his work. So far as I am aware, Bernal rarely, if ever, used the expression 'science of science' in connection with his own work, although it had already been used in connection with other social studies of science.

Personally, I am unhappy with the expression, although this is partly a simple question of semantics. The Anglo-Saxon usage of 'science' to mean natural science is so ingrained that the legitimate claims of the social sciences to be included (as they are in the Continental conception of *Wissenschaft* or *Nauk*) have never made much headway. In the English language, 'social studies of science' is still a far more intelligible description of what Bernal was about than 'science of science'. (Presumably the Science of Science Foundation itself took the same view as the name was soon changed to Science Policy Foundation.)

Perhaps it was the fragmentation of the social sciences that prevented social scientists from attempting to do what Bernal did in his book – to analyse how the science subsystem actually worked in contemporary society. Or perhaps, as Jewkes suggested when he came to study 'the sources of invention' twenty years later, they were put off by the seemingly intractable measurement problems. Or, as he again suggested, they may have thought that their other preoccupations were more urgent. A physicist or a biologist might be less bogged down by the traditional preoccupations of social science disciplines and bolder in charting a new course.

Bernal certainly made errors in economics and in politics. He had the physicist's determination to understand the behaviour of the system as a whole and to build order from a complex mass of detail by generating simplified fundamental laws and abstract models as a first approximation of the system's behaviour. He remained throughout his life first and foremost a natural scientist, but in relation to social systems this was a strength rather than a weakness for, whereas many social scientists never saw the wood for the trees, Bernal did see more clearly than any of them some features of social change that they largely or completely ignored.

Whatever the reason or combination of reasons, it was in fact a zoologist (Julian Huxley) and a physicist (Bernal), and not economists or sociologists, who made the first attempts to measure the scale of resources committed to scientific research in Britain in the 1930s: to examine the sources of financial support and to point out the political, social and economic implications of the pattern that emerged from this analysis. Bernal's was one of the first good maps of science as a social system and the first good guidebook too.

Before Huxley and Bernal, of course, it was not a blank sheet of paper. Many little pieces of information existed and parts of the map had been drawn. The situation was rather like maps of the world in the fifteenth century. Bernal's achievement was to establish a global perspective, to put the continents more or less in the right places and to give some valuable advice about the problems of navigation. As he knew only too well, much of his detail was inaccurate, being based on incomplete survey data and in some areas on unreliable observations. He did not have the resources of a national or an international organization to commission full-scale survey work. Nor was he able to resolve the definitional problems that abound in the field. Nevertheless, he made remarkably good use of the sources at his disposal and used intelligent guesswork to fill in some of the gaps.

After reviewing the available statistics he remarked:

> From what has already been said it can be seen that the difficulties in assessing the precise sum annually expended on scientific research are practically insurmountable. It could only be done by changing the method of accounting of Universities, Government Departments and industrial firms. . . . Nevertheless it is necessary to obtain some idea, however rough, inside which expenditure on research is likely to lie, in order to see the position of research in the national economy. (p. 62)

The estimate he came up with was that in 1934 Britain was spending about £4 million on scientific research. His upper limit, which he called 'gross', was £6.7 million and included applied research, and in some cases routine testing work. Definitions differed between government and industry. His lower limit, which he called 'net' was £1.9 million and corresponded fairly well to what would now be termed 'fundamental research'. He estimated that his figure of £4 million was about 0.1 per cent of the national income at that time (then being estimated unofficially by almost equally amateurish techniques). Comparing this with what was known about research in the USA and the USSR he came to the conclusion that both these countries were spending a significantly higher fraction of their national income on scientific research than Britain – 0.6 per cent in the case of the USA and 0.8 per cent in the Soviet Union (p. 65).

It was not until nearly twenty years after Bernal's original work that the first official survey of research and development expenditure was undertaken by the British government, in 1955–56. A few years earlier the National Science Foundation had initiated annual R&D surveys in the USA. Since the 1950s almost all industrialized countries have adopted R&D surveys as part of their regular statistical procedures, and many developing countries have done so too, so that the type of statistics that Bernal pioneered in the 1930s has now become an almost routine business in most parts of the world. The Organization for Economic Cooperation and Development has acted as a focus for standardization of definitions and international comparisons.

The major difference between Bernal's work and that of the OECD surveys lies in the approach to development (nowadays more precisely described as 'experimental development'). Bernal did not use the expression R&D in *The Social Function of Science*, nor did other scientists at that time. Even in the very incomplete surveys in British industry in the 1930s information was collected about 'industrial research' or 'scientific research', and 'development' was not mentioned. This was still true even in the early post-war surveys. Bernal's estimates therefore should be regarded as estimates of R rather than of R&D. The definitions were imprecise, in any case, so it is impossible to know the degree of underestimation which this involved if we are thinking of R&D on contemporary definitions. But probably D in the 1930s was about twice as big as R so that Bernal's figures would have to be trebled to make them comparable with current R&D survey data.

The same point affects Bernal's international comparisons. The US industrial surveys of the 1920s and 1930s were more complete than the British and took more account of D. The combined R&D Department was more characteristic of US industrial practice and the expenditure figures available in firms reflected this. The Soviet surveys were even more comprehensive and included many scientific and technical services (STS) as well as both research and development, such as geophysical exploration, project design and feasibility studies, scientific information services and so forth. The USSR led the world in measurement of all these activities but any reliable international comparison would have to take account of these differences in the scope of measurements. Thus a closer approximation to the true comparable figures of R&D as a fraction of national income would probably be 0.3 to 0.4 per cent in 1934 in the UK, 0.6 to 0.7 per cent in the USA in the same year, and 0.2 to 0.3 per cent in the early 1930s in the USSR rising to 0.4 to 0.5 per cent by the late 1930s. Bernal's upper estimate for Britain probably included all R, a little D, and some miscellaneous STS, while his central estimate was fairly near the mark for R alone. His figures for university research and government research were more reliable than those for industry, where he had to rely on very sketchy and incomplete data.

It might be tempting to infer that Bernal shared the disdain for technology or development that is often supposed to be a characteristic of British university scientists. But this would be remote from the truth. It is true that he does quote from Huxley's earlier report which uses the expression 'mere development', but Bernal was aware both of the importance and of the difficulties of development. He always insisted strongly on the importance of the interactions between science and technology in the historical development of science, and he was of course intimately involved in development work and applications research during World War Two. The exclusion of D from his measurement system was a largely pragmatic issue.

However, it probably would be true to say that he always regarded basic research as the heart of the whole science–technology system and that he saw the advance of fundamental science as the most dynamic element. After discussing the limitations of incremental rule-of-thumb improvements in traditional techniques he goes on to remark:

> The most complete integration of industry and science is, however, only reached when the knowledge of the fundamental nature of the processes is so extensive that it is able to lead to the development of entirely new processes unthought of, or indeed unthinkable, by traditional methods; as for example in the chemical synthesis of new dye-stuffs or specific drugs. The same result follows even more directly when a purely scientific discovery of a new effect is turned to some industrial use as, for instance, in the telegraph or the electric light. In these cases, we have an industry scientific through and through, an industry which owes its inception as well as its development to science. (p. 129)

Bernal's vision of the future was that this tendency towards industries that were 'scientific through and through' would predominate but would only reach its ultimate fruition in a socialist world. Much of Part I of *The Social Function of Science* is concerned with demonstrating the inadequacies and shortcomings of the organization and management of science in a capitalist society. Two of his most telling points are as important today as they were when he wrote: the diversion of a great deal of scientific and technical effort into armaments development and production, and the tendency for industrial R&D to achieve much better results in capital goods than in consumer goods and services.

Bernal's indictment of the failings of science policy in the 1930s thus comprised three main elements: first, the inadequate scale of resources for research and other scientific activities; second, the misdirection of those resources that were available; and, third, the inefficient use of the resources committed. Much changed with respect to the first point after he wrote his book, both in the capitalist world and the socialist world. The period from 1940 to 1970 was one of extremely rapid growth in the R&D system and in other STS. A part of this change must be credited to the influence of Bernal both in Britain and outside.

The second and third problems remain the focus of intense debate and controversy, and many of the issues of misallocation and inefficiency that Bernal identified are still very much with us today. In particular, the problem of a satisfactory scale, direction and effectiveness of consumer-oriented R&D is more crucial than ever, and the related problem of military R&D is far more serious and far larger in scale than it was even in the 1930s.

The socialist countries were not able to improve matters much in either of these areas and were not able to fulfil Bernal's hopes that the planned use of science and technology for social welfare would enable these societies to advance much more rapidly than the capitalist countries. Bernal did not discuss the limitations and problems of science and technology policy in the socialist world, but this became an extremely important issue both within these countries and outside them. He had an idealized vision of what science could do, which has not necessarily been falsified by the experience of the last forty years, but which obviously requires a great deal of re-examination and debate. As a result of Bernal's work and the subsequent developments there is fairly widespread agreement now on what science does but there is even greater controversy on what science could do, to which I now turn.

What Science Could Do

The first part of Bernal's book was analytical and descriptive, the second was mainly prescriptive. Having measured roughly the scale of resources committed to scientific research, their approximate distribution in terms of objectives and sector of performance, and the inefficiency of the system as he saw it, he went on to propose a series of radical changes and reforms which he believed would enable science to fulfil its true potential in terms of human welfare.

This method of tackling the subject was entirely characteristic of Bernal. He insisted always, both in this book and in all his other work, on the importance of the social implications and applications of any branch of knowledge. He believed profoundly that the interactions between science and the economy were mutually beneficial, and his main criticism of contemporary social science was that it was too much of an abstract spectator subject and not sufficiently involved in policy making and real-world testing.

It was equally characteristic of Bernal that he never took up the sterile position that nothing could be done in a capitalist society. Although he argues in 'What Science Could Do' that most of the reforms and changes he is proposing depend upon a socialist transformation to society, this does not stop him from advocating a great many reforms and improvements straightaway. Whether he was looking at air-raid precautions or building technology or policies for fundamental research, he was always ready to make constructive practical proposals for immediate action as well as fundamental criticisms and proposals for long-term structural change.

Unlike Keynes he was always a committed socialist but, like Keynes, he was ready to offer advice to governments and various political parties and even to work for governments both in official and unofficial capacities. Like Keynes, of course, he was often *persona non grata*, but especially during World War Two he was very much *persona grata*. For this reason, many aspects of the programme of reform that Bernal proposed in the second part of *The Social Function of Science* have had an appeal not only to Marxists but to many others who shared his concern for using science more effectively for human welfare but did not necessarily agree with him that the Soviet Union represented a good model for such policies, or even that socialist planning of the economy was essential at all.

This aspect of *The Social Function of Science* has led to some criticism by contemporary radicals who regard it as a manifestation of 'techno-economism' or 'reformism'. They are entitled to their view but, in my own, there is no justification either in *The Social Function of Science* or elsewhere in his life and work for the claim that there were two Bernals – one a radical revolutionary whom they would like to embrace, and the other a Fabian technocrat whom they would like to disavow. I shall return to this topic at the end of this chapter but here I want only to make the point that 'What Science Could Do' represents both a programme of short-term reforms and improvements and a long-term programme of more fundamental social changes related to deeper changes in the organization of society as a whole. Bernal never saw any contradiction in this manner of approaching the problem and neither do I. That his critics have problems of comprehending this reflects much more on them than on Bernal, and relates to their special interpretation of 'science' to which I shall return.

The first major point Bernal makes in the second part of his book is that there is a need for a massive increase in the scale of commitment of resources to research and other scientific activities. At the time, it appeared an incredibly bold proposal to suggest that support for research should be increased tenfold, as he himself acknowledged (p. 242). But such an increase was in fact achieved over the next twenty-five years both in Britain and in a good many other countries. As we have seen, Bernal overestimated the scale of Soviet R&D in the mid-1930s and, as part of his justification for a tenfold increase in British research, he argued that the USSR was already spending 1 per cent of national income, whereas Britain according to his estimate was spending only 0.1 per cent. Consequently he maintained that there was already a precedent for the scale of commitment he was proposing.

Although he was wrong about the precise figures and understandably did not resolve the problems of international comparability of definitions, he was right about two fundamental points. It was possible to expand the scale of R&D at approximately the rate he suggested for a considerable period, and this was already happening in the 1930s in the USSR. Other countries did indeed follow the Soviet example in this respect, in the first place because of the impetus of World War Two, but later on a more sustained basis in the 1950s and 1960s, largely because the kind of arguments that Bernal evoked in the 1930s carried widespread conviction far beyond Marxist or socialist circles.

As a result of this rapid and sustained expansion the leading industrial countries were spending about 2–3 per cent of national income on R&D by the end of the 1960s, and on a comparable basis the USSR was spending 3–4 per cent of national income. Typically, about one quarter or one third of this total was R and the remainder was D. In Bernal's terms, therefore, R had increased from his estimate of about 0.1 per cent of national income to about 0.7 per cent in the UK in the 1960s, whilst combined R&D had increased from about 0.3 per cent to about 2.3 per cent of a much larger national income. In Britain national income had doubled in this period and elsewhere it had grown even more. In absolute terms, therefore, the increase over twenty-five years may have been rather more than tenfold.

Since the 1960s the rate of increase has slowed down substantially and even gone into reverse for some years, especially in the UK. Bernal expressed alarm when this tendency first became apparent and he described the arguments then advanced for zero growth in the science budget as 'voluntary underdevelopment'. He continued to believe that still further expansion of scientific activities was desirable throughout the world but at an especially high rate in the developing countries. In *World without War* (1958) he visualized a long-term change in the distribution of employment over the next century in which scientific and technical occupations would come to employ about one third of the total labour force. Many of these would be in STS and only a minority in R&D but, nevertheless, in Bernal's view it would be desirable to expand R&D well beyond the 2–3 per cent fraction of national income that had been realized during his own lifetime and of which he had been the principal prophet and instigator. Long before Daniel Bell, Marc Porat and other apostles of the information revolution, Bernal had recognized the probable future evolution of occupational structure in advanced industrial societies and had projected the probable decline of manufacturing employment to a level comparable to that of agriculture in these societies today.

However, although the tenfold expansion of research that Bernal proposed was realized on a time scale not much longer than that which he projected, neither in Britain nor in the USSR, nor in any other countries did the consequences of this expansion quite live up to his expectations and hopes. He had believed that there would be a very high rate of return to the massive increase in R&D that he advocated, that is, that national income would grow more rapidly and that other welfare benefits would accrue not necessarily measured in national income statistics. For this reason he estimated that a tenfold increase in the real resources devoted to R&D would raise the fraction of national income devoted to research not from 0.1 per cent to 1 per cent, but from 0.1 per cent to 0.5 per cent (p. 242). This means that he calculated that national income would double during the period that R&D was expanding, largely or entirely because of this investment in R&D.

If we continue to regard Bernal's estimate of 0.1 per cent as referring to R rather than R and D, then he was not too far out in this rough calculation but, of course, as he would have certainly agreed, many other factors were affecting the growth of national income as well as the increased scale of investment in research, development and other scientific activities. Nevertheless during the 1950s and 1960s the evidence of empirical research by both economists and engineers has increasingly favoured Bernal's view that technical change associated with professionalized scientific and technical activities is the single most important element in the complex process of economic growth.

So far as R&D is concerned this is not quite so straightforward as it might seem. Clearly, rapid growth is quite possible simply on the basis of the diffusion of existing best-practice techniques. The differences in productivity between best-practice and worst-practice enterprises are so great that a doubling of national income would be feasible anywhere without new knowledge. Moreover, it is also true that a very important part of the technical improvements that are continually introduced in the production process do not depend on formalized professional research or even on R&D, but result directly from the activities of workers and engineers more directly involved in production. These considerations are especially relevant to the circumstances of developing countries with very limited scientific resources, but they are as important everywhere. In practice it has proved possible for a good many countries to achieve high growth rates of national income with a relatively low fraction of that national income devoted to professional R&D, even as low as 0.1 to 0.3 per cent.

When growth is based to a considerable extent on the assimilation of best-practice techniques first developed elsewhere in the world (and this must be true of most technical change almost everywhere) then local R&D may still be very important, as the experience of Japan has shown, since it enables the adaptation and improvement of imported technology to be carried through with maximum efficiency. Moreover, as a country industrializes and catches up with best-practice techniques, it will increasingly face the need to make original contributions to industrial R&D if it is involved in international trade competition. Consequently, the fact that other countries have achieved higher growth rates than Britain with a much lower commitment to R&D than Britain since the war certainly cannot be taken as invalidating Bernal's arguments for a massive increase from the pre-war level of British R&D. There is no country successfully competing on a significant scale in world trade in manufactured goods (as opposed to oil or primary commodities) that spends a lower fraction of industrial net output on nonmilitary R&D than Britain.

However, these considerations do serve to emphasize some aspects of the 'returns to R&D investment' problem that were relatively neglected in *The Social Function of Science*. Bernal recognized of course that the dissemination of scientific and technical knowledge was extraordinarily important. There is very great emphasis throughout this book both on education and on what is now called STINFO (scientific and technical information services). Some of the best chapters are concerned with reforms in science education, the science curriculum, and scientific publication and documentation. Moreover he went further than most contemporary reformers in his proposals for the popularization of science and technology on an enormous scale. He cannot, therefore, be charged with any failure to appreciate these aspects of the problem of widespread introduction of technical change. Where he was perhaps open to criticism was in the neglect of other complementary aspects of the process which are embraced in the concept of 'STS' but not of the narrower concept of R&D, still less in R by itself.

In particular, *The Social Function of Science* has little or nothing to say about some aspects of engineering and investment which are extremely important both for the effective implementation of much R&D work and for the efficient adaptation and exploitation of imported technology. Even forty years after Bernal, we are still very far from measuring (or even satisfactorily defining) the full range of STS, although we have made some progress with R&D measurement thanks largely to his original impulse. But we do know enough about STS as a result of the work of Martin Bell and others to know that project engineering activities, design engineering, production engineering and similar functions are extraordinarily important complementary activities in the process of efficient technical change. A fuller understanding of the efficient management of technical change will have to embrace the entire range of STS and not only those aspects on which Bernal focused in his pioneering work.

This leads on to a wider consideration of 'What Science Could Do'. Whereas Bernal has been at least partly vindicated in his belief in the importance of a massive increase in research and research-related activities for the future growth of national income, neither the extent of the increase nor its composition and distribution have been satisfactory. I do not share the view that the improvement in living standards in Britain since the war is trivial or unimportant. On the contrary, even though it was less than that achieved in most other European countries, and even though its fruits have been unevenly distributed, I do believe that it was a very important social gain just as it was in other countries both in Eastern and in Western Europe.

However, the returns to the vastly increased investment in R&D and other scientific and technical activities have been less than Bernal had hoped for and this has been a source of disappointment to many, both those who shared his political ideals and those who did not. The simple explanation of this disappointment is, of course, that whereas the *scale* of R&D was increased in the way which he suggested, the fundamental social changes he advocated and believed to be essential to realize the full benefits of the increased scale of research have not been made. However, in the Soviet Union and other socialist countries which increased their commitment to research and related activities on an even greater scale than Britain, and which also implemented many of the other reforms and changes that Bernal suggested, there was also some disappointment with the results. The plain fact is that in the 1970s growth rates for national income declined *both* in Eastern and in Western Europe and that this decline in growth rates occurred *after* the most massive expansion in scientific and technical activities that the world had ever known. Moreover, it is also the case that the bias in technical change in the direction of capital goods and materials, rather than consumer goods and services, has persisted worldwide.

Socialists, including Bernal, would have no difficulty in explaining the situation in capitalist economies. They would argue, as Bernal did in the 1930s, that the decline in growth rates, or the absolute fall in output which then occurred, was a failure of the economic system which could not make satisfactory use of the fruits of science and technology. The decline in the rate of growth in many of the socialist countries and their failure to surpass the growth rate of many capitalist countries in the 1960s are more difficult phenomena to explain in Bernalian terms. A satisfactory explanation would have to involve either a much more fundamental critique of the social, economic and scientific systems of the Soviet Union than Bernal was willing to make in the 1930s, or a hypothesis of diminishing returns to R&D investment, at least temporarily, both in capitalist and in socialist economies.

Such a phenomenon of diminishing returns is of course quite consistent with the strong emphasis on the two other major aspects of Bernal's programme stressed in 'What Science Could Do', respectively the misallocation of scientific resources and the inefficient use of those resources even where appropriately allocated. Although the scale of total resource commitment has increased much as he advocated, the major misallocations that he identified have not been rectified and may even have worsened, and the inefficiencies he identified may well have persisted despite some attempts to improve matters.

The greatest single misallocation of resources that Bernal identified in terms of human welfare was of course the diversion of scientific and technical effort into military R&D. The scale of this misallocation has vastly increased since the 1930s both in Britain and in the USSR and the USA, but the *proportion* of total scientific activities devoted to military objectives may not be so very different now from then – perhaps about one third of total R&D. There were periods during and after World War Two when this proportion was even higher – probably more than a half at one time in all three countries – and there are alarming signs that it is climbing once again after a period in the 1960s and early 1970s of slow decline. However this may be, the enormous waste of resources involved in the scale of this misallocation is even more evident today than it was in Bernal's time. He himself returned to this theme again and again after World War Two, most notably in *World without War* (1958) in which he argued cogently that the most straightforward way rapidly to increase the scientific resources available for global development was through world disarmament and the re-allocation of military R&D resources.

From time to time, voices have been raised in support of the spin-off hypothesis – that the civil economy benefits to a great extent from the technical advances won through military R&D. But the weight of the evidence supports the view that the effect of spin-off is slight, that by far the greater part of military R&D has little or no civil application, and that insofar as there is any genuine spin-off, this is an extraordinarily inefficient way to make technical progress, quite apart from all the other overwhelming arguments about the dangers of a technological arms race. A particularly alarming feature of the present massive misallocation of global R&D resources is the tendency for a growing number of poor developing countries to invest a high proportion of their extremely limited scientific resources in military areas.

Those who naïvely imagine that this process may ultimately enhance their economic performance should contemplate the experience of Britain and also that of the USA and the USSR. The evidence is now very strong that the opportunity cost of the diversion of a high proportion of the best industrial scientific and engineering talent into military projects in the 1950s was quite unacceptably high. The two strongest countries in world trade in manufactures since World War Two, West Germany and Japan, did not suffer from this disability. They were able to concentrate their scientific and technical efforts almost entirely on civil objectives until recently, and the consequences are there for all to see. In those industries that really matter for international competition, such as vehicles, engineering and chemicals, they have outstripped not only Britain but increasingly also the United States, too, both in new product design and in productivity. As Pavitt and Mary Kaldor have shown, the relative strength of the USA and Britain in a very limited area of defence-related industries, such as aircraft, has not compensated for this relative deterioration in competitive power across a much wider range of industries. The same point applies *a fortiori* to the USSR, although it is perhaps not one that Bernal would have made himself.

Bernal's analysis of this major resource misallocation must therefore be accepted as a large part of the explanation of the relatively disappointing results of the increase in sheer scale of R&D, particularly so far as Britain is concerned. Other types of misallocation identified in *The Social Function of Science* pale into relative insignificance by comparison with the military R&D problem, but undoubtedly some of them have also continued to be important. Among these are the relative scale of effort in consumer goods R&D compared with capital goods, the relative neglect of the life sciences and the social sciences compared with physics and chemistry, and the substantial effort devoted to product differentiation in industrial R&D as opposed to more fundamental technical advance.

Planning and Science

The central argument of 'What Science Could Do' is not merely one of gross underinvestment in science and technology, and serious misallocation of resources. It is also the more fundamental one that these deficiencies, together with many other sources of inefficiency in the system, really require some form of socialist planning if they are to be rectified. It was this argument which provoked the main attack on Bernal both at the time and since, on the grounds that planning would destroy freedom and would be even more inefficient. Since *The Social Function of Science* was published in 1939, the crude anti-planning standpoint has lost ground, although it remains a very important influence. But the view that identifies socialist science planning with authoritarian orthodoxy and bureaucratic repression has made considerable headway not because of anything in *The Social Function of Science*, but because of events since that time, especially the Lysenko controversy and its aftermath.

Hardly any government in the world, whatever its formal ideological complexion, has accepted the simple view of abdicating central government responsibility for science and technology and leaving these areas to the free play of market forces to determine resource allocation and scale of investment. This is for the simple reason that even the most extreme advocates of the virtues of the free market have recognized that, in terms of the pure principles of a *laissez-faire* world, there has to be public investment in basic science, science education and scientific and technical information systems. Most of those classical and neoclassical economists who have examined the issues at all closely, from Adam Smith onwards to Kenneth Arrow, would add to this list various other parts of the science-technology system such as the promotion of new technologies, technical advisory services in agriculture and other branches of the economy, military-related aspects of science and technology, and so forth.

Even if only the first group of activities is accepted as an area of public responsibility, primarily financed from central government funds (as it is in almost every country in the world), there is already a major government involvement, which requires some form of science policy. It is partly a matter of semantics as to whether this policy is described as 'planning'. Some countries which do not describe their system as a planned one pay a good deal of attention to aspects of policy making and execution normally associated with planning.

The rationalization and idealization of 'muddling through' is certainly one of the continuing sources of opposition to planning, both in theory and in practice. But a more sophisticated theoretical justification of *laissez-faire* in science was developed as part of the critique of Bernal. When *The Social Function of Science* first appeared, it was strongly attacked by J. R. Baker on the grounds that planning meant the suppression of freedom in science and the regimentation of scientists. Later, Baker was associated with various attempts to establish a theory of 'anti-planning' and give it greater credibility. The best-known of such attempts was that of Polanyi after World War Two.

In his article on the 'Republic of Science' in the very first issue of the journal *Minerva*, Michael Polanyi attempted to develop a doctrine of pure *laissez-faire* in relation to fundamental scientific research. He did this by making a direct analogy between a pure competitive economy and the fundamental research system. Just as the individual entrepreneurs were the only ones qualified to respond to market signals of price and demand so, Polanyi argued, only the scientists directly involved had the specialized expertise to make decisions in relation to a particular field of science. Unfortunately for Polanyi and his supporters, there were three fundamental flaws in this argument insofar as it was designed to limit government involvement in science or exclude it altogether.

The first and most important was that Polanyi himself (and everyone else involved in the argument) conceded at the outset that government must provide most or all of the money for fundamental research, however it might subsequently be distributed. This inevitably means that somehow or other government must develop a policy on the total budget for science and its growth (or reduction), just as Bernal had insisted. This was and still is often an unwelcome responsibility for some governments and one they usually attempt to devolve to various scientific advisory bodies, but few have been able to escape it altogether.

The second basic difficulty with the Polanyi approach is that – unlike the market economy, where the criterion of profitability can at least superficially be applied to every form of activity, quite irrespective of the actual products or services – the criterion of scientific merit has no such interchangeable recognition. Polanyi may well be right (and Bernal thought so too) that the specialists in a particular field of high-energy physics may be the only ones competent to decide on the relative merits of alternative research projects in that field, and the same may go for any other specialized branch of science. But there is absolutely no reason to expect that the specialists in high-energy physics can decide on the relative merits of competing claims in, say, cell biology, geophysics, medieval history, and colloidal chemistry. In fact, the contrary is the case on Polanyi's own theoretical arguments. This means that the Polanyi model breaks down at the most important level – the subdivision of a global budget between various disciplines and subdisciplines. So long as the total science budget was both small and steadily expanding, this was not too serious and the Polanyi 'autonomy of science' system appeared to work. But once budgets became large and more seriously constrained, this problem became much more acute and is a critical one in all science policy bodies today, irrespective of whether they talk about 'planning' or not.

Moreover, the problem has become accentuated by the development of 'big science' – areas of investigation where the capital cost of instrumentation or other types of equipment has become so high that it may pre-empt a large part of the total funds available for science. The extreme example is space exploration, but there are many other cases where the issue is important – accelerators and telescopes are two obvious ones. This problem may partly be circumvented by separating these decisions and funding them under separate budget headings, for example, NASA or military or nuclear power budgets. But this is only a pseudo-solution since the development of science is then affected by many different budget decisions and 'coordination' or some form of mutual consultation becomes necessary.

Finally, a third difficulty is that government involvement inevitably includes responsibility not only for the total level of funding of science (and some areas of technology) and for the principal subdivisions within that total, but also for the 'health' or 'efficiency' of substantial parts of that system. Even under highly decentralized management systems, such as the universities, this remains true, and it is a much more direct responsibility in the case of government laboratories, research stations, libraries, patent offices and the like.

In the 1930s even the idea of 'science policy' was still unfamiliar and the notion of a long-term plan for science was regarded as far-fetched and utopian. As with the Keynesian heresy in economics, with which Bernal's ideas have many parallels, it was World War Two that made possible the more widespread acceptance of the notion of explicit overall central government responsibility for the healthy development of science and technology as for a healthy state of employment, investment and growth in the economy more generally. Partly for this reason, Baker's initial attack was never very effective and its revival by Polanyi after the war was hardly more successful.

If the area of public responsibility is more widely defined (and in almost every country central governments are as a matter of fact involved much more widely), then government policy extends to technology policy as well as science policy more narrowly conceived as policy for fundamental research. In these areas too, it is much easier to point to the shortcomings and inconsistencies of government policies or 'plans' (when they are described as such), than it is to destroy the argument for some degree of government responsibility. Even if planning is regarded with great distaste and reluctance as an inevitable residual obligation, in practice it has been found difficult to dodge the responsibility entirely. This means of course that, as Bernal insisted, the job is often done half-heartedly and incompetently.

This can be confirmed by simply examining the history of public administrative and legislative activity in the USA or any other industrialized society during and since World War Two. There has been a bewildering succession of different committees, councils, centres, departments and administrations but the underlying trend has clearly been towards a more extended involvement not only with respect to the funding of basic science through the National Science Foundation, but also in many other directions. The disavowal by the Rothschild Report of any central system of priorities for science or technology in the UK was accompanied by a substantial increase in government departmental 'planning' of science and technology and the re-entry of *de facto* central priorities by the back door.

All these problems were clearly recognized by Bernal in the 1930s and it is hardly surprising that in subsequent issues of the journal *Minerva* and elsewhere, the Polanyi school had distinctly the worst of the argument. They really did not have a leg to stand on and could only score points by pointing to the inefficiency and incompetence of government policy making and 'planning'. Here they were on relatively strong ground as progress in developing a more efficient way of managing the science–technology system has not been conspicuous since 1939, whether under socialist 'planned' systems or others.

Thus, although it is true that since *The Social Function of Science* government responsibility for scientific activities has been greatly extended, both in scale and in range, it remains true that many of the sources of inefficiency identified by Bernal are still evident. Some of these relate to the management and planning of work in individual institutions, others to the coordination of groups of institutions and their joint action, and finally some relate to the management, coordination and planning of the entire system or network of institutions.

It is by no means the case that, when he identified the sources of inefficiency, waste and misallocation in British science in the 1930s, Bernal confined his strictures to the influence of the market or its failings. On the contrary he stressed, very strongly, bureaucratic government mismanagement and waste as well as the deficiencies of the private market system. He was particularly insistent that the whole style of work in government laboratories was often inimical to creative work, and he emerged as an early and extreme advocate of what would now be called Flexitime:

> The special difficulties which beset scientific research carried out under government control are largely due to bureaucratic methods. Civil Service and Army methods of administration are essentially unsuitable for the carrying on of research. Research is always an exploration of the unknown, and its value is not to be measured by the amount of time spent on it but by the output of new ideas thought of and tested out. Regularity of hours or days, clocking in and clocking out, with an annual fortnight's holiday, is not conducive to original thought. The work of a scientist requires the most irregular hours. Sometimes it may be a matter of wanting to work sixteen or twenty-four hours a day for weeks on end; at others all the time spent in the

laboratory is useless and the best results would come from going to parties or climbing a mountain. (p. 106)

He also realized clearly the dangers of an authoritarian or autocratic style of management both at the laboratory level and higher up. All the post-war discussion among sociologists about the most efficient way to manage laboratories and the empirical investigations of management techniques have only served to confirm the conclusions reached by Bernal. He pointed out that:

> We have to balance, as in all political affairs, the dangers of arbitrary or incompetent action of a single individual against the possible lack of unity, inconsistency and general obstructiveness of any corporate body. According to the temperament and abilities of the people available for the work at any particular time or place, one or the other method may be the more suitable. In a scientific laboratory an individual who, with his guiding ideas, markedly leads the way, will be willingly chosen as a personal director; in other cases it may be a small group who combine in a harmonious way a set of ideas which they can realise effectively only by working in close collaboration. (p. 268)

Whether the responsibilities of the director were exercised by an individual (which he thought generally the most effective way) or by a small group, he saw the need to guard against arbitrary power, ageing and incompetence by involving the whole staff of an institution in discussion of major issues: 'Any new organisation of science, if it is to be vital as well as effective, must bring with it the democratic principle which will ensure adequate participation in responsible control by scientific workers of every grade of seniority' (p. 116). At higher levels of management and planning he thought the same principle was valid but recognized the need for some delegation of power and responsibility. However, it is at this level of decision making that Bernal's analysis becomes more vulnerable and he sometimes underestimates the problems involved, in terms both of the political power structure and of the content of the decisions themselves. Nevertheless he can hardly be faulted for his clear recognition that *some* method of determining central government priorities for science and technology must be established, and that there was merit in deliberately debating these priorities and encouraging new initiatives, especially in the areas between established disciplines. The idea of a publicly discussed flexible plan remains the most viable approach despite our greater awareness of all the pitfalls involved. Experience so far suggests that it is more difficult than Bernal supposed to eliminate inefficiency and waste and to achieve optimal use of resources through any known planning system, but this does not destroy his basic argument.

Moreover, the very difficulties that have been encountered in all types of economy surely strengthen his case for public criticism and debate and for not concealing these problems by a conspiracy of silence:

> If we try to examine more closely the inefficiency of science as a method of discovery, we find that it originates in two major defects. The first is the totally inadequate scale of finance ... the second is the inefficiency of organisation which ensures that these small resources shall be to a large extent wasted. This last remark may seem to the scientist something approaching treason. Even if it were true it should not be publicly stated, for the little that science gets now it gets as the result of the belief in its effectiveness. Once it is suspected that scientists waste the money that is given to them, they will not be able to get even that much. Yet the gentleman's agreement to gloss over the internal inefficiencies of science is bound to be disastrous in the long run. However carefully concealed such things are, they are always suspected and give rise to an attitude of vague distrust on the part of possible benefactors and the public at large, which is far more damaging to science than openly brought charges. (p. 99)

Ironically, since the wasteful misuse of resources reached a much vaster scale in government military and prestige projects, scientists often actually found it easier to get more money but Bernal was nevertheless right about the long-term damage brought about by this type of situation, whether publicly or privately financed.

Thus, in his critique of science under capitalism, he was not simply contrasting 'public' and 'private' and arguing that public operation and responsibility was somehow always better, he was rather contrasting an idealized model of socialist planning with the private *and* public decision making of a captialist or mixed economy. Unfortunately for his case, he chose to identify such an idealized model of socialist planning very largely with the actual workings of the science–technology system in the USSR. This was the most vulnerable part of the whole argument of *The Social Function of Science* and the remainder of this chapter discusses the reasons for Bernal getting himself into this trap, if trap it were.

The extent to which Bernal idealized the operations of the Soviet science–technology system can be seen especially clearly in his discussion of the relationship between basic research and industrial technology. Since World War Two, and especially in the 1960s and 1970s, a lot of the internal debate in the USSR concentrated on the weaknesses of this relationship and the relative failure to achieve a satisfactory two-way flow between academy (basic research) institutes, industrial research institutes and factory-level development and implementation of technical change. It has been widely recognized, both inside and outside the USSR, that the organizational separation of the various institutions led to major communication failures and other deep-seated problems in the system, including much duplication and waste of resources. Yet Bernal wrote in *The Social Function of Science*:

> From the scientific point of view the research institutions in industry and agriculture are closely linked with the Academy, and the separation which exists in Britain between academic and industrial science is largely non-existent. The idea behind the organisation is that there should be a two-way flow of problems and solutions. The problems of industry put in a precise form by works laboratories are passed on to the technical institutes. In so far as their solution falls within the scope of existing technical knowledge they are solved there. But if some more fundamental ignorance of the working of nature is revealed, this is passed on to the Academy. Thus industry serves to present science with new and original problems. At the same time any fundamental discoveries made in the universities or the Academy are immediately transmitted to the industrial laboratories so that anything which can be turned to useful purposes may be used in practice as soon as possible. (p. 227)

The schematic idealized description here contrasts strongly with his great sense of realism, balance and sophistication in discussing those political and organizational structures with which he was familiar from first-hand experience. The naïvety of the phrases 'the problems of industry put in a precise form' and 'fundamental discoveries ... are immediately transmitted' is quite extraordinary for someone who perceived so clearly in his chapter on scientific information and documentation that direct personal contact was the most important form of communication, and who advocated interchange of people as essential to achieve efficiency and avoid bureaucracy.

One of the reasons he misjudged the real situation in the USSR was that he relied on two main sources of information, neither of which was reliable. One was the official propaganda accounts of the system and the other was the opinion of colleagues who had more direct knowledge of the workings of the system than he had, especially Martin Ruhemann and J. G. Crowther. In addition he was inclined to distrust the accounts of more critically minded colleagues and outright opponents of the Soviet system because of the very large amount of prejudicial comment on Soviet affairs which was characteristic of the interwar period and which he suspected biased the general view in an anti-Soviet direction.

It is important to remember too that the USSR really was the first political system in the world to accord science high priority in its economic policies in a deliberate and systematic way. It was also the first to give the highest priority to science and technology in its education system, and promoted the growth of the science system and the science education system more rapidly than any backward country had previously ever attempted. It is not surprising that these very positive features of Soviet policy made a considerable impression on Bernal and many other scientists at a time, in the 1930s, when science was being cut back in many other European countries, or mutilated and militarized as in fascist Germany.

Nevertheless, there was clearly something more fundamental in the relative suspension of his critical judgement in relation to Soviet affairs. In part it was a failure that was common to the whole Communist movement and sympathizers like the Webbs. Bernal was certainly no more gullible than I was myself, and he was less gullible than many others. Even though the general tenor of his remarks about Soviet science in *The Social Function of Science* was one of adulation he did point out in an extremely important passage:

> ... the great disadvantage is the lack of sufficiently rigid criticism, but this again is to be expected. A critical attitude is the fruit of long experience and well-established schools, its absence one of the faults of youthful enthusiasm which only time and experience can correct. A certain part is played here by the long period in which Soviet science was cut off from the rest of the world and the degree in which political, financial and language barriers still cut it off today. It is only by the comparison of the work of a very large number of scientists in different places that a fully critical attitude can be developed. (p. 230)

This, together with other passages which have already been quoted, shows that Bernal was certainly not entirely unmindful of the importance of open debate and criticism and restraints on the exercise of political power in any social system. However, he was sufficiently confident that the socialist countries were developing in general on what he regarded as the right lines that he was almost always inclined to give them the benefit of the doubt and to abstain from any public criticism of their shortcomings. It was only after this approach led to the tragic episode of his support for the official Soviet line on Lysenko that he came to realize that some more fundamental problems were involved than simply those of 'youthful enthusiasm' and immaturity.

In addition to the more general political problem of the uncritical approach of part of the left to the USSR in the 1930s and 1940s, there was clearly a specific reason for Bernal's identification with official Soviet views in this period. After all, both Bertrand Russell and Rosa Luxemburg, both greatly respected figures among left-wing intellectuals, had clearly recognized as early as 1918 the fundamental political problems of the Soviet system: the monopoly concentration of power in the party as opposed to the soviets, trade unions, or parliament; the concentration of power within the central committee of the party, and the tendency towards an even further concentration of power within a small group or one single individual. Both Russell and Luxemburg predicted clearly that the emergence of an authoritarian 'socialist' orthodoxy was just as inevitable within the confines of the Soviet system as it was in similar political systems of power concentration throughout the ages, whatever the initial ideals and scientific intentions of the founders.

Although Bernal clearly recognized the analogous problem in capitalist countries and in his brilliant indictment of fascism, there was some mental block which prevented him thinking equally clearheadedly about the USSR and later about other socialist countries. The additional peculiarity in Bernal's case was his identification of 'science' with 'socialism' and his readiness to accept one version of Marxism as 'social science'. He idealized science not just as knowledge but in a political sense too, believing that the management of human affairs could also be more scientific by virtue of being socialist. He was thus particularly inclined to accept the claims of Soviet Marxism to represent science in general, and to accord it the same degree of respect. This is a strong claim and it can only be substantiated by considering not just *The Social Function of Science* but also some of his earlier work.

The Social Function of Science was not his first book. It was preceded by his extremely daring, imaginative and provocative essay *The World, the Flesh and the Devil* written ten years earlier and published when he was only twenty-eight years old. This first book is important in giving us an insight into some basic values and ideas which affected his approach to science and society throughout his life. As he himself said forty years later in writing a foreword to the second edition: 'This short book was the first I ever wrote. I have a great attachment to it because it contains many of the seeds of ideas which I have been elaborating throughout my scientific life. It still seems to me to have validity in its own right.' In the final section of this chapter I therefore turn to the consideration of this book in relation to the central weakness of *The Social Function of Science*.

The World, the Flesh and the Devil

Enthusiasm for science is evident throughout Bernal's entire life and work. Sometimes, even in *The Social Function of Science*, he seems to forget the wider social and political system and subconsciously to identify the interests of science with the interests of society. It is not too much to say that his position sometimes came very close to 'what's good for science is good for the world'. This must be qualified by the recognition that he saw science as the instrument for overcoming poverty and war and for promoting welfare on a vast scale.

Many years later, in some notes for the Russian novelist Boris Polevoi, Bernal traced his enthusiasm for science to his boyhood in Ireland. When he was only five years old his sister was taken to be X-rayed in the hospital in Limerick thirteen miles away. Tremendously impressed by the talk in the family about the X-ray machine, he tried to simulate it in his bedroom with a paraffin lamp, as recounted by Peter Trent in Chapter 4. The lamp was knocked over and his father gave him a very bad beating which, Bernal said, he had remembered ever since. 'But it fixed firmly in my mind the determination to find out what those x-rays were and what to do about it.' When he was seven he read Faraday's *The Chemistry of the Candle* and he was excited by the prospect of making oxygen and hydrogen. He succeeded in persuading his mother to write a letter to the local chemist requesting various chemicals for experiments:

> My mother, who knew less science than I did, because she had not even read Faraday on *The Chemistry of the Candle* obediently wrote it all down and the chemist handed the stuff over, which of course if he had had any sense, he would not have done, because what he gave me in fact was a small bottle of concentrated sulphuric acid – oil of vitriol – with which I could have done myself and the house very much damage.

The experiments (conducted just outside the house) ended with a magnificent explosion. Bernal recalled:

> I was more frightened than hurt because in fact I kept my hand fairly far away. . . . But I was absolutely convinced of the truth of science. So you see, these two stories together were sufficient to start me off on a scientific career, though it was a difficult career to follow in Ireland, and it was not for many years that I was to learn about the chemical actions of sulphuric acid, and to find that my life's work lay in the use of x-rays for determining crystal structures. Strangely enough I was later to take on Faraday's old job, Assistant at the Royal Institution in London, and to work in the cellar he worked in.

Psychologists may make of these stories what they will, but what is certain is that Bernal's boyhood enthusiasm for science was to continue undiminished throughout his life. The most outstanding characteristic of *The World, the Flesh and the Devil* is the total commitment to science, and the subtitle 'The Three Enemies of the Rational Soul' indicate the basic approach of this brilliant essay in futurology. It is also better written and gives the impression of greater spontaneity than anything he wrote later.

The extent of his faith in science can best be described as religious devotion.[1] This can be seen most of all in those passages in which he looks beyond the short-term problems of scarcity and poverty (which he expects to be overcome at the latest within a few centuries) to the long-term perspectives for the human race. Here he makes clear his own preference for a society of interlinked disembodied minds devoted to the pursuit of research and control of the universe. He accurately foresees space travel and contemplates not merely reaching the stars but using them as 'efficient heat engines'. A characteristic comment is that the 'stars cannot be allowed to continue in their old way'. He is fully prepared to make the ultimate Faustian bargain, although he saw no need for bargaining.

Looking back forty years later, he was entitled to some self-congratulation for the way in which he anticipated developments in rockets and space travel, as well as other, more mundane, developments in synthetic materials and industrial technology. Indeed the prophetic technological forecasting is both bolder and more accurate than that in *The Social Function of Science*. Even his vision of space colonies does not sound anything like so implausible as when it was written half a century ago.

The boldness of his speculation about technology is, however, far exceeded by his speculation about the future of human beings and their social systems. The book should be regarded, of course, as to some extent a form of science fiction but he is ready to ask the question 'What is to be the future of feeling? Is it to be perverted or superseded altogether?' And to give the answer 'Feeling, or at any rate, feeling tones, will almost certainly be under conscious control: a feeling tone will be induced in order to favour the performance of a particular kind of operation' (p. 80). Yet he himself is obviously not satisfied with cold rationality for he comments that one of the advantages of the interconnected super-brains of the future that he envisages would be that 'feeling would truly communicate itself, memories would be held in common, and yet in all this, identity and continuity of individual development would not be lost' (p. 54). Moreover he recognizes the strength of the revulsion against technical changes that might be considered dehumanizing and confesses to sharing it himself.

> It is not sufficient, however, to consider the absence or presence of desire for progress, because that desire itself will not make itself effective until it can overcome the quite real distaste and hatred which mechanization has already brought into being. This distaste is nothing to what the bulk of present humanity would feel about even the milder of the changes which are suggested here. The reader may have already felt that distaste, especially in relation to the bodily changes: I have felt it myself in imagining them. (p. 55)

He is not so intolerant as to wish to compel humanity to follow the scientists to the stars. Indeed he takes seriously the possibility of 'dimorphism' for future humanity, that is, a division into one section which colonizes space and pursues the scientific enterprise to the ultimate, and the other – 'the old mankind – which would be left in undisputed possession of the earth, to be regarded by the inhabitants of the celestial spheres with a curious reverence' (p. 73). One is reminded irresistibly of the biblical prophecy 'Blessed are the meek for they shall inherit the earth.'

Bernal comments that this prospect should satisfy both the scientists with their unending quest for knowledge and experience, including observation of the 'human zoo' on earth, and the humanists with their desire for idyllic happiness on earth. The 'zoo' would be so well-managed that the inhabitants would not even be aware of it.

The point of this brief consideration of some of the basic ideas in The World, the Flesh and the Devil is not to dispute Bernal's vision of the future, which he was the first to recognize was inevitably speculative and highly uncertain. Characteristically he rejected the 'dimorphic' perspective outlined above because it was too probable, and 'we do not really expect or want the probable'. The point is rather to demonstrate some basic features of Bernal's way of looking at the world which shaped also his approach to The Social Function of Science. These may be summarized as follows:

1. Science is the most important activity of human beings both in the short term and in the long term. It is its own justification.
2. Science is the specialized activity of a social group – the scientific community – which has a dual function: 'to keep the world going as an efficient food and comfort machine, and to worry out the secrets of nature for themselves'.
3. The scientific community has already increased in number and compactness, so that the world may be run more and more by the

scientific experts. Real sovereignty will tend to shift to advisory bodies composed of scientists, as the influence of force diminishes and the world becomes more rational.

4. The problems of overcoming poverty could be relatively quickly solved on a historic time scale to leave humanity free to pursue other goals. Scarcely anything would ultimately be beyond the powers of organized human scientific intelligence: 'the motions of stars and living things could be directed'. God may not exist, but science has become God.

When Bernal came to reconsider his book years later, he did not retract his vision of brains interlinked to form a complex hierarchical system, but he did allow that the book was overinfluenced by the Freudian ideas of the 1920s. Apparently this criticism refers to the notion of sublimation of sexual desires into science and art. But, in considering the significance of *The World, the Flesh and the Devil* for *The Social Function of Science*, it is not so much the small dose of Freudianism in the book that is of interest as the faith in planned scientific social organization on a vast scale.

This dream was, of course, not a new one. Armytage has demonstrated that it has a long intellectual lineage and has been a recurring theme of utopian and science fiction literature for centuries. It is not surprising that a brilliant young scientist like Bernal should be attracted by this vision in the atmosphere of the 1920s and 1930s. In his own work and in that of his colleagues he could see the frontiers of science changing so rapidly as to appear almost limitless. Perhaps more than any other scientist of his generation he was aware of the new implications of scientific discoveries in relation to biology as well as to physics. Nor is it surprising again that he should share the dissatisfaction of his generation with the time-serving politicians of the 1920s and the depression and poverty of post-war Europe and contrast this with the achievements and the potential of science. It was an obvious step to link this utopian idealism about the future with what appeared as the one hopeful attempt to apply science systematically for human betterment – the young Soviet state.

As Bernal himself says, he remained in some ways true throughout his life to his utopian dream of the future. Personally I do not criticize him for this, rather the reverse. But it would be ridiculous and unjust to him to make no distinction between brilliant speculative semi-fiction and the political programmes that he actually advocated in his numerous later books. I am not suggesting that he did not distinguish between long-term speculation and everyday politics, only that there was an important connection. He came down to earth from the stars but he never lost sight of the stars.

Whatever he may have said in *The World, the Flesh and the Devil*, in *The Social Function of Science* he unequivocally dismisses the notion of government by scientists:

> The solution that is most often and most persuasively put forward, from Plato to H. G. Wells, has been the ideal one of placing the management of affairs, in general, in the hands of the philosophers or scientists. Unfortunately it suffers from two radical objections: first, that no-one can think of any way of transferring control into their hands; and, second, that most existing scientists are manifestly totally unfitted to exercise such control. The reluctance of democracies to choose people who appear, at least to themselves, so eminently suitable for controlling the community has led most of the proposers of these schemes to turn to authoritarian or, in modern language, fascist solutions. But in fascist states it happens that the scientists are used merely as tools for war preparation and propaganda. Nevertheless, though we may dismiss as fantasy the prospect of the scientist ruler, the scientist will certainly have a large and critically important part to play in the formation and development of the social organization of the future. (p. 398)

However, in the light of *The World, the Flesh and the Devil* one can far more readily understand why in *The Social Function of Science* he so easily accepts the notion of the perfectibility of organizational systems and does not see the need for pluralism and toleration of organized groups of dissent and disagreement in the way that Laski, Luxemburg and Russell so clearly did.

In trying to ask the question: 'Need science be organized?' he writes:

> A quite opposite objection to any reorganisation of science is based on the recognition of this very danger of control by elder scientists. The existing anarchic state of science gives many opportunities to evade particularly obnoxious control. If objections are taken to the policy of one committee, another can be formed to do the same work under different auspices. It is felt that organisation might put an end to these possibilities and perhaps more effectively than ever block unorthodox developments in science, through the danger of carrying over to that organisation the principles of autocratic control. But this is not so much an objection to organisation as one against existing abuse of such organisation. (p. 116)

This passage is particularly interesting because it shows that Bernal was missing the main point of the Russell–Luxemburg–Laski argument, which was precisely that a 'margin of contingent anarchy' was absolutely essential to avoid the 'abuse of such organisation', and that pluralistic political and social arrangements that deliberately tolerate organized criticism and opposition and the expression of alternative approaches were not evidence of irrational waste but essential features of a civilized community, and of an efficient system of developing and testing new ideas.

Bernal saw 'participation' as the answer to the danger of intellectual and political orthodoxy, but he failed to see how easily 'participation' could be manipulated in centralized power structures. Some of Bernal's contemporary 'radical' critics still miss the point. Whilst highly critical of Bernal's behaviour over the Lysenko issue they have not recognized how easily the insistence on a 'socialist science' in opposition to a 'capitalist science' can degenerate into the suppression of unorthodox and unwelcome ideas from whatever source. The best way to criticize and expose reactionary ideas in science surely remains to demonstrate that they are unscientific, not to rely on political labels to discredit them, or fail to perceive the common human basis of most of our science and art.

The lesson I personally draw from Bernal's *The Social Function of Science* is not that he was wrong in his enthusiasm for the ambitious and well-organized use of science and technology for human welfare. On the contrary I still share this ideal, and believe that in varying degrees it is being realized in many parts of the world, partly as a result of Bernal's impulse. But this ideal needs to be complemented by an equally explicit commitment to political and social institutions that promote open critical debate appealing to fact, experiment, experience and logical argument. Bernal did recognize this in *The Social Function of Science*, but as I have tried to show he mis-specified the basis on which it could be realized. The most important debt that I believe we owe to Bernal is to try to realize both these ideals simultaneously, not one without the other.

Note

This essay was written in 1980. A slightly different version of it was first published in Chris Freeman, *The Economics of Hope: Essays on Technical Change, Economic Growth and the Environment*, London, 1992.

Red Scientist:
Two Strands from a Life in Three Colours

Hilary Rose and Steven Rose

The Colour Purple

Francis Aprahamian, who worked with Desmond Bernal for many years, closely overseeing the editing of much of his later writing, including the final editions of *Science in History*, once described to us Bernal's characteristically innovative plan for a biography.[1] It was to be written in three parts and in three colours: red for his politics, blue for his science and purple for his sexual life. That life in three colours was never written, and the biography by Maurice Goldsmith published in 1976 was written without access to the archives or the cooperation of those close to him.[2] Thus although the biography touched on all three parts of his life it is less than adequately grounded.[3] Bernal and his literary executors – mainly, but not only, the women who were central in his life – ensured that much of the material relating to his personal life was to be held back from the archives for fifty years. Science and politics, as two overlapping, mutually reinforcing public discourses, were to constitute the archive, the private only to be marginally acknowledged through the ephemera and writings that recorded the cognitive and political development of the child, youth and public man.

The archival record of the public man, with sexual politics and even the body itself cast into the silences of the private, ironically reproduces Bernal's own vision of the future. In *The World, the Flesh and the Devil*, written before and while he was becoming a Marxist,[4] his science-based speculation saw human beings as evolving into nothing other than disembodied intelligences. Only the blue and red strands were to show; the purple was to be more or less concealed from scrutiny. As a public record we have only the account by Margaret Gardiner,[5] life-long pacifist and sponsor of the arts, of her visit to the young Soviet Union with J. D. B. That Gardiner had a child with Bernal (the historian of science Martin Bernal) or that after World War Two Jane Bernal (the psychiatrist) was born to Margot Heinemann and Desmond Bernal, and that the two children felt themselves to be brother and sister, is somehow outside the story. Only the children of the marriage to Eileen Bernal were officially recognized.[6]

Some of the silence can be explained pragmatically in that at the time of Bernal's death there were still too many living people to be hurt by an account insensitive to the fusion of private and public politics. Those of his own generation knew of his extensive sexual activities, but like many on the left had learned during the Cold War years that it was unwise to share such knowledge with outsiders. The commitment of thirties left intellectuals to sexual liberation found little support in the chilly climate of sexual hypocrisy associated with both Stalin's USSR and post-war Britain. This struggle to live personal life as new women and new men during the early sixties so scrupulously analysed in Doris Lessing's novel *The Golden Notebook* was not on the agenda for the historian of social relations in science until the arrival of an entirely new generation of feminist science historians claiming the personal as political. In consequence even Gary Werskey's account of the left scientists in the interwar period, while sympathetic to the women whom he saw as doing much of the organizational work of the social relations of the science movement in which Bernal was central, still leaves them as insubstantial figures.[7]

Joseph Needham once described Bernal to us as an extraordinarily attractive person, observing, 'Desmond was wonderfully...' then he paused, and smiled with pleasure as he found the word ... 'polyvalent'. Some of the women among Bernal's contemporaries were well aware that these affairs had not infrequently cost the woman more than the man. Bernal held strongly (and conveniently, a subsequent generation would come to think) to a view of sexual liberation as a progressive ideology to be shared equally by men and women. In practical terms this meant multiple sexual partners and no possessiveness. He had, for example, one lover in the USA who was supposed to share in a passionate relationship as and when they could get together but not to mind the separations. Unfortunately for her she did mind, and it hurt. To be told this by another of Bernal's ex-lovers showed something of the complexity of such situations. Yet despite all this he inspired tremendous and enduring love from women and men alike. As a committed Communist, above all during the Cold War, he was also to inspire hatred – the price of the life of the public intellectual.

Bernal's sexuality and its linkages with his Communism and his science is mostly a story of silences and tacit knowledges. But if we try to use the colour purple there are important clues concerning his understanding of both sexuality and science, in that his relationships to women scientists were privileged. They were almost entirely excluded from his heterosexual enthusiasm.[8] By contrast, left women were a particular focus of passion. This privileging of women scientists casts light on why, after her conflict with Maurice Wilkins at King's College, London, the X-ray crystallographer Rosalind Franklin, whose pictures were stolen to provide the crucial clue as to the helical structure of DNA, went to Bernal's laboratory at Birkbeck. We know, from the account by Franklin's close friend and biographer Anne Sayre, that his lab was indeed a refuge.[9] In this laboratory, to be a woman scientist was to enter in an honorary capacity the brotherhood of men and thus to be (almost) above heterosexual invitation. As many women who have worked in almost entirely male enclaves know, becoming this kind of honorary man is one strategy to desexualize the environment and make work relations manageable.[10] The sexual self is denied and the woman scientist is that much less a whole person, but at least it is no longer quite so easy to have one's creative work appropriated, as the King's lab culture had tolerated, or to be offensively written off as sexually unattractive as in James Watson's story of the double helix.

Bernal's ideology of sexual freedom, however challenging to the ideology of bourgeois monogamy, was none the less unthreatening to the accepted construction of scientists' masculine heterosexuality. Even at Cambridge, where a certain toleration extended to the homosexuality of arts dons (and even economists), it seemed that for natural scientists heterosexuality ruled. It was not that natural scientists were or are seen as glamorous, sexually desirable figures; they had, however dully and conventionally, to be *real* men.[11] Those who were caught transgressing this rigid construction of an acceptable sexuality came under intense social pressure; the highly publicized suicide in 1954 of the brilliant mathematician and code cracker Alan Turing, arrested for cottaging, speaks of the price gay men scientists paid.

Perhaps today it would be at last possible to be able to reflect on the silences and to write a critical biography of Bernal, from his belief in human liberation through class struggle, in abundance and peace through science, and in personal happiness through sexual freedom, through to the accommodations made with both Stalinism and the Cold War. Today, sixty years after the high optimism of *The Social Function of Science*, when whatever was both good and bad about the

Desmond's parents, Samuel Bernal (left) and Elizabeth (Bessie) Bernal, née Miller.

In the drawing room at Brookwatson, 1906. Desmond (seated, left) aged 5, with his younger brother Kevin, baby sister Geraldine and their nurse.

Photograph courtesy of Margaret Gardiner

Desmond Bernal, 1932   (Photo: Lettice Ramsay)

Bernal in the Physics Department at Birkbeck College with a Weissenberg Camera on a fixed-tube X-ray generator, used for collecting X-ray diffraction patterns from crystals.

From Felix Topolski's *Conference Sketchbook*: International Congress of Intellectuals for Peace (Wrocław, Poland, 1948). Left to right: J.D. Bernal, Hyman Levy, Ivor Montagu, Hewlett Johnson (behind him, Julian Huxley), J.B.S. Haldane.

Bernal at work on a model of the molecular structure of a liquid.

Khrushchev addressing the World Peace Council, with Desmond Bernal seated to his left. Also present: Liu Shao Qui (third from right); Paul Robeson (fourth from right); Mme Blum (sixth from right).

Bernal (left) in India in 1950 with Irène Joliot-Curie, Nobel prizewinner C.V. Raman and Frédéric Joliot-Curie.

Bernal with György Lukács at the Hungarian Parliament, April 1954.

Bernal (left) with T.D. Lysenko (far right), director of the Soviet Institute of Genetics.

With Nobel prizewinner Dorothy Crowfoot Hodgkin, a former research assistant of Bernal's.

Professor of Crystallography, Birkbeck College, 1965.

dream of a Communist utopia, or even 'actually existing socialism', has been at least temporarily lost, and when feminism has taught us something of the complexity and pain of gender relations and their implications for the production of knowledge, we could in principle begin to talk seriously about a life in three colours. Here however our focus is on two, and on how during Bernal's lifetime the red faded down to pink with red patches, and the science went from blue to red and then to unequivocal blue – but harnessed to red goals.

Bernal in Red and Blue

We explore Bernal's contribution to the politicization of science and scientists, above all the development of the Social Relations of Science movement. Inevitably there is a periodicity to Bernal's activities, with his early social concerns in the 1920s, the distress of the Depression, the evident threat of Nazism and Fascism, and his unequivocal political engagement during the 1930s forming one distinct arc, and the greatly changed political situation after World War Two another. The earlier period was an immensely productive one for Bernal. In his laboratories, first in Cambridge then at Birkbeck, he was laying the foundations of crystallography and the work that would make the new molecular biology possible.[12] He was preoccupied with the struggles to build the Association of Scientific Workers as an effective trade union, and more generally against the Nazism and Fascism that were appearing within science as elsewhere. Bernal loathed their obscurantism and volkish mysticism: these were the enemies of reason.[13] His science and his politics were integral to his Enlightenment universalism. Science served Marxism, and Marxism served humanity, justice and freedom. It was for a later generation to learn painfully that this 'universalism' had served men more than women and the interests of white people rather than black. This is not to diminish the historic role of the left's ideological struggle against Fascism and Nazism; rather it is to locate it in context and, above all, to understand the very different ideological positioning of science.

Today antiracism and opposition to eugenics go hand in hand; at that time, numerous British left scientists (particularly the geneticists rather than crystallographers like Bernal)[14] and, above all, welfare reformers, from Marie Stopes to Richard Titmuss and William Beveridge, were very attracted to eugenicist ideas. Improving the 'race' scientifically was seductive to intellectuals of the time. What in the USA had led to the compulsory sterilization of 20,000 poor women was in Britain an intellectual and political flirtation with intermittently negative practical outcomes; but in most of the Nordic countries eugenics was integral to their systematically constructed welfare states.[15] Outside this Nordic social-democratic consensus, however, which was only broken by growing wealth and the less authoritarian state of the seventies, the left had become swiftly disenchanted with eugenics. As early as 1936, eugenics was officially denounced in the Soviet Union as the evil accomplice of Nazism and Fascism.[16] There has been some remarkably underinformed comment recently by the journalist Jonathan Freedland on eugenics, racism and the British left scientists of the thirties, not least Bernal,[17] that has attempted to make them all out as a pack of eugenicists and racists. It is important to set the record straight, warts and all. There were rather few liberals of the Freedland model at the battle of Cable Street, but numbers of Communists and trade unionists prepared to fight alongside the Jews against Mosley's Fascists.[18]

This crisis-ridden period also saw the production in 1939 of Bernal's germinal text *The Social Function of Science*. This book, together with his towering intellectual presence within the British Communist Party, gave British Marxism a peculiarly natural scientific slant.[19] Francis Bacon, that seventeenth-century founding father of modern science, with his highly empiricist theory of natural philosophy, was recruited by Bernal into his Marxist analysis of science. Bacon's belief that science could be produced by 'any man'[20] who followed the test of empirical practice found echo in the experimental scientist. For Bernal, the democracy of science and the democracy of Communism reinforced one another;[21] at its best, science was Communism. *The Social Function of Science* immediately made Bernal the founding figure in the now huge field of science, technology and society studies. Even today, one of the most prestigious awards of the US Society for the Social Studies of Science is an annual Bernal prize; that all too many of its recipients have been the narrow professionals of whom his life as a public intellectual stands in contempt is just one of life's ironies.[22]

With the post-1945 Cold War came Stalin's 'two camps' view of the world and the Lysenko affair, with its especial challenge for the Social Relations of Science Movement. By the 1950s there was, on the one hand, an increasing acceptance by political orthodoxy, especially within the Labour Party, of a Bernalist view of the social functions of science, but with the class politics replaced by a nation-state corporatism. As depoliticized techno-economism, this revisionist Bernalism became the powerful orthodoxy of both national and international science policy, above all through that influential rich nations' club the Organization for Economic Cooperation and Development (OECD).[23] Meanwhile in the difficult context of the Cold War, Bernal turned his energies towards building internationalism through the peace movement. His writing, moving from his contribution to *Britain without Capitalists* (1936) to *World without War* (1958) is expressive of the change in direction.

But despite the discontinuities imposed by world events, if one rereads today the corpus of Bernal's writing, from *The World, the Flesh and the Devil* (1929) through to the valedictory interviews and the essay 'After Twenty-five Years' (1964), there is a remarkable homogeneity and coherence of themes: a continuing, unshakeable optimism about what science, once free from the shackles of capitalism, could do for human liberation; the belief in the essential possibility of a rational world, a world planned for the benefit of humanity; and a certainty that this was a vision that natural scientists, if only they could cast off the constraints on their thinking imposed by an irrational capitalist research system, would see and embrace. Today few of these propositions are so self-evident; doubt and concern about the very nature of the scientific enterprise inhibit such optimism. For that matter, where Bernal saw an undifferentiated humanity as potentially freed by Communism, today's radical intellectuals are more aware of the taken-for-granted masculinity and ethnocentrism of Marxist thought. Where Bernal saw Communism as an inevitable consequence of the scientific and technological revolution, today's radical intellectuals see pollution, chemical and biological warfare, and genetic reductionism. Where Bernal lauded an abstract rationality, today's radical intellectuals see the instrumental rationality of capitalism with science as its uncritical servitor and seek to rebuild both science and rationality as socially and environmentally responsible. Lastly, where Bernal saw scientists as potential organic intellectuals, today's radical intellectuals see them as a divided group of elite managers on the one hand, and casualized research workers on the other.[24]

The Radicalization of the Thirties Scientists

Bernal's optimism and enthusiasm – essential ingredients of any successful revolutionary change – stem in part from a powerful positivistic tradition in early-twentieth-century English socialism, a tradition common, for example, to G.B. Shaw, H.G. Wells and the Webbs. Recruiting

Bacon, both Queen Elizabeth's Lord Chancellor and modern science's founding father, into this left tradition was an extraordinary example of Bernal's capacity for creative synthesis. Like his socialist peers, Bernal shared a belief in the importance of intellectuals, the need for men (and rather rarely women) of exceptional talent to bring their services to bear on the making of a beautiful world. The relationship to the Shaw of *Back to Methuselah* and the struggle to overcome death through a disembodied intelligence or world brain is there in *The World, the Flesh and the Devil*, and it is present in the 1960s when he writes, more poignantly now, of the struggle against debility and death.[25] With Wells, he saw the scientists and engineers as having the 'future in their bones' (to use C.P. Snow's often-quoted phrase); they were to be amongst the leaders, the planners, the theoreticians.

As Benjamin Farrington has remarked, Bernal's strengths were not those of the philosopher; his Marxism was almost an extension of his natural science. For him science and Marxism formed the ultimate happy marriage.

> My impression was that at least half of the Marxists whom I met [in the 1930s] were scientists. But I had the impression also that their Marxism had originated from scientific sources – I mean the physical sciences – and not to be so much aware of the social and philosophical background. In England, chiefly also among scientists, I found a complete optimism about Marxism and science. It seemed to them, and I heard the actual words from them, that Marxism was the theory which gave science its opportunity. And it seemed as if science and Marxism had absolutely been married to one another.[26]

Yet it was precisely this which was Bernal's strength; the strength of a positivist tradition taken and made over, simultaneously claiming not only historical but dialectical materialism, in the interests of working people and of socialism. And, above all, the understanding that making a revolution was not merely a business of getting one's theory right, or influencing the '2000 who really mattered' in the elitist Fabian conception, but of practice, of struggle at all levels, in the streets and factories as well as the laboratories and committees. It is this personal commitment to public political action that must weaken the conspiratorial view, constructed by Werskey,[27] of Bernal as merely a Communist Party tool for winning over the intellectuals.

Bernal's emergence as one of the central figures among the left scientists of the 1930s must be seen in the context both of his own development over the period and of the general political and social situation. The global background is well known: the economic depression and massive unemployment in all the capitalist countries; the emergence of Nazism in Germany and fascism in Britain, and the then apparently clearly shining beacon that the Soviet Union represented. In this context, and with the most recent of a long line of capitulations of the Labour Party leadership – that of Ramsay MacDonald – there for all to see, there was only one place for socially aware intellectuals as well as workers to be, and that was the Communist Party.

Bernal himself saw this plainly. He had joined the Communist Party in the 1920s. A long draft of an unpublished article written in 1931, 'What Is an Intellectual?' – a modified version of which was published in *Cambridge Left* in 1933[28] – argues that intellectuals must be politically engaged, that they cannot withdraw into their laboratories or studios or into the fragmentation of bourgeois life. (Oddly, in the light of their later roles, he lists Einstein and Picasso amongst the retreatist intellectuals[29] – but then his relationship to Einstein, even when writing an obituary, was always somewhat ambivalent. The relationship to Picasso, not least through the peace movement, was to become more cordial.) An intellectual must mix as a citizen in the life of the time. Intellectuals who oppose fascism must ally themselves with the working class and support the Communists as offering the only possible alternative.

And many, of course, did. As Perry Anderson points out, for some of the artists and literary figures it was to prove a transient, romantic attachment – an attachment that admittedly led many to their deaths in Spain.[30] But Anderson, in his splenetic dismissal of Bernal, fails to recognize that for many of the British left scientists the commitment was to last much longer, even if their socialism was to become translated into more uncritical enthusiasm for the Soviet Union. Why? Perhaps Farrington is right and 'scientific socialism' spoke particularly to them; perhaps the influence of the leading figures was more powerful. But also, socialism and Communism seemed to offer something concrete to scientists that it could not so straightforwardly provide for other intellectuals. The enormous growth of and respect for science in the Soviet Union, its full incorporation into the political structure of the state, enthusiastically proclaimed by Bernal in article after article throughout the 1930s (and of course, long beyond), seemed to be a glimpse of the possibility of a promised land.

For the situation of science in Britain was truly impoverished. Jobs were scarce and poorly paid (£150–£450 a year for a researcher or lecturer) at least until the professorial level (£1,000 to £2,000 a year). Research funds were scarce in universities, almost nonexistent in industry. In both industry and government, the isolation of the researchers from their academic counterparts guaranteed their mediocrity. In the aftermath of the 1914–18 war and the Haldane *Machinery of Government Report* of 1916, the state had begun to support basic science by way of the research councils, and industrial science by way of the research associations. But the funding was reluctant and, during the 1930s and the Depression, almost stagnant. Even by 1936, when Bernal was to make the first detailed estimates of science spending, he calculated it at no more than 0.1 per cent of GNP – compared with 0.6 per cent in the USA and more in the USSR. In a nonscientific society, science education was correspondingly poor, with anachronistic syllabuses (Bernal pointed out that science syllabuses in schools mainly concentrated on work that had been done prior to 1810, with a brief excursion in chemistry towards the 1890s), and, with the exception of biology (then usually called 'nature study'), was virtually nonexistent for girls.

It was against this background that critical voices in science had begun to be heard from the 1914–18 war on, demanding the professional recognition of scientists. Hence the emergence, in 1917, of the National Union of Scientific Workers (which, to help recruitment, changed its name in 1927 to the Association of Scientific Workers (AScW)) with both the professional and critical concerns firmly at its masthead. Membership was always small – mostly less than 1,000 – and fluctuated with its relative success or lack of success in local industrial negotiations, but it provided a focus for the activity of the left scientists.[31] Bernal himself was active in the AScW throughout the 1930s, and by 1938–39 was chairman of the union executive.

In addition to work in the AScW there was a determined effort by left scientists to capture the main ground of scientific opinion in the country. Dismissing the Royal Society as virtually impregnable (despite the fact that the leading left scientists, Haldane, Levy and Needham, were members, as was Bernal from 1937), a determined effort was made to insist that the British Association for the Advancement of Science (BA) put the question of the social functions and responsibilities of science onto its agenda. This was easier ground, partly because the BA, after all, had itself something of a critical past (having been established by Babbage and others in 1831 in response to the then irrelevance of the Royal Society whose membership was largely of fashionable amateurs and aristocrats and indifferent to the needs of industrial science), and partly because the structure and membership of the BA made it more open to democratic demands. In the battle to win over the BA, the left scientists had a major ally in Richard Gregory who, as editor of *Nature*, used its columns to open up the science and society debate. (It was Gregory who said of himself, 'My grandfather preached the gospel of Christ, my father the gospel of socialism. I preach the gospel of science.'[32] Gregory was no Marxist, but for much of this period there was a relatively easy alliance between liberal reformists such as Gregory and the biologist Julian Huxley and the left scientists. It was this combination that resulted, eventually, in a characteristic concession by the BA. It opened a special section, Section X, for the discussion of the social context of science, thereby neatly encapsulating and potentially incorporating the radicals. The alliance with the liberals, and the openness of the BA, were both to be strictly limited. (In due course those, such as Bernal, who supported the Soviet Union and Lysenko in the post-1945 'Two Camp' period and the Cold War would find themselves excluded.)

A crucial mobilizing issue for the left scientists was the growing threat of fascism. The tasks confronting them ranged from the ideological to the practical. Within the ideological there was an increasing necessity, as the 1930s wore on, to expose and demolish the claims of Nazi science and its Aryan racist genetics. The 'perversion of science under Fascism' was a subject Bernal increasingly returned to in his writing, though he himself was less actively involved in the ideological struggle within biology than geneticists and social biologists such as Haldane and Hogben. Practically, there was the struggle to rescue scientific victims of Nazi persecution, a task that, like so much of the work of the left and liberal scientists, was Cambridge-based, centring on the laboratory of the biochemist Frederick Gowland Hopkins.[33]

From 1934, Bernal was actively involved with the Cambridge Scientists' Anti-War Group (CSAWG). Initially concerned with protesting publicly about the 'prostitution of science for war purposes', the group went on to mobilize, organizing marches, pickets of military bases and petitions. As the military situation became more threatening, the group became more directly concerned with the consequences of war for the civilian population in Britain, and, in particular, the question of defence against air attacks. The group employed a combination of direct scientific experimentation – documenting, for example, the inadequacies of the government's advice on protection against gas attack and of the gas masks provided – with publicity by way of pamphlets, Left Book Club books[34] and a campaign of public meetings across Britain which again involved many of the central figures among the left scientists. This campaign, too, began to involve the development of the subsequent generation amongst the left scientists: Eric Burhop, Dorothy Hodgkin and Maurice Wilkins[35] amongst others have described their mobilization through the CSAWG.

The Social Function of Science

These concerns, then, the organization of scientists into the AScW, the raising of their political consciousness particularly regarding fascism and the danger of war, and work in the peace movement, formed the major elements of Bernal's political practice over the decade. But his outstanding achievement during the thirties was theoretical: the summary statement of the problematic state of science under capitalism and a prophetic attempt to describe the goals of a science under socialism, represented, in its final form, by *The Social Function of Science.*

The Social Function of Science seems to have taken the best part of the decade of the thirties to mature. Bernal wrote several shorter versions of its main themes early on. Some of the ground is covered in a manuscript book synopsis entitled 'The Scientist in the Contemporary World' (1934) (of which only four chapters were written) and Bernal's contribution to *Britain without Capitalists,*[36] a *tour de force* written by an 'anonymous' collective of economists, scientists, and technicians. Bernal was the main author of a chapter entitled 'Science and Education', which rehearses many of the main themes of *The Social Function of Science.* In some ways the optimistic belief in the imminence of revolution that the collective work expresses permeates Bernal's chapter and makes it more direct and explicit than his later book. 'It is not usual to think of science as an industry,' the chapter unequivocally proclaims before going on to show that it is such, although 'unlike other industries that are concerned in keeping a certain state of production going, science is concerned in changing that state' (p. 407). Later the chapter goes on to consider the potential effects of a socialist revolution in Britain on science. Bernal observes that some scientists, of course, will oppose it, but they are likely to be the older ones with little further contribution to make; some may get killed in a civil war. There will, however, be a tremendous pool of talent and most will go along with the revolution either out of conviction or because they are cramped under the present system. The revolution offers scientists the possibility of working under more attractive conditions because of the expansion of science that will come about as a result of it.

There were two major additional theoretical inputs into the thinking of the left scientists during the 1930s. The first was the new perspective on the theory of the growth of science introduced to Western scholars at the International Congress on the History of Science and Technology in 1931. The Soviet Union sent a major delegation headed by Bukharin, and including the physicist Hessen. Hessen's path-breaking paper on the 'Social and Economic Roots of Newton's *Principia*'[37] made a profound impact on the left scientists attending the meeting.[38] Bernal was at the conference, and acted as Bukharin's London guide during the meeting. Bernal's subsequent writing on science was to show the influence of both Bukharin and Hessen. Hessen's method of relating scientific development to the economic and social order and the demands of the rising bourgeoisie shaped Bernal's later writings. Subsequently Bukharin was to become one of the leading Communists to be tried and executed in the purges, and as a nonperson his ideas could not be directly discussed. Bernal's high prestige within the Soviet Union meant that Buhkarinist theories could be explored indirectly through Bernal's writing but, despite the evident indebtedness of the author of *The Social Function of Science* to Bukharin, there is no public record of Bernal defending him.

The second theoretical input was the rediscovery of dialectics. Engels's major attempt to set out the meaning of dialectical materialism for the natural sciences, *The Dialectics of Nature*, was not published until 1927 and only became available in English in 1940 when Lawrence and Wishart published it with a preface by the biologist and Marxist J. B. S. Haldane. But the earlier *Feuerbach* and *Anti-Dühring* were available as, of course, was the much cruder *Materialism and Empiriocriticism* of Lenin. Bukharin's contribution to the 1931 conference emphasized the importance of dialectical materialism, and Western Marxists were soon to respond. The French biologist Prenant, the Scottish mathematician Levy, and Haldane were to write what they claimed to be 'dialectical' accounts of science and the scientific method.[39] Bernal's attempts in this direction were fewer: Farrington's comment quoted earlier applies here, too, and in general Bernal seems, as we suggested earlier, to have been happier developing a positivistic Marxism. However, he did contribute to a book on dialectical materialism which appeared in 1934,[40] and write an article on 'Dialectical Materialism and Modern Science' in the Marxist journal *Science and Society* in 1937.[41]

The 1937 article represents an advance on the earlier piece and eschews the repetitious churning out of the 'laws of the dialectic' and their exemplars which disfigure many of the writings of the period (the melting of ice and boiling of water as representing transformation of quantity into quality, north and south poles as interpenetration of opposites, et cetera) in favour of a more sensitive description of Marxism not as a method of natural science, but as one that shows the limits of science, enabling radically new things to be known, pointing to new solutions and enabling an ideological critique of existing science to be made. Hence dialectical materialism shows the economic determinants of science and makes possible a science for the people. The theses of this article, whilst they are submerged in *The Social Function of Science*, are expressed in their final and clearest form in Bernal's pamphlet *Marx and Science* which appeared in 1952 and to which we will refer later.[42]

These theoretical inputs of Hessen's economic determinants of science and of dialectics of nature, together with the practical experiences gained through the political organizing of the 1930s, crystallized the earlier drafts into *The Social Function of Science*. Chris Freeman's chapter in this volume discusses the impact of this book in 'science policy' terms. Here we want only to point to some of the essential political issues it raises. Bernal emphasizes the critique of science, its consequences in unemployment caused by new technology, in weaponry, in the sense of loss of individual security. Faced with these threats from science, there has been a revolt from reason, a retreat into irrationality and idealism of which fascism is the extreme case. Yet this idealism is no substitute for the creative use of science; science is power, the scientists as workers must use it in the interests of humanity. Bernal's emphasis on technological unemployment foreshadowed the dramatic restructuring of capital and production in the eighties.

The parallels should not be pushed too far. Largely missing from Bernal's account are today's knowledge of the threats of a science that generates industrial pollution, health hazards, nuclear, chemical and biological weaponry. His enemy was too little science; we and our environment all too often suffer from too much capitalist science. His optimism at the inevitably happy marriage of science and socialism led him to be silent on the ideological role of science, to say nothing of the problems of ideology in science. These silences are no more than the constraints of context: what is impressive is how often he moves beyond them.

The first section of *The Social Function of Science* is a pioneering attempt to describe the existing state and frustrations of science in Britain and throughout the developed world.[43] It was this strand within the book that was to provide the basis for the techno-economist interpretation of Bernal in the 1950s and 1960s. It is, therefore, important to recognize that Bernal (like Marx) could – and indeed did – say, 'I am no Bernalist'.[44] None the less, the critique of the inefficiencies of capital's use of science and the need to rationalize it, to plan, was shared by many sections of advanced capital, if not in Britain then certainly in Germany and the United States. David Noble, in his book *America by Design*,[45] time and again quotes passages from leading industrialists in the United States in the 1920s and 1930s urging scientific reform that could have come straight out of *The Social Function of Science*.

The second section of the book is programmatic – what science could do. It calls, even under capitalism, for the massive expansion of spending on science to 1 per cent of GNP (by the 1960s, developed countries were spending 2–3 per cent of GNP on science). It calls for the restructuring of the science curriculum at all levels, for the reorganization of the structure of research. This was to involve not merely the development of scientific plans for the country and for the laboratory – the call that was to evoke so much hostility from the right – but also the reorganization of laboratory practice itself. Part of the reorganization envisaged is pragmatic and determined by criteria of efficiency, for instance the appointment of administrators and librarians; but part touches on a major political objective of the later New Left: that of laboratory democracy. Laboratories should neither be autocratically run by a director nor left to anarchy, but run by a regularly meeting laboratory council involving all the scientific staff, including technicians. Authoritarian management of science should be prevented not by denying expertise but by making progress through the ranks (from lab assistants upwards) much easier and more flexible.

The rest of the book is a more euphoric account of what science could achieve for humanity once the shackles of capitalism had been thrown off, a return to the straightforward optimism of *The World, the Flesh and the Devil* and indeed the theme that was to be reiterated again and again, ultimately in *World without War*. It concludes with the claim that science is Communism, Communism is science. But where Marx's science was 'proletarian science', Bernal's science had unequivocal roots in natural rather than social philosophy. If in this chapter we wish to rescue the liberatory dimension of Bernal's agenda, we do so conscious always of the shadow of that other that prioritized the happy marriage of *natural* science and Communism.

No account of *The Social Function of Science* would be complete without reference to the response that greeted the publication of the book. It was, of course, widely seen as the manifesto of the left scientists, the most fully developed statement of their position. The issue of 'planning in science' became a battleground between right and left. In particular the cytologist J. R. Baker (who re-emerged in the 1970s from relative Oxford obscurity with a book entitled *Race* which threw its support behind the scientific racism associated with Eysenck, Jensen, Shockley and Herrnstein) issued what he called a 'Counterblast to Bernalism' in which he attacked the idea of any planning or 'dictation' in science.[46] His polemic against Bernal was vivid, if off-target: 'I know what he's after: he's going to tell me that I dare not ever again use gentian violet as a stain for my sections, but from now on I shall have to restrict myself to methylene blue'.[47] Bernal responded by pointing out that planning meant no such thing. By contrast if science were to progress, it demanded a freedom of thought at the bench and in the library; what planning was about was the creation of the conditions for such freedom. 'If Bernalists existed,' Bernal said, 'they certainly didn't believe what Baker claimed for them.'[48]

But the debate had a deeper political agenda, an agenda that resulted in Baker, the philosopher of science Michael Polanyi and the botanist A. J. Tansley establishing a small and brief-lived Society for Freedom in Science, which in the aftermath of the 1939–45 war seized on the Lysenko question as precisely exemplifying its fears, and issued a series of pamphlets exposing the consequences of Stalinism for Soviet science. The society seems never to have been much more than a paper organization, and there were tensions between those high liberals for whom even to sign a manifesto was in contradiction with their belief in freedom, and those for whom the cloak of liberalism hid other Cold War goals.[49] Despite its continued existence throughout the war, the reality of the transformation of the British science scene after 1939 made the Society for Freedom in Science irrelevant.

Bernal's immense activity in the 1939–45 war is discussed by Ritchie Calder in Chapter 7. The mobilization of the scientists in Britain is a story that has been told many times.[50] As 'boffins' they worked on schemes ranging from civil defence to the treatment of wounds, the atomic bomb, radar, the Mulberry harbour, and a vast project to tow icebergs south to act as aircraft carriers – we do not propose to discuss them here. The scientific mobilization of the war effort rendered many of the themes of the frustration of science raised in *The Social Function of Science* temporarily irrelevant. By the time the war ended, the tenfold increase in science spending that Bernal had called for in 1939 had been achieved; science was on the political agenda as never before, and the time should have been right to move both the theoretical analysis and political struggles to a higher level. What happened, as we know, was rather different.

Science in Two Camps

To understand the fate of the left scientists' movement in the postwar world we have to recognize the changed historical context which suddenly denied space for left politics. Whereas in 1945 it had seemed that a significant move to the left was about to take place and that social justice was about to break out in the wake of the suffering of the war, international capital, spearheaded by the US, had no intention of simply letting go. A combination of military repression (Greece), financial inducements (the Marshall Plan), and political manipulation (France), to say nothing of the Soviet army's role in the conversion to Communism of Eastern European countries, led inexorably to the Iron Curtain. The result was that the world was split into what Stalin was to call the 'two camps'.

For left scientists the two camps meant that political space was denied to them. Section X of the British Association, which in a less ideologically split world had provided an arena for left/liberal discussion, became closed to the Marxists. Liberal scientists who had hitherto quite often worked with the left, such as Julian Huxley, were no longer willing to work with Communist scientists. Bernal himself was excluded from the Council of the BA following reports of a speech he made in Moscow attacking the nature and control of science in the capitalist West. In the Cold War it was no longer possible to be a (public) revolutionary[51] and keep a toehold in the British establishment. To get indignant about the exclusion of Bernal from positions of power in postwar Britain would be not to take seriously what he was saying and to fail to recognize the anti-Communist hysteria of the first Labour government.[52] The options for left scientists in Britain in the late forties were cooption or isolation. Despite the potentiality of people to make history, they do so, as Marx reminds us, in circumstances not of their own choosing. And the circumstances of the Cold War were indeed singularly unpropitious for the optimism of the thirties scientists.

For Marxist scientists, and particularly Marxist biologists, the crisis that engulfed the world Communist movement in 1956 with Khrushchev's denunciation of Stalin had a serious foretaste in the Lysenko affair. Culminating in 1948 with the publication of Lysenko's triumphalist *Soviet Biology*,[53] it had a long history, beginning in the 1930s when Trofim Lysenko, a plant breeder of peasant origin, began to challenge the theories and findings of conventional genetics and especially the new plant genetics which was strongly represented in the Soviet Union by Nicolai Vavilov. The crisis in genetics was part of a general assault in the Soviet Union on the claims for the universality of science. Hessen's theoretical case for the possibility of two sciences, bourgeois and proletarian, found its first concrete embodiment: Vavilov's bourgeois science was to be replaced by Lysenko's proletarian science.

Lysenko claimed that Vavilov's genetics, despite – or rather because of – their world (that is, bourgeois) fame were in actuality merely a combination of the idealist formulations of Mendel, Morgan and Weismann. As such they were self-evidently reactionary, anti-Darwinist and antisocial. The thrust of Lysenko's argument was that one philosophy in science and another in politics is *a priori* a lie. In part what Lysenko was doing has to be seen as an aspect of the questioning of authority and the attempts to build a socialist science and overcome bourgeois ideology that the late 1920s and early 1930s movement had opened up in the Soviet Union – an incipient cultural revolution.[54] Whatever the revolutionary intent, however, Lysenko's administrative and political methods of winning the debate, and indeed his 'science' itself were both disastrously awry. Lysenko's challenge to the confused and often mystical underpinnings of much contemporary genetics could not be dismissed, but his own claims were based on singularly slender and even forged evidence (although, as Levins and Lewontin point out,[55] they were based on a practical appreciation of the irrelevance of the contemporary statistical procedures in genetics in the climatic conditions of the USSR).

Lysenko's claims were initially treated with scepticism by both Marxist and non-Marxist biologists inside[56] and outside[57] the Soviet Union. However, this scientific and ideological debate was set in the specific context of Stalinism and the crude 'partisanship' thesis of Zhdanov. Vavilov, along with so many other professionals and workers loyal to the hopes of a socialist society, was dismissed from his post and arrested; he was eventually to die in transit to a labour camp in 1943. Despite the dangers, Lysenko's position was none the less systematically criticized within the Soviet Union by both geneticists and philosophers of science right up to 1948. It was then, at the height of Zhdanovism, that there took place the crucial meeting of the All-Union Lenin Academy of Agricultural Science. Lysenko, despite the inadequacy of his scientific position, crushed all opposition by revealing that Stalin had administratively ruled in favour of 'proletarian science' and was backing him. Soviet geneticists, faced with the administrative decision, enforced by terror, that genes have no material existence, found themselves committed either to acquiescence or to persecution. Several sought shelter within nuclear physics departments, as radiation biologists. All were denied the conceptual apparatus of genetics which was to lead in the West within the next five years to the Crick and Watson era of molecular biology.

As we know, the Communist Party of Great Britain, along with other Western Communist parties, was, in the context of the Two Camps thesis, to rule that there were only two possible positions: either support for or hostility towards socialism, and thereby Lysenko. Debates within the Engels Society, the Marxist natural scientists' grouping within the CPGB, revealed a range of positions not so unlike those of their Soviet counterparts.

The crudest response was that of uncritical support for the Soviet Union, in which the task of CPGB was seen as 'to study the conclusions of the leading communist authority in order to understand them'. For Emile Burns, who had the task of summarizing the meetings, it was a simple matter of loyalty. 'We meet here as communists. Hence we shall never tolerate any expressions hostile to the USSR. We regard the attitude to the USSR as fundamental to loyalty to the party and so to the working class.'[58] The Communist geneticists, on the other hand, were disturbed by the lack of experimental details reported by Lysenko. Their responses ranged from a robust rebuttal of his claims to a more modest request for research papers that were longer than one or two pages. It was, however, the responses of the theoreticians Haldane, Levy and Bernal that were most interesting and important.

Haldane, as a geneticist who had himself written on dialectics in biology, argued a measured case for Mendel–Morganism while entirely agreeing that Weismann was an idealist. He claimed that his own work demonstrated how temperature alone could alter the chromosomes of Drosophila and hence that environment interacted with the genes. Genetics, he argued, was causal not deterministic. But at the same time he felt that Michurinism (the Lysenko doctrine that there could be a direct influence of the environment on heritable characteristics) had much to offer. He gave as an example extra-nuclear inheritance – that is, inheritance dependent not on the cell nucleus but on factors in the cytoplasm of the maternal egg – and, as a second example, how the presence of certain substances in the milk of females results first in hypertrophy of the mammary glands in their offspring, and subsequently in cancer. He concluded by saying that, having supported Lysenko as far as he could, he would go no further without fuller data.[59] As is known, in the aftermath of this affair, unwilling as a Marxist to denounce the Communist position, Haldane was to allow his membership of the Communist Party to lapse. He eventually left active politics in Britain for a continuation of the nonviolent theories that had underlain much of his research, in the more congenial context of India.[60]

Levy began by observing that the Mendel–Morganists present in the Engels Society seemed to be different to those that Lysenko was attacking in the Soviet Union. He proceeded as a nonbiologist to ask some penetrating questions which are worth quoting in full:

1. 'Did geneticists hold that there is a special substance which alone conveyed the hereditary qualities?'
2. 'Did they really maintain it is independent of the rest of the organism, and, if not, what is the nature of the interdependence?'
3. 'Did the Lysenkoists deny that there is a special substance which is necessary to start off the development of hereditary characteristics?'
4. 'Is it true that Mendelians are mainly interested in the transmission of characters that are the same from one generation to the next, whilst Lysenkoists are concerned with the creation of differences between parent and offspring?'[61]

Levy's contribution thus sought to clarify the scientific and philosophical issues. Clearly endeavouring to find a synthetic and general solution he concluded that what was going on was something like the debate over field and particle in physical sciences, which he saw as a struggle against the permanence of artificial categories.

Bernal's contribution centrally faced the issues within the philosophy of biology but for him, like Burns, the choice was ultimately political. First and foremost the Lysenko controversy was being used against the Soviet Union as part of the Cold War. Second, he saw Lysenko as a Marxist scientist standing in opposition to bourgeois science:

> In the past there has been one science. Because modern science was part of the origin and development of capitalism, it was necessarily the production of bourgeois thinkers and steeped in bourgeois ideology. It is only now in the Soviet Union with the new generation of scientific workers that it is possible to build a socialist science.

So far as the defence of genetics was concerned he argued that despite their claim that they allow for the reaction between genes and the environment . . . to make this allowance goes only part of the way. It still places the emphasis on the autonomous gene and treats the action of the organism as mere perturbation. The important point is that heredity is a process involving the whole organism rather than the substance depending on permanency. The recognition of this fundamental Marxist fact – that the world consists of processes not things – is the essence of Lysenko's approach.[62]

These debates over Lysenko are not of mere historic interest; they go to the heart of contemporary biological theory, expressed in the conflict around the new molecular biology and genetic determinism. Just how far is DNA, as Watson claims, the 'master molecule' and how far is it part of a biological process? To discuss them in detail would take us outside our brief here, but the questions that were then posed as the difference between mechanical and dialectical materialism are as vital today as they were in 1948.[63] The closure of the debate by administrative means and terror in the Soviet Union, and by Cold War mechanisms in the West, caused profound damage to Marxist thought about natural philosophy, a disaster so great that many would argue that the profound scar that it has inflicted on biological theory remains even today.

Culturally and politically constrained in the British context, Bernal turned his attention to the building of international bridges and to circumventing the constraints and threats of the Cold War. His work in the AScW had already led to the foundation of the World Federation of Scientific Workers (WFSW) in 1946, in whose activities he became increasingly involved. The successor to the CSAWG was Science for Peace, founded in 1951, many of whose activists were in turn to become involved in the Campaign for Nuclear Disarmament in the mid-1950s. With the distancing from Stalinism following Khrushchev's speech in 1956, there was little energy to grapple with the theoretical issues of what a socialist science could be. Instead Bernal's response was essentially pragmatic, leaving the past to bury the past. From the 1950s on he never refers again to Lysenko except in the context of *Science in History*,[64] whose successive editions in the 1950s and 1960s gradually reduce the prominence given to Lysenko's claims.

Bernal's uncritical eulogy of 'Stalin the scientist', which appeared as his part of the Stalin obituaries of 1953 is the last reference to what he then described as 'the greatest figure of contemporary history', who was 'at the same time a great scientist', who 'combined as no man had before his time a deep theoretical understanding with unfailing mastery of practice . . . his wonderful combination of a deeply scientific approach to all problems with his capacity for feeling and expressing himself in simple and direct human terms'.[65] Of course Bernal should not have abdicated critical thought – any more than those of us who were subsequently to speak of the Chinese Cultural Revolution or the Vietnamese revolution in similar, if less grandiloquent terms, should have done. But if failing to appreciate at the time what was happening was understandable, failure to speak at all subsequently was not. Nor was it appropriate to talk, as Bernal did, about the post-1953 successes of molecular biology and the materiality of the gene as if they were ideology-free, without a backward glance at 1948.

It is of course one of the ironies of the history of molecular biology that Bernal's scientific techniques, and indeed the pioneering research of his own laboratory, were centrally concerned with the elucidation of the structure of DNA, the very genetic material of which Lysenko had denied the existence. Bernal himself in 1968, reviewing the period and Watson's book *The Double Helix*, is mainly concerned to describe his own lab's involvement in the work and how they finally missed it.[66] He never took the opportunity to reflect on his own previous emphasis on process rather than permanence. The retreat from a socialist science into bourgeois neutrality meant that the ideological framework of the new molecular biology went unchallenged until the 1970s and the rise of the New Left radical science movement.

These silences meant that the critical questions, of whether and under what circumstances there can be a socialist science, were set aside. It may well be that the barbarous constraints imposed by the context of the Cold War made it impossible for such difficult and potentially threatening theoretical work to be continued. Left scientists turned to issues of world peace and the dangers of nuclear holocaust as both a necessary and in some ways a less problematic activity. Bernal himself abandoned wrestling with the ideological questions to go on to the writing of the optimistic *World without War* and the massive project of *Science in History* (see Chapter 12). For the 1970s radical science movement, the closure of these questions and the long silence over Lysenko meant that the theory of the two sciences had to be rediscovered rather than developed from an ongoing debate. Only in 1976 was the discussion, for example, of science and ideology thoroughly reopened.[67]

Even without a cold war denying us intellectual and political space (although there has been a major ideological shift to the right from the 1980s onwards), the context of the current debate is difficult. The problems criticized are those of silence, a silence that meant that it was

Bernal's legatee as president of the World Federation of Scientific Workers, the physicist and Lenin Prize winner Eric Burhop, who most prominently defended the position that 'science is neutral' (that is, that there is only one modern science) within the radical science movement of the early 1970s.

Back in the immediate post-war period of the late 1940s, however, the positivist strand of Bernal's pre-war writings was to remain and flourish whatever the theoretical and political difficulties his dialectical materialism encountered. He had revealed the inefficiencies of British science and had inspired a generation of younger scientists to think in the same vein. The war absorbed all their talents and by the end of it many had tasted the delights of power. Had a Tory government been returned immediately, many might have gone back into opposition, but Labour's mandate for change and its continuing belief in the need to reconstruct British science attracted many. At any rate, scientists touched by the *Social Function of Science* debate, like the physicist P. M. S. Blackett and especially Solly Zuckerman, moved closer to the state scientific advisory apparatus, carrying with them many of Bernal's views on how to organize science, but without his commitment to class politics – which meant that he himself was soon excluded.

Bernal's thesis that capitalism was incompatible with the development of science and technology – a thesis he shared both with orthodox Marxist theorists in the Soviet Union[68] and with influential Trotskyists such as Ernest Mandel,[69] was about to be put to the test – and found wanting. The subsequent thirty-five years were to show the unabated growth of science in all advanced capitalist countries – a point Bernal himself makes when he considers the transformation of the post-1945 world in his retrospective 'After Twenty-Five Years'. However, what could not be seen so clearly prior to its collapse was the devastating consequences of the imbalance of research towards the military which played such a significant part in the decay of the entire Soviet system. By contrast in the West it was the critique of the instrumental rationality of science by the neo-Marxist Frankfurt School of Adorno, Horkheimer, Marcuse and, later, Habermas that was to inform the new challenge to science.[70]

In Britain it was above all the Labour Party that was open to the technoeconomist argument. Over the twentieth century Labour has favoured industrial capital, unlike the Tories, who favour finance capital. The period of 'thirteen wasted years' of Tory power between 1951 and 1964 briefly invested the techno-economic argument with some of the glamour of opposition. In *Science and Society* we have shown how techno-economism was adopted by Harold Wilson in the run-up to the 1964 election and given an almost messianic quality, of the building of socialism 'in the white heat of the scientific and technological revolution'.[71] Despite the hostility of the Labour Party to Communists, Bernal was too talented to ignore and he was involved in the key science policy discussions that the Labour Party held in the run-up to the 1959 and 1964 elections. The failure of the Wilsonian promise was also the failure of a revisionist Bernalism, a failure that no British government in the post-1964 period has been able to overcome.

This cooption of Bernalism was not confined to the political sphere. The later 1950s and early 1960s saw the development of a sort of academic 'Bernalism without Bernal', the growth of university and para-university units devoted to the agenda of *The Social Function of Science*, now variously named 'Science of Science' or 'Science Policy'. Bernal had empirically demonstrated the 'state of science' and this approach, suitably sanitized in the Cold War aftermath, was taken to the US by a Bernal disciple and historian of science, Derek de Solla Price. Price, using abstract numerical techniques alone, showed the irreversible transformation that had taken place as the production system of knowledge changed from little science to big science.[72] The philosophical origins of this brilliant analysis of the qualitative effect of the quantitative transition in the scale of science was never recognized in the US, where Price's work was acclaimed.

Nearer home the period saw the growth of university science study units in Sussex (Chris Freeman), Edinburgh and Manchester, all of whose initial debts to Bernal were apparent. The independent Science of Science Foundation (later the Science Policy Foundation) was founded by Maurice Goldsmith, with a residue of Bernalian charisma.[73] Had Bernal died before the birth of the Radical Science movement in 1969, it would have been as if the revolutionary flames he had helped to light in the 1930s had burned down to a very dull ember. It was for a new generation to attempt to capture that spark and fan it to life.

What should critical thinking about the sciences see as its legacy from Bernal? While Bernal confronted class antagonism in science, the analysis of the social relations of science has been extended over the last three decades to questions of ethnocentricity and androcentricity. Racist and patriarchal science, above all in sociobiology, has been interrogated and confronted. Yet although these dimensions of social relations were underrecognized in Bernal's time, it was he who gave us the essential concept of the social relations of science that allowed these questions to be raised. Nowhere in the corpus of Bernal's writings are these ideas expressed better than in his *Marx and Science*, a short pamphlet published in 1952: Farrington's introduction is brief: 'I do not think Bernal has ever written better than here,' he says, 'and to those who know only the Bernal of *The Social Function of Science* it will prove a startling experience.' In describing Marx's approach to science and dialectical materialism, Bernal opens many of the themes the later radical science, feminist and environmental movements came to make their own.

The task of making a new science, he concludes 'is not a simple or an easy one. It involves very great struggles and contradictions, because the whole ideology of science itself, an ideology implicit in all scientific theory, is derived from that of capitalism.' Science must become 'the property of the whole people, firstly by ensuring that most scientists are drawn from the working people, and then by directly involving working people in scientific research relevant to their own problems ... the philosopher has finally started to change the world, and what we have seen now is but a small foretaste of things to come. The struggle is still in front of us but we can be confident of the future.' That future, in Bernal's own phrase, is of 'a science for the people'.[74]

Acknowledgements

This chapter builds upon two earlier accounts: Hilary Rose and Steven Rose, 'The Two Bernals: Revolutionary and Revisionist in Science', *Fundamentia Scientia* 2, 3/4, 1981; and Hilary Rose, 'Talking About Science in Three Colours: Bernal and Gender Politics in the Social Studies of Science', *Science Studies* 1, 49–65, 1990. This latter paper was given at the Finnish Society for Science Studies 1989 meeting celebrating the 50th anniversary of the publication of *The Social Function of Science*.

Notes

1. John Desmond Bernal was variously called Bernal, J. D. B. and Desmond Bernal. Those intimates who knew him as a very young man were happy to call the young polymath Sage.
2. Maurice Goldsmith, *Sage: A Life of J. D. Bernal*, London, 1976. For his science see Dorothy Crowfoot Hodgkin, *John Desmond Bernal 1901–71*, Biographical Memoirs of Fellows of the Royal Society of London, Vol. 26, London, 1980; see also the Finnish journal *Science Studies* special issue celebrating the fiftieth anniversary of *The Social Function of Science* (1990) 1.

3. Gary Werskey's study of five left scientists had similar problems in that only Joseph Needham was willing to be interviewed. G. Werskey, *The Visible College*, London, Allen Lane, 1978.

4. J. D. Bernal, *The World, the Flesh and the Devil*, London, Kegan Paul, 1929.

5. Margaret Gardiner, 'Moscow Winter 1934', *New Left Review* 98, July/August 1976.

6. Thus when in the 1990s the ICA sought to sell Bernal's gift of the mural Picasso had sketched on Bernal's apartment wall at the time of the aborted London 1950 peace conference, only Michael, as Eileen's elder son, was consulted.

7. Werskey, *Visible College*.

8. Dorothy Hodgkin was probably the single exception, but she was very clear about the dangers of a woman being married to someone in the same field (interview April 1991).

9. Anne Sayre, *Rosalind Franklin and DNA: A Vivid View of What It Is Like to Be a Gifted Woman in an Especially Male Profession*, New York: Norton, 1975.

10. This does not mean that rampant sexism was not present in the Birkbeck laboratory environment. In the early seventies while waiting for a meeting to start, one of Bernal's close colleagues showed a group of radical science activists some pornographic photographs. The evident embarrassment of the group persuaded him to put them away. As the one woman present I acutely recall my feelings of rage and humiliation. – HR.

11. Sharon Traweek's 1989 anthropological study of physicists (*Beamtimes and Life Times: The World of High Energy Physics*, Cambridge MA: Harvard University Press) meticulously documents this compulsory heterosexuality for men physicists and a sort of androgenous sexuality for women.

12. His PhD student, the immensely distinguished crystallographer Dorothy Hodgkin FRS, OM, described her feeling when she learned that she had been awarded the Nobel Prize: intense pleasure at the honour modified by sadness as she felt it should have been shared with Bernal. Interview, April 1991.

13. Tor Wennerberg (1997), in an otherwise excellent piece on the recent eugenic 'scandal' in Sweden, mistakenly seeks to recruit Bernal's hatred of unreason into today's language of antiracism. He misreads Edwin Roberts, *The Anglomarxists*, New York, 1996. See p. 146 where Roberts describes Bernal's passionate commitment to reason and his hatred of irrationalism. See Tor Wennerberg, 'Nazism, Sterilisation and Propaganda', *New Left Review* 226, pp. 146–54.

14. The Marxist geneticist Herman Muller, working in the USSR in the early years of the Revolution, was an outstanding supporter of socialist eugenics. His science fiction speculations echoed his deeply androcentric dreams, cloning lots of Lenins, Einsteins and other important men in science and left politics. Among the biologist leftists speculating on the future of reproduction, only the geneticist J. B. S. Haldane – with a feminist sister, Naomi Mitchison, and married to a feminist, Charlotte Haldane – thought of cloning women as well.

15. Nils Roll-Hansen, Gunnar Broberg and Mattias Tydén, *Eugenics and the Welfare State: Sterilization Policy in Norway, Sweden, Denmark and Finland*, East Lansing, Michigan, 1996.

16. Nils Roll-Hansen, 'The Practice Criterion and the Rise of Lysenkoism', *Science Studies*, 1, 1989, pp. 3–16.

17. Jonathan Freedland, 'Master Race of the Left', *Guardian*, 30 August 1997.

18. For all his eminence in scientific circles, Bernal was entirely capable of taking part in direct confrontation. The story of his going to an antifascist demonstration with half a brick in his pocket told to HR by a local Camden Communist Party militant was part of that. While I am in no position to judge the veracity of the story it does tell us how Professor Bernal FRS was seen by his fellow militants. – HR.

19. While this is very evident retrospectively, at the time only the Marxist historian Benjamin Farrington understood this particular British development.

20. Feminist critics of science have long pointed out that this apparently democratic and universal 'any man' was historically restricted to men of a certain class. See also

Steven Shapin, *A Social History of Truth, Civility and Science in Seventeenth-century England* London, Chicago, Chicago University Press, 1994.

21. Bernal's utopian view of science as necessarily Communist and democratic found echo in an immensely influential text by the pioneering sociologist of science Robert K. Merton, 'The Normative Structure of Science' Chicago: Chicago University Press (1973). Between the first publication of the paper – pre Cold War – and later versions, the political Merton changed a key norm, which had initially echoed Bernal, from 'Communism' to 'Community'.

22. Bernal might well have been pleased, however, that the sociologist and critic of science Dorothy Nelkin was a recent recipient.

23. The English Bernalist Alexander King pioneered the growth of science policy studies at the OECD.

24. See Michael Gibbons et al., *The New Production of Knowledge: The Dynamics of Science and Research in Contemporary Societies*, London, Sage, 1994.

25. J. D. Bernal, 'The Struggle with Death', *New Scientist* 12 January 1967, pp. 86–8.

26. B. Farrington, transcript, BBC interview with G. Werskey, 'A Generation for Progress', broadcast 27 September 1972.

27. P. G. Werskey (1975) 'Making Socialists of Scientists: Whose Side is History on?' *Radical Science Journal*, 2/3 pp. 13–50: 'the CPGB was able to formulate the ideological and organisational basis of the radical intelligentsia's politics in the 1930s. . . . the Party encouraged its bourgeois members and "fellow-travellers" to concentrate their energies on organising and writing for their own political subgroups. With the intelligentsia so divided, the Party's officials could then rationalise their total control . . . This rationale for a privileged elite within the vanguard was already quite familiar to Soviet intellectuals' (pp. 30–31).

28. J. D. Bernal, 'The Scientist and the World Today: The End of a Political Delusion', *Cambridge Left*, Winter 1933–34, pp. 36–45.

29. J. D. Bernal, 'Albert Einstein: An Appreciation', *Labour Monthly*, 37(6), 1955, pp. 268–73; 'Picasso 80th Birthday Tribute', *La Nouvelle Critique*, Paris, 23 November 1961.

30. P. Anderson, 'Components of the National Culture', *New Left Review*, 50, 1968, pp. 3–57.

31. For accounts of the AScW see Werskey, 'Making Socialists of Scientists'; H. Rose and S. Rose, *Science and Society*, London, Allen Lane, 1969; R. Macleod and K. Macleod, 'The Contradictions of Professionalism: Scientists, Trade Unionism and the First World War', *Soc.Stud. Sci.* 9, 1979, pp. 1–32.

32. See W. K. G. Armitage, *Sir Richard Gregory, His Life and Work*, London, Macmillan, 1957; also Werskey 'Making Socialists of Scientists'.

33. Bernal organized the first conference on academic freedom in Oxford in 1935.

34. CSAWG, *The Protection of the Public from Aerial Attack*, London, Gollancz, 1937. The science writer Ritchie Calder, who took part in this campaign, has described how, to make the demonstration of the inadequacies of the gas masks more apparent to their audiences, the wearer was instructed to smoke a cigarette inside the mask; the smoke would appear on the outside. To ensure that the demonstration was theatrically effective, they removed the filters from the masks!

35. The physicist Eric Burhop, later to work on the A-bomb, later still to become president of the WFSW (as well as an FRS), was a postgraduate student member of CSAWG; Wilkins was also a student member of CSAWG, involved in the gas protection experiments. He became the president of the British Society for Social Responsibility in Science in 1969, and continued in this role for the lifetime of the organization.

36. 'A group of economists, scientists and technicians', *Britain without Capitalists: A Study of What Industry in a Socialist Britain Could Achieve*, London, Lawrence and Wishart, 1936. See 'Science and Education', pp. 407–68.

37. B. Hessen, in *Science at the Crossroads* (no editors) London, 1931; Kniga, pp. 149–212.

38. Joseph Needham describes this in his foreword to a new edition of *Science at the Crossroads*, 1971, London, Cass, 1971.

39. See M. Prenant, *Biologie et Marxisme*, Paris, Editions Société Internationale, 1935; H. Levy, *The University of Science*, London, Watts, 1932; and *A Philosophy for Modern Man*, London, Watts, 1938; J. B. S. Haldane, *The Marxist Philosophy and the Sciences*, London, Allen and Unwin, 1938. Lancelot Hogben, the social biologist, loathed dialectical materialism while endorsing historical materialism, and refused to join the CPGB until it embraced British empiricism.

40. J. D. Bernal, 'Dialectical Materialism', in *Aspects of Dialectical Materialism*, London, Watts, 1934, pp. 89–122.

41. See text 2(1), *Science and Society*, Winter 1937, pp. 58–66.

42. J. D. Bernal, *Marx and Science* (edited by B. Farrington) Marxism Today series no. 9, London, Lawrence and Wishart, 1952.

43. Our own ambivalent relationship with the Bernalian tradition during the 1960s in, for example, *Science and Society* is neatly captured by Ravetz. He writes, 'it is interesting that throughout their historical study, the authors of the (latter) book show impatience with the slow growth of the industrialisation (of science); but in the final chapter they find themselves making a radical criticism of the state of affairs that has finally been achieved.' J. Ravetz (1971) *Scientific Knowlege and its Social Problems*, Oxford: Clarendon Press, 1971.

44. J. D. Bernal, 'Professor Bernal Replies', *New Statesman and Nation*, 18, 441, 5 August 1939, pp. 210–11.

45. D. Noble, *America by Design*, Cambridge, MA, MIT Press, 1979.

46. J. R. Baker, 'Counterblast to Bernalism', *New Statesman and Nation*, 18, 440, 29 July 1939, pp. 174–5.

47. Quote by an unnamed researcher, believed to be Baker, in S. Zuckerman, *Science at War*, London: Hamish Hamilton, 1966.

48. 'Professor Bernal Replies'.

49. See W. McGucken, 'On Freedom and Planning in Science: the Society for Freedom in Science, 1940–1946', *Isis*, 1979.

50. There is a huge and often repetitive literature here which we do not cite.

51. There was of course no problem for the secret revolutionaries, the spies Burgess, Maclean, Philby and Blunt.

52. The Royal Society has not, since the 1939–45 war, admitted publicly engaged leftists to membership – unlike the National Academy of Science in the US – and perhaps we should not expect it to.

53. T. D. Lysenko, *Soviet Biology: a Report to the Lenin Academy of Agricultural Sciences*, Moscow, 1948; London, Birch Books.

54. As president of the Lenin Academy of Agricultural Science, Vavilov tried to recruit Lysenko but was blocked by senior plant physiologists who found him uneducated and unwilling to learn. Despite the debate between historians concerning the later relationship between the two men, Vavilov began as Lysenko's patron. N. Roll-Hansen, (1989) 'The practice criterion and the rise of Lysenkoism', *Science Studies*, 2, 1989, pp. 3–16.

55. R. Levins and R. C. Lewontin, 'The Problem of Lysenkoism', in H. Rose and S. Rose (eds.), *The Radicalisation of Science*, London: Macmillan, 1976, pp. 32–64.

56. L. R. Graham, *Science and Philosophy in the Soviet Union*, New York, Knopf, 1972; D. Joravsky, *Soviet Marxism and Natural Science*, London, 1961: Routledge and Kegan Paul; D. Lecourt, *Proletarian Science? The Case of Lysenko*, London: New Left Books, 1977; Z. A. Medvedev, *The Rise and Fall of T. D. Lysenko*, New York, Columbia University Press, 1969; Roll-Hansen; A. Vucinich, 'Soviet Physicists and Philosophers in the 1930s: Dynamics of a Conflict', *Isis*, 71, 1980, pp. 236–50.

57. For responses in Britain see J. Huxley, *Soviet Genetics and World Science*, London: Chatto and Windus, 1949; E. Ashby, *A Scientist in Russia*, London; Penguin, 1947; J. Langdon-Davies, *Russia Puts the Clock Back*, London: Gollancz, 1949; M. Polanyi, 'The Autonomy of Science', *Science Monthly*, 60, 1945, pp. 1410–50; C. D. Darlington, 'Retreat from Science in Soviet Russia', *The Nineteenth Century and After*, 142, 1947, pp. 157–68 and Society for Freedom in Science, *Papers on the Soviet Genetics Controversy*,

occasional pamphlets 9 and 10 (1949 and 1950). For later reflections see G. Jones, 'British Scientists, Lysenko and the Cold War', *Economy and Society*, 8, 1979, pp. 26–58; for France see Lecourt.

58. This and the following contributions of Levy and Bernal are to be found in the *Proceedings of the Engels Society*, nos 1, 2 and 3 (1949) and 7 (1952). Bernal's contribution is published as 'Biological Controversy in the Soviet Union', *Modern Quarterly*, 4, pp. 203–17.

59. J. B. S. Haldane, *Proc. Engels Soc.* 1, 1949, pp. 8–9.

60. R. W. Clark, *J. B. S.: The Life and Work of J. B. S. Haldane*, London, Hodder and Stoughton; 1968.

61. H. Levy, *Proc. Engels Soc.*, 1949, pp. 7–8.

62. J. D. Bernal, *Proc. Engels Soc.*, 1, 1949, pp. 11–12.

63. There is a vast literature here, but in its sharpest form the conflict can be seen between the gene-based writing of Richard Dawkins and the more organismic, process-based position of Steven Rose. See R. Dawkins, *The Selfish Gene*, Oxford, Oxford University Press, 1976; S. Rose, *Lifelines*, London, Allen Lane, 1997.

64. J. D. Bernal, *Science in History*, first, second, third editions (1954, 1957, 1965); London: Watts.

65. J. D. Bernal, 'Stalin as a Scientist', *Modern Quarterly*, 8, 1953, pp. 133–42.

66. J. D. Bernal, 'The Materialist Theory of Life: a Review of *The Double Helix* by James Watson', *Labour Monthly*, 50, 7, 1968, pp. 323–6.

67. See H. Rose and S. Rose (eds.) (1976) *Ideology of and in the Natural Sciences* (2 vols) London, Macmillan; R. M. Young (1977) 'Science *Is* Social Relations', *Radical Science Journal*, 5, pp. 65–131; H. Rose and S. Rose (1979) 'Radical Science and Its Enemies', in R. Miliband and J. Saville, eds., *Socialist Register*, London: Merlin, pp. 317–35; and the special issue of *Science Bulletin* (The CPGB's science discussion journal) 22, 1979.

68. See for example M. Millionshchikov, in *The Scientific and Technological Revolution: Social Effects and Prospects*, Moscow, Progress Publishers, 1972, pp. 13–38.

69. See E. Mandel, *Late Capitalism*, London, New Left Books, 1975.

70. J. D. Bernal, 'After Twenty-five Years' in M. Goldsmith and A. Mackay, *The Science of Science*, London, Penguin, 1964; H. Marcuse, *One Dimensional Man*, London, Routledge and Kegan Paul, 1964; J. Habermas, *Towards a Rational Society*, London: Heinemann, 1971.

71. Rose and Rose, *Science and Society*.

72. D. J. de Solla Price *Little Science Big Science*, New York, Columbia University Press, 1963.

73. This foundation was conspicuous for the number of industrialists and the shortage of science critics amongst its members.

74. J. D. Bernal, *Marx and Science*.

Bernal at War

Ritchie Calder

If Desmond Bernal, in order to get into the Second World War, had had to submit himself to a services' selection board, he would have rated the category 'definitely not officer material'. With his shaggy, finger-combed hair and his scholar's slouch he would have broken the drill sergeant's heart. With his unconcealed contempt for official stupidity he would have been booked for dumb insolence or worse. (One recalls his private exercise in operational research: his 'Theory of Combat' in which he claimed to have calculated every factor for infallible victory in the Desert War – except for one variable: the credulity of generals.) Sartorially he was hopelessly unsuited. The mind boggles at the idea of Bernal in mess kit. His chosen wartime uniform was a shapeles duffel coat with wooden toggles and a warden's tin hat in which he clambered over bomb sites. Even in the compulsory battle dress and forage cap, which he had to wear in the war zone, he could never be mistaken for a soldier. To get to the Normandy beaches with the invasion forces he had to don conventional uniform, and he himself wrote in his journal of that day how he set off in 'the very doubtful disguise of a naval officer'. He went on, 'The commodore appeared. I saluted – my first salute.' His first salute, after nearly five years of association with the services!

Then there was his MI5 record. His dossier was a bulky one as behoved the assiduous attentions of the Special Branch in the thirties. He had never concealed, or disavowed, his Communist affiliations. Politically, he was no mole: he was a chanticleer. The political police had kept tabs on his utterances and activities. What riled their masters and intensified their surveillance was when the Cambridge Scientists' Anti-War Group (CSAWG), of which he was president, began by actual experiments to expose and deride the government's air-raid precautions.

In a West German publication (Elga Kern, ed., *Wegweiser in der Zeitwende*, Munich/Basel 1956) Bernal wrote his account of the formation of the group:

> It was through the Spanish Civil War that the full dangers of Nazism and Fascism were revealed to the majority of the British people and formed the background of a general resistance which was to find expression when the World War came. It was the forms of violence it revealed, the bombardment of open towns in particular, a foretaste of what was to come to people in the rest of the world in a very short time indeed.
>
> And yet it would be an exaggeration to say that the majority of the people realised the full danger and the need to take action against it. We found, those of us who were active in trying to win support for the Spanish people, a general attitude of apathy and withdrawal, a feeling that one must attend to one's own affairs, that these things in distant countries were not our concern. This feeling went right on, through the seizure of Austria and Czechoslovakia right up to the beginning of the war. Indeed it needed the falling of bombs on Britain before it was finally shaken.
>
> Some scientists at Cambridge formed an anti-war group. We studied poison gases and protection against them. We saw that, in the event of war, science would have to play a big part in it. What we wanted to ensure was that, in the first place, the people would get a sufficiently clear grasp of the dangers they were running into and that they would unite to prevent war. But for that there was not enough goodwill in time.

As a matter of fact, the CSAWG had been in existence since 1934, coinciding with the rejuvenation of the local branch of the Association of Scientific Workers (AScW). In both, Bernal had been the moving spirit. The CSAWG had been basically pacifist, meeting for discussions on international tensions, taking part in demonstrations when the RAF had air displays at neighbouring airfields at Mildenhall or Duxford, supplying speakers and speakers' notes for sympathetic organizations, and writing letters to the press. What transformed the group, as it did for many other peace-seekers, was the Spanish Civil War, that dress rehearsal for Nazi aggression. If the fascist powers were to be contained, Great Britain, France and the Soviet Union must be seen to make common cause. This was the basis of the Popular Front, which the group supported. One form of preparation which the Popular Front accepted was the need for civil defence.

The CSAWG's intervention in the controversies raised about civil defence was to have immediate and durable effects. Scientists, as functional citizens, were not only expressing their opinions: they were backing them up by analysis, measurements and actual experiments, exposing themselves in some cases as human guinea pigs to personal danger. With predictable certainty (and with Guernica to prove it) Britain, in a war with the Nazis, would be exposed to air attack, with the civil population deliberately involved and bombarded with high explosives, incendiaries and probably with gas bombs. The Home Office's Air Raid Precautions (ARP) department had issued handbooks instructing the civil population, on a do-it-yourself basis, how to prop up their premises and gasproof them. The responsibility was on the citizens but if they followed the ARP guidebooks carefully (so they were told) they would survive the worst effects of blast, heat, splinters and poison gas. The government was resisting expenditure on public shelters both in terms of cost and of 'undermining the self-reliance of the people'. The brave people of Britain did not need 'funk holes'.

Official proposals were at best trivial and at worst cynical. Although it was pretty obvious that high explosives would be the most destructive element, the government (because little would be done about explosives) played instead on the risks of poison gases and then encouraged people to believe that the gases were not really as awful as had been made out.

In November 1936, the CSAWG, with W. A. (Peter) Wooster as chairman and his wife Nora as secretary, decided to test the claims being made by the Home Office for its gas protection measures. The testing labs were in a basement room on King's Parade, or the Wooster's sitting room, or in the college rooms of some enthusiastic research students, to the consternation of their 'bedders'. They were testing the gasproofing advocated by the Home Office for all and sundry, including the millions who, in overcrowded conditions, had no spare rooms to be sealed off, and the millions more who could not have afforded the expense necessary to make any room really gasproof.

The facts the group produced attracted public attention and questions in parliament. The government's predictable reaction was to attack the political motivation of the group and to get establishment scientists to repudiate their findings. They then staged demonstrations with lethal concentrations, using a solidly built and windswept cottage on Salisbury Plain that bore little relation to the substandard alley-locked city tenements where the gas would seep in and would not disperse quickly. And, of course, the Salisbury Plain cottage had not been rattled, much less shattered, by high explosives which would surely accompany the gas bombs.

In spite of being denounced as pursuing political ends and creating public panic, the CSAWG persisted. It caused genuine dismay among the government's own backroom boys who had already been trying to get the Cabinet to take ARP more seriously. One exercise of the group would have been funny if it had not been so much a matter of life and death. The government had plans to issue civilian gas masks (costing 2s. 6d. at pre-war prices) to every member of the public. The group managed to get possession of some ahead of issue. The investigation (carried out among others by the biochemist R. L. M. Synge, later to get the Nobel Prize for his work on partition chromatography) found that the canister snout containing carbon was no protection against arsenical smokes.

As a science reporter, I was in close touch with the group, and, in addition to writing in the *Daily Herald* and the *New Statesman*, I was speaking about their findings at meetings arranged by the Union of Democratic Control, by the Socialist Medical Association and, to the special annoyance of the government, by branches of the National Association of Local Government Officers (NALGO). Since NALGO members would be involved in the ARP activities this was 'subversion', which meant that I was shadowed by the Special Branch. My 'shadow', whom I got to know well, would turn up at my meetings and watch me do the trick that Bernal taught me. It was simple: one removed the valve from the gas mask, filled one's mouth with tobacco smoke, put on the mask, and breathed out the smoke through the canister. With a spotlight and a white background the smoke could be seen curling out.

That campaign was certainly successful. The shamefaced authorities produced as a smokeproof filter a wafer which was strapped on to millions of stocked snouts with sticking plaster before the masks were publicly issued.

These activities were not likely to endear Bernal to the authorities. The Munich Agreement had jolted many more scientists out of their political neutrality. Within two weeks of Munich, Bernal had contributed a signed editorial to *Nature* on 'Science and National Service'. This was to be a trailer for a memorandum on 'Science and National Defence', written in October/November 1938 by Bernal after discussion with Solly Zuckerman, a zoologist at Oxford, and sent first to Secretary of State for War Leslie Hore-Belisha and then, when Hore-Belisha ignored it, to Liddell Hart, the military correspondent of *The Times*. Solly Zuckerman, in *From Apes to Warlords* (Hamish Hamilton, 1978), gives an account of the memorandum and of the abortive efforts to bring it to offical attention. The opening paragraph stated that it was:

> . . . of the utmost importance that the special character of scientific work in the service of national defence should be realised, and that the proper steps

should be taken in time to see that the work of scientists is not wasted. Scientific workers have, in relation to national service, two special qualifications. In the first place, they are expert in the use of essential instruments and methods which are acquired only after long training, and it would be impossible under war conditions to replace them. But this is the least important of their potentialities. They also have a special training in the ability to deal with new or unexpected developments and to suggest new possibilities of action.

The memorandum, as Zuckerman summarizes it:

... went on to consider the need for scientific research under the six following headings: Maintenance of civil and military life; maintenance and extension of war production; defence against aerial attack; carrying out of military operations; caring for casualties; and maintenance of morale. In retrospect I find it interesting that what the memorandum did not consider was the use of scientists in the analysis and planning of actual operations, an omission to which Liddell Hart in fact drew attention, and which in the end was to be my main preoccupation.

Bernal's induction into the defence establishment, as one of the first of the 'boffins', was a freak of history. His anti-fascist record was impeccable but, in some high quarters, that was no recommendation. His public identification with the peace movement, including Lord Robert Cecil's peace campaign, with the Popular Front and with the CSAWG, and his membership of the Communist Party were 'security' objections. After the bumbling efforts at a Soviet–British entente, the signing of the Molotov–Ribbentrop Pact, the Soviet support of the invasion of Poland and the Russo–Finnish war, right up to the Nazi invasion of the USSR on 22 June 1941, any known Communist was automatically debarred from any security post. Yet Bernal was recruited at the very outset of the war and continued throughout in positions of high sensitivity. Indeed, he had been earmarked before the declaration of war by a most unlikely person.

In January 1939, Sir Arthur Salter, Fellow of All Souls, Oxford, gave a luncheon party for Sir John Anderson, Lord Privy Seal in the Chamberlain government, with responsibility for civil defence and likely to be Minister of Home Security. Salter thought he should be exposed to his most outspoken critic, Bernal, and asked Zuckerman to bring him along. Sir John Anderson (later Viscount Waverley) was an archetype of the establishment man. He had climbed the civil service ladder to become Permanent Secretary at the Home Office. He had been Chief Secretary for Ireland and Governor of Bengal. He had switched to parliamentary politics and, with a safe Commons seat, was in Neville Chamberlain's Cabinet. Sir Arthur Salter, a distinguished economist who had been elected independent MP for Oxford University in 1936 as a left-of-centre candidate, beating the official Conservative, Sir Farquhar Buzzard, must have had impish satisfaction in bringing together the pukka sahib, Anderson, and his antithesis, Bernal.

As it turned out, the two hit it off extremely well, for the very good reason that Anderson, before he opted for the civil service, had studied science at Edinburgh University and Leipzig. Like Ernest Rutherford he had been a meritorious '1851 Exhibition' scholar. Bernal was forthright in his criticism of civil defence but he could quantify his objections, and Anderson listened attentively. On his return to London he sent for Dr Reginald Stradling who, from being Director of Building Research, had been appointed Chief Scientific Adviser on civil defence and was setting up the Princes Risborough research station. Anderson told Stradling to get Bernal to join the Civil Defence Research Committee. This first met on 12 May 1939. Stradling, who was in no way a political animal, ventured to remind Anderson that Bernal was known as a 'red'. The minister, to his everlasting credit, replied, 'Even if he is as red as the flames of hell, I want him,' and, just as remarkably, Bernal agreed to serve.

From the outset of the war Bernal was involved in the Research and Experiments Department of the Ministry of Home Security, which had been installed at the Forest Products Research Station at Princes Risborough, twenty miles from Oxford. There he was able to help Stradling, who was wide open to ideas, to collect a team and produce from scratch a programme of research and inquiry. Great emphasis was laid on statistics and mathematical modelling. Completely new techniques had to be devised for handling a mass of data. Bernal was responsible for bringing in Frank Yates of the Rothamsted Experimental Station; a mathematics lecturer, Jacob Bronowski, from Hull University College; and Bradford Hill of the London School of Hygiene and Tropical Medicine. Yates was already a colleague of R. A. Fisher – with whom he had compiled the Statistical Tables for Biological, Medical and Agricultural Research – and later became a master of computer science. Bronowski was to become world-famous as a popular expositor of science and an outstanding television personality (*Ascent of Man*, et cetera). His success owed a good deal to his training at Princes Risborough, since he first made his mark through a brilliant radio broadcast on the bombing of Hiroshima and Nagasaki, where he had gone as a statistician on the Chiefs of Staff mission of inquiry. Bradford

Hill (with Richard Doll) was to be responsible for the post-war report that established the relationship of cigarette smoking to lung cancer.

With characteristic energy, Bernal threw himself into the work of civil defence, abandoning his academic activities.

In his very first weeks Bernal made a habit of driving into Oxford for consultations with Solly Zuckerman about the possible effects of blast, and the consultations became actual experiments. This was the theoretical phase, during the Phoney War. In spite of World War One with its millions of casualties, its intensive trench warfare, the strafing and the 16½-inch shells of Big Bertha, the nature of explosions was imperfectly understood. It was not just a matter of rethinking – new thinking was needed. Bernal was exasperated by the lack of valid information about the behaviour of explosives, in spite of the Ordnance Board which had been operating since the fifteenth century. Bernal immediately launched himself into the physics of explosions, the measurements of resistance of structures to various types of shock, and the effect of shock waves on the living body. His experience confirmed his contention that the British scientific establishment was ill-organized for war and, for that matter, for peace.

With his extraordinary capacity for reducing complexity to essentials and being able to explain, he quickly made himself an authority on explosions:

> High pressure waves have not the smooth wave form of ordinary sound waves. Instead, the pressure in front of the wave rises immediately to its highest value and then falls gradually, to be followed by a phase of less than atmospheric pressure, i.e. by suction. The generation of this steep-fronted or shock wave is very similar to that of the waves that break on the sea-shore: the high pressure phase of the disturbance, always travelling faster than the low pressure phase, gets to the front just as the top of the wave on the seashore, being held back less than the base that is being slowed by friction of the sand, moves forward and ultimately breaks the wave.
>
> . . . A shock wave behaves in a characteristic way towards limited obstacles. Ordinary sound waves throw shadows only from the largest obstacles, such as hills or high buildings, because their wave length is of the order of 10 to 100 feet. A shock wave has no single wave length but a number of wave lengths. The sharp high pressure part has a very small wave length and the tail suction part has a very long one. Consequently a shock wave passing through an aperture or round an obstacle is changed in character. Roughly speaking, the pressure part goes straight and casts shadows, whereas the suction part travels round a corner without any difficulty. This is very useful as it has been shown that it is the pressure part of the wave that is responsible for most physiological damage. To have even a small garden wall between oneself and

the bomb is practically to be secure from the direct effect of blast. On the other hand, an open doorway is a danger. (J. D. Bernal, 'Physics of Air Raids', Royal Institution Lecture, 3 December 1940)

To investigate the risk that people in underground shelters might suffer from concussion as a result of shock waves that passed through the earth when a bomb exploded nearby, Bernal called in Zuckerman, who had been investigating the behaviour of primates for over ten years. In October 1939, Zuckerman took along his monkeys for a series of tests on the effects of ground shock on various kinds of trench. In the first experiment the trench wall nearest the bomb had been fractured but the monkeys were unscathed. Reassured – but against the advice of the Home Office Explosions Inspector – Bernal and Zuckerman went into the next trench themselves and waited for the bomb to be fired. Nothing went amiss.

To the official indignation of the Medical Research Council (MRC), which regarded anything to do with bodies as its affair, Zuckerman, encouraged by Bernal and Stradling, persisted in experimenting with animals to test many officially accepted assumptions. These included the idea that blast exercised its effect by forcing its way down the mouth and nose and into the lungs, bursting them.

Before the fall of France and long before the air blitz on London, Bernal – through his re-examination and experimentation on the physical and structural effects of blast – and Zuckerman – with his investigation of the effects of blast on living organisms – had begun to revise drastically the assumptions on which air-raid precautions and anticipated casualties had been based. At the beginning of the war it was officially believed that human beings would be killed if struck by a blast pressure of 5 lb per square inch. Bernal and Zuckerman showed that blast was in fact one hundred times less dangerous than had been supposed. 'This discovery revolutionised conceptions of the effects of bombing,' says the official British history, *Science at War* prepared by J. G. Crowther and Professor R. Whiddington FRS (HM Stationery Office, 1974). And, at the same time, Bernal was challenging J. B. S. Haldane's views, derived from his Spanish Civil War experiences. Haldane had said:

> When a big bomb explodes the blast becomes translated into a wave of sound like that of the last trumpet which literally flattens everything in front of it. It is the last sound many people ever hear, even if they are not killed, because their ear drums are burst and they are deafened for life. It occasionally kills people outright without any obvious wound. (*ARP*, Gollancz, 1938)

These assumptions had not augured well for the garden bunkers – Anderson Shelters (dugouts, arched with corrugated steel and sand-bagged) – and the street shelters like flat-topped toilets that the government was promoting as against the idea of underground public shelters. On those same assumptions, however, they were preparing for 100,000 air-raid casualties per week, with corresponding stocks of cardboard coffins.

During the Phoney War, Bernal and his colleagues on the Civil Defence Research Committee and his team-mates at Princes Risborough were seeking facts and setting up their statistical systems and models. The Germans, having mopped up Poland, were preparing their Western *Blitzkrieg* by land and air. There was little practical material to work on. There were simulations and animal experiments, and a great deal of scepticism and active opposition. Bernal overcame the Medical Research Council's disapproval of Zuckerman's activities by taking the findings to the Civil Defence Research Committee, which included distinguished physicists who were not unduly impressed by the MRC's conventional wisdom. The result was that Zuckerman was given a civil defence research grant to support his Oxford Extra-Mural Unit and fieldwork facilities for blast experiments at Stewartby in the brickfields near Bedford.

This was the weird, twilight period of the war. Soviet forces were fighting the Finns in the winter snows. The French army, after a feint offensive, had immured itself in the Maginot Line. The British expeditionary force was building up only slowly. Belgium and the Netherlands were neutral. The Germans were regrouping for their *Blitzkrieg*. The Western Front stagnated for seven months.

Bernal's official position did not gag his criticisms. I was present at the dinner of the Tots and Quots on 23 November 1939 when he opened the discussion in forthright fashion. The Tots and Quots was a remarkable dining club started in 1931 by that social catalyst Solly Zuckerman, then a 26-year-old research anatomist at the London Zoo. Its title was a corruption of the Latin tag 'Quot homines, tot sententiae,' which can be freely translated as 'As many opinions as there are men.' The idea, apart from conviviality, was to bring scientists into touch with the 'outer world' (ultimately it acquired such a reputation that few men of affairs, however eminent, would turn down an invitation to be a guest). The Tots and Quots nurtured most of the scientific Young Turks who enlivened the 1930s, all of whom finished up famous. From an ebullient start, with Zuckerman as impresario and major-domo, and after its impact in the mid-thirties, it had been dormant until, at

Bernal's suggestion, Zuckerman decided to revive it in the autumn of 1939. Its second phase became historic. J. G. Crowther, then scientific correspondent of the *Manchester Guardian*, kept his notes of that November dinner, which he quotes in his *Fifty Years with Science* (Barrie and Jenkins, 1970).

> Bernal spoke of the disorganisation of science. All that we had feared before the war began would happen, had happened and was in fact worse. Those traditionally in charge had no influence although they had prevented anybody else from doing anything on the ground that they were busybodies.
>
> Funds for science were liable to end at any moment. This was a mistake from the point of view of those running the war. The Government research formed vested interests. The situation in the Services was lamentable. There was obstruction of research on things which should have been done years ago. He (Bernal) had heard three years before of an important device; research on it had started two weeks after the commencement of the war. We were not getting any effective utilization of science, and in addition to this, academic science was being stopped. The threat of absence of funds stopped research. This was happening because the Government and its servants were unfitted to use science. No plan for science could be expected of them.

That dinner, and Bernal's presentation and exhortation, gave the kiss of life to the Tots and Quots. It reared into action. Julian Huxley immediately produced a paper on the organization of scientific research in Britain, arguing that there should be more executive power, planning and coordination in government science, with direct access to the Cabinet through a senior minister. A month later, with the sodality reinforced by scientists from the fields concerned, the discussion took up where Bernal had left off, with detailed examples from others of where things were going wrong and what should be done about food and nutrition.

At the January 1940 dinner we were discussing the neglect of science in war industry and deciding to produce a series of pamphlets. In February the movement's anxieties were continued by a visit of leading French scientists. The French, with some exceptions like Auger and Langevin, alarmed even the orthodox British scientists by their complacent account of French military scientists. Something had to be done about our allies as well. Contact was made with the French embassy to discuss a plan of action, which later led to the formation of an Anglo-French Society of Sciences. In March, Tots and Quots had as guests Lord Melchett of ICI and Professor Lindemann (later Lord Cherwell), Churchill's grey eminence. The theme was science in wartime industry, and the input was again disquieting. Bernal brought the discussion fiercely back to the misuse of science, arguing that science was suffering from a process not so much of repression as of misdirection. Lindemann said pontifically that the social functions of science would have to wait until the war was won. This was hotly disputed by those who were maintaining that the war could not be won unless science was applied over a wider front and by people who knew what they were doing with it.

In April, Bernal, in his official capacity, contrived to go to France, taking Zuckerman with him. The mission was organized through the Ministry of Home Security with the object of seeing what civil defence research might learn from French experience. The cicerone to Bernal and Zuckerman was that legendary character the Earl of Suffolk, the British government's scientific liaison in Paris. Later, Suffolk was to organize the escape of Halban and Kowarski, complete with the result of Joliot-Curie's research on the chain reactions in uranium and the whole French stock of heavy water, which was the largest existing supply. He got them on board ship, under shellfire at Bordeaux, and brought them and their momentous cargo back to Britain and to the Cavendish Laboratory. A year later, engaged on bomb disposal work, Suffolk was blown to pieces: he was awarded a posthumous George Cross.

The French official contacts confirmed Bernal's misgivings: French science was no better organized than the British for making its contribution to victory – or survival – in the war against the Nazis. Behind the barbed wire of bureaucracy they found some who could be regarded as their opposite numbers in the work they were doing, and many who were like-minded and frustrated about how science could be organized. The latter included Jean Perrin, Paul Langevin, Frédéric Joliot-Curie, Pierre Auger, Henri Laugier and H. Longchambon. It was the familiar story: the military resisted the scientists 'muscling in on their territory', and the scientific departments were jealous of their privileges and afraid of revealing their incompetence.

For the work on hand, however, they gained insights. They learned a lot from a day at the Bellevue Ballistics Laboratory. They were taken by the director, Colonel Libessart, to Modane on the Maginot Line and were overawed by the sheer vastness of the underground fortifications and appalled by the smug reliance on them. It really was *folie de grandeur*. But they learned something from the mock explosions. On their way back to Paris they were taken to the artillery range at Bourges where they saw the firing of the biggest guns and high explosives that the French had in their armoury.

Comparing notes with French experts actually on the job, they discovered how little exchange of information there was between the allies. Inquiries they themselves had made had never reached their French opposite numbers, whose reciprocal inquiries about blast and bomb casualties had got lost in British in-trays.

Metadier, of the French Admiralty, had been invited to a Tots and Quots dinner to discuss how relations between the French and British scientists could be strengthened. The question of an Anglo-French society of science was pursued as a matter of urgency. Metadier was so impressed that on 8 April 1940 he took J. G. Crowther, as the emissary of the Tots and Quots, back with him in his plane to Paris to conduct an intensive series of high-level meetings which led to Joliot-Curie becoming the president, and Auger the secretary, of a prestigious French Committee. Crowther flew back to Britain and mustered a British Committee with Paul Dirac, the Nobel laureate, as president; Professor A. V. Hill, Sir Alfred Egerton and Sir Henry Tizard as vice-presidents, and a clutch of Tots and Quots as its executives – Bernal, Blackett, Cockcroft, Darlington, Waddington and Zuckerman – with Crowther as secretary.

It was a bit late in the day: in the week that it took to form the British Committee, the Nazis had occupied Norway and, incidentally, had overrun another of the Tots and Quots, Lancelot Hogben, who had been lecturing in Oslo. He opened his hotel room curtains to discover the Germans marching up the street. He and his daughter escaped in a milk float across the border into Sweden.

A month later Hitler struck through the Netherlands and Belgium, bypassing the Maginot Line and paralysing the French general staff. The British evacuated 233,000 of their troops and 112,500 French from Dunkirk. The French government capitulated. The Anglo-French Society of Science became, in effect, a rescue operation. It not only arranged the escape of distinguished French scientists but re-rescued them from humiliation at the hands of the British authorities who, obsessed with their suspicions of the infiltration of the French evacuees by German agents, had incarcerated some of the most eminent of them. This crassness was later redeemed by the hospitality and vigorous intellectual exercise provided by the graciously housed Society of Visiting Scientists which burgeoned from the Anglo-French Society of Science.

The violent end of the Phoney War gave urgency and impetus to the demands for reorganization of science for war that Bernal had spelled out at the November 1939 dinner of the Tots and Quots. Allen Lane, the publisher of Penguin Books, was a guest at their dinner on 12 June, a week after Dunkirk. So was Kenneth Clark (later Lord Clark, OM) from the Ministry of Information. After a rousing discussion among the scientists present, Allen Lane said it was a pity that a verbatim record had not been made because it would have provided a powerful Penguin Special. At the table Zuckerman immediately nailed him. If Lane could have a typescript within a fortnight, could he produce a book within a month? Lane accepted the dare, and before the meeting broke up, Zuckerman, briskly as ever, had allocated specific tasks to those present and had collected suggestions as to others who could provide further documentation.

The insistent Bernal and the indefatigable Zuckerman got down to work immediately. They drafted an outline of the book, reflecting very much the heads of Bernal's 1938 memorandum to Hore-Belisha. By telephone they chivvied their contributors into action and pestered them into delivery. Within days, the material was coming in from the work of many hands, and the editing began. The compilation and rewriting were carried out with newspaper deadline efficiency. By 23 June the typescript was ready. Allen Lane kept his promise: the typesetting began and Zuckerman and Crowther started the proofreading. The presses rolled. As Allen Lane was rightly to boast in his publisher's note: 'Science in War crystallizes the discussion of a group of distinguished scientists who met round a dinner-table less than a month ago.'

Science in War was a publishing feat. It was also a historic event. The twenty-five contributors were anonymous which, in a curious way, increased its impact. It was realized that they were, in their frustration, appealing to the court of last resort, the general public. The book was widely and influentially reviewed, with the ironic twist that anonymous contributors (such as Julian Huxley) were called upon to review it and reinforce it. Its circulation ran into tens of thousands of copies and its royalties helped to subsidize subsequent dinners of the Tots and Quots. It also provided food for thought for the policy makers, and substance for questions in Parliament.

It is always difficult to establish post hoc ergo propter hoc but it cannot be without significance that within a year every 'Quotentot' (Hogben's version) was in a key position in the war effort.

Hitler's onslaught on Britain still did not happen. The air-raid wardens were on the alert. The Home Guard (mustered as the Local

Defence Volunteers in May) were awaiting the paratroopers and drilling with broomsticks while awaiting their rifles. Road signs, even street signs, had been removed to bamboozle the enemy. The blackout was rigorously enforced. Between 13 and 18 June nearly one hundred thousand children were evacuated from the London area, many for the second time. But there was little to test the air-raid precautions until August, when the Luftwaffe tried to neutralize RAF airfields and destroy radar masts, and an attack on Croydon airport gave London its first taste of bombing.

Bernal was assiduously chasing 'incidents', trying to identify the characteristics and, with his statistical colleagues, trying to estimate possible damage and casualties if serious raids happened. Stradling had organized a field staff to collect statistics of damage when raids occurred. There were 120 observers with 40 headquarters to analyse the data when collected. The nucleus was provided by the Cement and Concrete Association, whose advisers and salesmen could use their local connections. The official history, *Science at War*, states:

> In June 1940 Bernal and Dr F. Garwood forecast the results of a raid by 500 enemy bombers on a typical English town. They happened to choose Coventry. They worked out what the effects would be from their new data on the destructiveness of bombs and their probable distribution on the town as determined by statistical experiments. Some time afterwards Coventry was attacked by about 500 bombers: the forecast by Bernal and Garwood of the amount of damage and casualties was exactly confirmed when the results of this serious attack were surveyed.

This, it should be noted, was a theoretical prediction and not to be confused with the Hull and Birmingham survey that Bernal and Zuckerman prepared considerably later, in March 1942, based on the actual raids on those cities. This became a central issue in the arguments about the area-bombing strategy against Germany that Lord Cherwell successfully advocated.

Cherwell – a member of the War Cabinet and Churchill's most influential adviser – was seeking to show, by simple arithmetic that the Prime Minister could grasp, that if Bomber Command were given the necessary aircraft, Germany could be bombed into submission. However, he needed plausible figures. He knew the work of Bernal and Zuckerman and the insights and information they had accumulated from the raids on London and the provinces during the first real air blitz starting in September 1940. He approached Sir John Anderson to get them to do a survey of the overall effects of the bombing of some

British cities. Birmingham and Hull were chosen because the Bomb Census had an almost complete record of the bombs that had fallen on them and because they could be regarded as typical of manufacturing and port towns.

Bernal and Zuckerman organized an inquiry covering physical damage, casualties, effects on production, absenteeism, evacuation and morale. They assembled the staff for the various investigations and began to cope with the masses of reports that came in. It was certainly the most thorough and informative study of the effects of bombing that had ever been made anywhere.

Zuckerman was keeping Cherwell posted about interim findings, on the understanding that he would not anticipate the results of the survey that he himself had initiated. Cherwell jumped the gun. A week before receiving a full assessment of the report he sent a minute to Churchill (30 March 1942):

> Careful analysis of the effects of raids on Birmingham, Hull and elsewhere have shown that on the average one ton of bombs dropped on a built-up area demolishes 20/40 dwellings and turns 100/200 people out of house and home. We know from our experience that we can count on nearly 14 operational sorties per bomber produced. The average lift of bombers we are going to produce over the next 15 months will be about 3 tons. It follows that each of these bombers will in its lifetime drop about 40 tons of bombs. If these are dropped on built-up areas they will make 4,000 to 8,000 people homeless.
>
> In 1938 over 22 million Germans lived in 58 towns of over 100,000 inhabitants which with modern equipment should be easy to find and to hit. Our forecast output of heavy bombers (including Wellingtons) between now and the middle of 1943 is about 10,000. If even half the total load of 10,000 bombers was dropped on the built-up areas of these German towns the great majority of their inhabitants (about one-third of the German population) would be turned out of house and home.
>
> Investigation seems to show that having one's house demolished is most damaging to morale. People seem to mind it more than having their friends or even relatives killed. At Hull, signs of strain were evident although only one-tenth of the houses were demolished. On the above figures we should be able to do ten times as much harm to the 58 principal German towns. There seems little doubt that this would break the spirit of the people.
>
> Our calculation assumes, of course, that we really get one half of our bombs into built-up areas. On the other hand no account is taken of the large promised American production (6,000 bombers in the period in question). Nor has regard been paid to the inevitable damage to factories, communications etc. in these towns and damage by fire probably accentuated by breakdown of public services.

This was too simple arithmetic. Sir Henry Tizard, chairman of the Aeronautical Research Committee and Rector of Imperial College, London, warned the Secretary of State for Air that Cherwell had overestimated the probable effects of using the entire heavy bombing force by a factor of at least four. Blackett at the Admiralty said bluntly, 'Lord Cherwell's estimate of what can be achieved is at least 600 per cent too high.'

Anyway, the assumptions on which the minute was based – that bombing people's homes would demoralize them and disrupt war production – flew in the face of the findings of the survey he had commissioned. The report, for which he had not waited, said categorically, 'There is no evidence of breakdown of morale for the intensities of the raids experienced by Hull and Birmingham (maximum intensity of bombing 40 tons per square mile).' Other findings showed that temporary evacuation did not interfere with the work of the town. Loss of production was due almost entirely to direct damage to factories. Indirect effects of raids on labour, turnover, health and efficiency were insignificant. The direct loss of production in Birmingham due to raids was about 5 per cent and the loss of production potential was very small.

Cherwell did not get his full *Delenda est Carthago* policy through, but the area-bombing strategy he encouraged – indiscriminate destruction as distinct from selective bombing of military installations or productive plants – led inexorably to the firestorms of Hamburg, the destruction of Dresden and the holocaust of Hiroshima. When C. P. Snow's Godkin Lectures, 'Science and Government', in 1960 revived the issues and led to an acrimonious debate that still goes on, Bernal wanted to intervene and suggested to Zuckerman that they write a joint letter to *The Times*. Zuckerman demurred and discouraged Bernal from sending the following letter to the editor:

> No one who had worked with Lord Cherwell in the war could doubt his intense and detailed interest in the scientific aspect of operations: what was in doubt was, rather, his judgment and, in particular the way in which his treatment of scientific data was distorted by his preconceptions and aims. This was most evident in his preparation for the bombing offensive against German cities in 1942. To justify this course he initiated a survey of the effects of German bombing on a British industrial city and munitions centre. Birmingham was chosen as the principal target and Hull as the control. It was undertaken by the Ministry of Home Security under the direction of Solly Zuckerman and myself.
>
> It was the most elaborate combined military, economic and social study of a large city ever carried out, for we had to measure the degree of bombing,

the casualties and destruction, and to compare them, day by day, with the effects on munitions production and morale in the city.

We had two teams of about 40 workers each which went into everything in great detail, down to the number of pints drunk and aspirins bought in the chemists. The factories were accurately sampled. The conclusion which might well have been reached without this fuss and expense was that the bombing of Birmingham had only a small effect on production, quite outweighed by the positive effects of higher piece-rates, so that more workers actually came into the city during the period of bombing.

To the ordinary scientists, this report would hardly have been taken to support the thesis of the strategical value of area-bombing, but Lord Cherwell claimed that it proved that as German bombing had produced little effect, a very much heavier bombing of German cities must produce a decisive one, and he took action accordingly with the results that we all know now. It was this kind of wild extrapolation that made most scientists who served under both prefer the careful logic of Tizard's approach to such problems. For Tizard never neglected – as Cherwell nearly always did – the operational factor: the ratio between what is found by experience to happen in the field and what is deduced from trials carried out under the most favourable conditions. It is this factor, rarely less than five and often greater than ten, that makes many schemes which seem effective on paper fail altogether in practice. (27 April 1961)

Within a few weeks of the completion of the Birmingham/Hull survey, Bernal moved from the defensive to the offensive. He received a letter from the Chief of Combined Operations, Lord Louis Mountbatten, saying that Tizard had proposed him as Scientific Adviser to 'Combined Ops'. Bernal consulted Zuckerman and replied to Mountbatten – yes, if he could bring Zuckerman with him. Mountbatten had, in fact, asked Tizard for two first-rate scientists who could be put on to the operational analyses of devices and equipment and techniques for landings. Bernal was delivering both.

It was a fateful event for both scientists. It was to take Bernal (*mirabile dictu*) into the Forbidden City of military security and to the Quebec Conference of 1943, where Roosevelt and Churchill and the Combined Chiefs of Staff took the decision on Overlord (the code name for the landings in Europe). It took Solly Zuckerman into the innermost councils of the Supreme Commander who executed these decisions.

The irony of Mountbatten's bold and unregretted choice is illustrated by Lord Blackett's account in his Chairman's Address at the J. D. Bernal Lecture at Birkbeck College (23 October 1969):

On one occasion Bernal urgently asked for an assistant to help him on his study of the Normandy beaches for the invasion of Europe, and suggested an ex-colleague from his bombing days to take this post at the Headquarters of Mountbatten's Command. Endless delays ensued and eventually the Security people prohibited the appointment of this young man. When asked why, the Security people explained that before the war the man they wanted had been associated with a noted Communist. What was his name? they were asked. The answer came – J. D. Bernal!

Mountbatten and Bernal made an incongruous pair: the Almanac de Gotha and the Communist Party; Buckingham Palace and King Street; the acknowledged authority on ceremonial pomp and the iconoclast. Bernal was the exception to any rule in Mountbatten's book. Pomp gave way to circumstance as is illustrated by Zuckerman's account (in *From Apes to War Lords*) of how King George VI came to review the staff of Combined Operations: it was a quarter-deck ceremony on the terrace of Montagu House:

> Bernal and I had been told to be on parade, but when I arrived, after most of the staff had already formed up, Des was not there. Mountbatten greeted me by saying that he was worried lest Bernal, who usually wore his hair very long, might have gone to get it cut. Fortunately, Des turned up in time with his hair in its pristine state, and the King, who had been told what to expect, was not disappointed.

Bernal was Mountbatten's tame magician – or licensed court jester. The C-in-C issued instructions to his top brass:

> I have decided that the two Scientific Liaison Officers, Professor Bernal and Professor Zuckerman, should be allowed to act as scientific observers of the Meetings of the Planning Syndicates, each taking a different operation in turn.
> This is an experiment to which I attach great importance as I am anxious to link up the scientists from the very beginning of operational planning so that when their scientific knowledge is required, they may be completely in the picture.

Bernal responded to Mountbatten's avid interest in science, his open-mindedness, his vitality, his interest in problems and people *per se* and his grasp of the principles of operational research. A genuine affection developed between them and a mutual respect for each other's exceptional qualities.

There can be no better description of the essentials of Bernal's role at Combined Operations than the Earl Mountbatten of Burma's own account in this volume (Chapter 8). And the author's glosses on those essentials are peculiarly illuminating of Bernal's qualities. He was, as Mountbatten so generously acknowledges, identified with Mulberry, the artificial harbours that made the invasion possible, and with that entrancing nonstarter Habbakuk. Behind both were the characteristics that made Bernal very special. His real friends will recognize the quintessential Bernal in Mountbatten's statement, 'His most pleasant quality was his generosity: he never minded slaving away on other people's ideas, helping to decide what could or could not be done, without him being the originator of any of the ideas on which he actually worked. This may be why his great contribution to the war effort has not been properly appreciated.'

The Normandy landings owed much to that other characteristic of Bernal, which earned him as a young scientist the nickname Sage. This was his encyclopaedic knowledge, or rather his insatiable curiosity, coupled with a retentive memory, a data bank for serendipic trifles that a problem or an occasion would make relevant.

When I asked Bernal after D-Day what had happened at Quebec in the discussion of landing points, he said, 'I just happened to remember a holiday in Normandy when I had been bathing off the beach at Arromanches and I had noticed that I was swimming through peat . . .'. That observation made history not only in determining the nature of the beaches and the military traffic they would have to carry, but in the development of the science of waves; for it was a remarkable fact that while so much was known about tides that nautical almanacs could give local information with minute precision, comparatively little was known scientifically about the behaviour of waves, which had tossed and battered ships through millennia.

Typically, Bernal chose the occasion of the Caen Congress of the French Association for the Advancement of Science in July 1955 to tell the full story for the first time, with a graceful acknowledgement to the Linnaean Society of Caen, the proceedings of which, over its century of existence, he had studied in connection with the landings.

> I came to learn something of the plan of the landings as early as the summer of 1943 when I took part in the Conference at Quebec. It so happened that I was about the only person concerned with the operations who had actually been to the place of landing, namely Arromanches. For that reason I was given the task of examining some of the physiographical information that might be useful for the landings – nothing to do with military dispositions, at any rate until the last phase.

He had started, he said, by having a look at the guide books and found in *Le Guide Bleu* the description, 'The beach at Arromanches is very sloping, indeed the sea goes out a long way and at extreme low tides the peasants go out to the end of the beach and pick up a material they call "gourban" which they use to manure their fields.' Bernal reported:

> It was immediately apparent to me that this kind of beach might well be treacherous and that it was probably laid down on a basis of clay and peat – on an old forest of glacial times. . . . The problem therefore was to find something of the character of the sand deposits on such beaches, how thick they were and how safe for the passage of military vehicles.

He went on to explain the great amount of research carried out in numerous laboratories, but especially by Brigadier R. A. Bagnold who, a medallist of the Royal Geographical Society for his explorations in Libya, had raised and commanded the Long Range Desert Group in 1940–44. Bagnold had made himself a leading scientific expert on the movement of sediments by wind and water, for which he was elected FRS in 1944. Bernal, speaking at Caen, said:

> Bagnold showed, by beautiful experiments, that on a normal gentle day the waves coming on to the beach do not produce simple circular movements of the water but have a residual progressive movement. In other words the water at the top moves forward and so does the water at the bottom. The excess water is carried away by a current flowing between the top and bottom – the undertow which is often fatal to swimmers.
>
> In gentle weather, the current along the bottom carries grains of sand and builds up the beach and normally a beach will continually grow or, wind-blown, form dunes. However, in a storm, an entirely opposite phenomenon occurs. The sand is stirred up by the waves and lifted into the returning stream, to be carried right out to sea and deposited in deep water. A beach may be built up as much as three to four metres in a season and then swept clean in a single night of storm in winter. Fishermen know this very well and so do people living by the sea, but the ordinary seaside visitor only sees the beach at its best, when it is fully built up in the summer.

Although later, in January 1944, after-dark landings would be made from a midget submarine by a specially instructed recce unit to get confirmatory tangible evidence and samples, Bernal had to set about reconstructing and charting the far shore from evidence remote in distance and in time. It was scholarly detective work and cloak-and-daggerish at that. It was essential that his interest in a particular part of the French coast should not raise questions. Mountbatten's head-quarters provided him with a warrant that gave him the run of the British Museum Library to seek out and borrow books without their nature being revealed to any official of the library. He had to remember where to put them back because there was no record of their removal.

One might think that the sheer variety and abstruseness of his reading would have been cover enough: it ranged from the Anglo-Norman poem 'Roman de Rou' to a fourteenth-century lawsuit between le Sieur de Courseulles and the King of France. It covered all the volumes of the Archives of the Linnaean Society of Caen. Discarding existing charts of the Bay of the Seine he went back to the charts of the French hydrographer Beautemps Beaupre and found that he had taken them from charts prepared on royal orders in 1776. He found the first charts admirable but that in subsequent copying a number of rocks had been omitted because the copyists were paid by the chart and not by the number of rocks. He found a map prepared by Abbé Huc who had been interested in the distribution of glacial erratic rocks and had gone around with the local fishermen identifying them by names such as Dos de l'Ane and La Vieille Femme and, by taking cross bearings on them at high and low tide, had determined the safe channels through them. (On a post-D-Day landing Bernal was to navigate a boat through those rocks from memory of Abbé Huc's soundings.) With aerial photography, Bernal's deductive charts were refined and were validated by the fact that the whole landing force crossed the reefs without serious loss. This was the more remarkable because one of the reasons why the German general staff had considered the landings on that coast unlikely was that the reefs were too dangerous for shipping, as indeed the record of wrecks, including ships of the Spanish Armada, could testify.

Tides were crucially important. The precise points of debarkation from the various types of vessel had to be established. By updating tide tables and the use of tide analysis machines it was possible to predict the tide of all the places of landing within 5 centimetres. At Caen, Bernal told his audience:

> If that degree of accuracy seems absurd, consider that with a beach-slope often as low as one part in 200 an error of 5 centimetres meant an error of 10 metres along the beach. The tides also proved a way of measuring the depth of water down to the low water mark. For deeper water one had to rely on indications given by the reflection of the waves – a wave travels more slowly in shallow water, and by using appropriate formulae it is possible from an air photograph to calculate the depth of water.

Bernal's investigations were not confined to getting troops ashore. They and their massive equipment had to be deployed and moved into action in the interior. Apart from anything which the enemy might dream up, the terrain itself might be teacherous. Here, his reading of the 'Roman de Rou' had yielded an interesting piece of information. That romantic account of the escape of Duke William (before he became the Conqueror) from a castle prison near Cherbourg gives a very good description of the countryside, including an account of William's crossing on foot of 'le Grand Vey', a muddy bay. An air reconnaissance in fact established to Bernal's satisfaction the existence of a ridge of rock and gravel which could have enabled the fugitive to avoid a detour of some 40 kilometres.

Another of his studies of old place names revealed a no-longer-extant 'Village du Marais' indicating an old marsh where there were now fields, and 'Hable de Heurtot', which Bernal decided meant there had been a harbour in what was now countryside. He investigated further and found there had indeed been a harbour in the Middle Ages that had been the cause of the lawsuit between Le Sieur and the King because the latter alleged that the lord, with his 'Hable de Heurtot', was undercutting the royal harbour dues elsewhere. Bernal then traced the storm in the sixteenth century that had sanded up, beyond dredging, the mouth of the harbour. Satisfied that the whole of the area of what had been the harbour was most unreliable ground, he declared it 'unfit for tanks'.

On 30 May 1942 Churchill sent a minute to Mountbatten:

> Piers for use on beaches: they must float up and down with the tide. The anchor problem must be mastered. Let me have the best solution worked out. Don't argue the matter. The difficulties will argue for themselves.

Actually the minute referred to a pet project of Churchill, who had wanted to replace the lost Singapore base by an artificial harbour on an island in the Indian Ocean. Mountbatten put his experts, including Bernal, on to the assignment. The Indian Ocean project was abandoned, but two years later it materialized on the Normandy coast as Mulberry.

In the interval the proposal had been processed through Mountbatten's Department of Wild Ideas and examined by Bernal, whose function was to tame those wild ideas by testing their scientific validity. To secure an area of sheltered water, a haven, the relentless rhythm of the wave motion has to be interrupted. A wave will break on a solid obstacle such as a breakwater, but it can also expend its energy in compressing air. That was what interested Bernal in a Russian device for laying a perforated pipe on the sea floor, ejecting jets of air upwards. Physically it was plausible (indeed this pneumatic method can be used for 'fencing' bays and inlets for fish farming, checking the incoming waves but also discouraging the fish from migrating). For invasion purposes it was militarily impractical because it would involve a pretty massive compressor close inshore, an easy target for enemy attack. Another pneumatic breakwater he considered was the Lilo – an air-filled 'sausage' on which the waves would expend their energy like a pugilist punching a punchball. The problem was how to anchor it securely. A 200-foot long Lilo needed a 700-ton sinker.

Then there was the Swiss Roll. This device appealed to Mountbatten's boyish sense of novelty. It consisted of a great drum on which was wound a flexible roadway (like the steel-slatted 'carpets' laid down as temporary air-landing strips). This would provide the grip for army vehicles unloaded from landing craft and a track across the beaches. The novelty was that it was propelled by catherine wheels (the coiled explosives of fireworks) on each side of the drum, A prototype was built but the trial was a fiasco: the catherine wheels, electrically detonated, were not properly synchronized and, instead of heading towards the cliffs, the contraption zigzagged all over the beach to the peril of the troops taking part in the exercise. That was one wild idea that the conventional services had no difficulty in blocking.

Bernal's role in the Mulberry operation was mainly that of troubleshooter for Mountbatten. In March 1942, at the age of forty-one, Mountbatten had been made Chief of Combined Operations, with the rank of Vice-Admiral in the Royal Navy, Lieutenant-General in the Army, and Air Marshal in the Royal Air Force. He was made a full member of the Chiefs of Staff and he was under Churchill's special protection. He was the embodiment of amphibious warfare, and therefore very much concerned with Mulberry. The Admiralty and the War Office, jealous of their antique privileges, both staked their claims for primary responsibility and both had their own pet schemes. To the British inter-service rivalries were presently added the say-so of the US allies.

One thing could not be gainsaid: the decision to land in Normandy would never have been approved without the Mulberry concept. The Dieppe raid on 19 August 1942 (among other dire and salutary lessons) had made it plain that capturing an existing port (for example, Cherbourg) was out because the all-out assault that would have been necessary would have entailed the destruction of the port facilities beyond immediate repair. But on open beaches an improvised haven and off-loading quays were indispensable to the consolidation of a bridgehead.

A prerequisite of any such enterprise was a knowledge of the characteristics of 'the far shore', and Bernal's ingenuity in determining the nature of the Normandy beaches, of the terrain over which the troops and armour would have to advance, and of the tides and wave formations, had made him the authority. His ability to convert erudite and recondite information into simple and relevant terms made him Mountbatten's trusted adviser, watchdog and salesman.

By an adroit manoeuvre Mountbatten broke loose from the intrigues and the in-trays of rival departments. At the behest of Lieutenant-General Frederick Morgan, who had been appointed Chief of Staff to the still-unappointed Supreme Commander of the invasion forces, Mountbatten called a conference, code-named Rattle. It was held at Largs and brought together the top brass of all the services to examine and expedite the plans for the Normandy invasion. They were to decide between the competing schemes for the artificial harbourage.

There are still arguments as to who should get the credit for the Mulberry project, and even as to the origin of the code name itself. (It is nice to think of it as a Churchillian glottal slur of 'Marlborough'.) In the outcome, Mulberry was a combination of many components, but Mountbatten says in his Bernal memoir (see page 194) that the idea of sinking ships and concrete caissons (the gigantic Phoenix units that so baffled the shipbuilders who had to prefabricate them) was first put forward by Commodore John Hughes-Hallet, Naval Adviser to Mountbatten at Combined Operations Headquarters; when he made his proposal he was Naval Chief of Staff at COSSAC.

As Mountbatten recalled, 'I called in Bernal who quickly proved to be a powerful exponent.' As Mountbatten's scientific observer in the discussions between the services, which were pushing their pet components, he could bring quantitative reasoning to rival concepts. One concession to the Admiralty's pressures, which neither he nor Mountbatten liked, were the bombardons. These were 200-feet-long cylindrical tanks with four steel fins set longitudinally at 90-degree intervals. These fins were supposed to stabilize the bombardons which, anchored on the perimeter, would protect the *Mulberry* structures from the full force of Channel storms. During the invasion, in the storms of June 1944, these steel tanks broke loose from their anchorings, smashing up everything they encountered, including the American harbour, *Mulberry A*, at Omaha Beach.

The War Office's strong bids for Churchill's 'Piers for the Use on Beaches' had been Spud Pierhead and the Whale. The pierhead, 200 feet by 60 feet, had four legs ('spuds') standing on the sea bottom. The loading platforms slid, with the help of a diesel engine for lifting and lowering, up and down the legs with the variation of the tides. Although a prototype had been successfully demonstrated on the Solway Firth, there were reservations about its robustness under Channel conditions. The Whale was a flexible floating roadway which would link the pierhead to the beach. Their limitations were overcome by protecting them within the breakwaters of the artificial port.

The *Mulberry* concept and its components had to have the approval of the Prime Minister and the British Chiefs of Staff, and be sold to President Roosevelt and the US Chiefs of Staff. *Overlord* was at stake because any decision had to take into account the landing and sustained advance of armies of a million rising to 2 million men with all their modern equipment – a throughput of at least 12,000 tons per day. What was being proposed was the assembly of two artificial harbours as big as Dover from components that had to be towed across the Channel. A second bold and successful enterprise was the laying of the oil pipeline Pluto across the Channel.

Bernal was an unlikely 'salesman' for such dealings with the military high command, but Mountbatten was brazen enough to take him along to the Quadrant conference at Quebec in August 1943 where the site of the invasion was to be decided. At Combined Operations headquarters Mountbatten called him in to discuss the Canadian conference. He said, 'I think you ought to go fairly soon. In fact, I've arranged for you to go this afternoon.' (Bernard Fergusson, *The Watery Maze.*) He ordered Bernal to pack his bag and join the *Queen Mary* boat train leaving in secret from a siding at Olympia. On this occasion Mountbatten did have misgivings about the impression Bernal's appearance would make on the service brass hats and sent him to the barber on the *Queen Mary* to have an inch cut off his hair.

Mountbatten presented Bernal to the Prime Minister and the Chiefs of Staff in the large stateroom where they were waiting for a presentation of the technical details of the artificial harbour. For a practical demonstration they adjourned to the bathroom. The bath was filled with water and twenty boats were made out of folded newspapers. Captain T. Hussey RN, the head of Bernal's department at Combined Operations, went to one end of the bath and Bernal took the paper flotilla to the other. Hussey used a back brush as a wave-making machine and the 'fleet' sank. Bernal then used an inflatable lifebelt to represent *Mulberry*, produced another twenty paper ships, called for another wave storm, and showed how the energy of the waves was frustrated and all the ships remained afloat. This child's play was watched by the Prime Minister and the chiefs of the three services (the

First Sea Lord, Admiral Pound, stood on the lavatory seat to get a better view). They were duly impressed and were made responsive to the technical expositions of the details that the other experts were to provide with diagrams, reports on trials, and maps. Churchill went into Quadrant enthusiastically and argued the merits of *Mulberry* with a conviction that persuaded the Americans.

Churchill, however, had another ploy – Habbakuk (*sic*) – in which Bernal was even more involved. In his history of the war, Churchill wrote:

> There was another associated problem on which my mind dwelt, namely the maintenance of fighting air superiority over the battle area. If we could create a floating airfield we could refuel our fighter aircraft within striking distance of the landing points and thus multiply our air power at the decisive moment. Among the numerous devices discussed on this busy voyage [that is, the journey of the *Queen Mary* to Canada – R.C.] was one called *HABAKKUK* [*sic*]. This project was conceived by a Mr Pyke on Mountbatten's staff. His idea was to form a structure of ice, large enough to serve as a runway for aircraft.
>
> It would be of ship-like construction displacing a million tons, self-propelled at slow speed, with its own anti-aircraft defence, with workshops and repair facilities and with a surprisingly small refrigerating plant for preserving its own existence. (Winston Churchill, *The Second World War*, vol. 9) (Churchill spelt it right: Habakkuk). The code name got it wrong. It derived from the Old Testament prophet who said, 'I shall work a work in your days which you will not believe though it be told you'. (Habakkuk 1:5)

The Mr Pyke referred to was Geoffrey Pyke, an ingenious eccentric and an old friend of Bernal, who had been wished on to Mountbatten by the Cabinet minister Leopold Amery. Even Bernal was conventional compared with Pyke, who was already installed at Combined Operations headquarters when Bernal and Zuckerman were appointed. Pyke had already been responsible for several by no means cranky devices, and Mountbatten had regarded him as a genius, to be protected and encouraged. Typically, he turned him over to the care of Bernal, who 'never minded slaving away on other people's ideas'. And slave he did on *Habbakuk*, to the extent of eventually breaking his partnership with Zuckerman, who later wrote, 'Habbakuk, or rather Pyke, proved the beginning of the parting of the ways for Bernal and me' (*From Apes to Warlords*, p. 161).

Pyke had got the idea of Habbakuk while in Canada working on his design of a snow vehicle called the Plough, a premature prototype of the Weasel, which was later to play an important part in polar exploration.

He sent back a voluminous report interlarded with remarks for Mountbatten's special benefit. This was not just an idea for adapting a natural iceberg or an ice floe (as was later done for providing ice stations for polar observations). It was no less than a prefabricated carrier to be constructed of what became known as pykerete, a form of ice produced by freezing an aqueous solution of sea water and fibrous wood pulp.

The frozen hull would be thirty feet thick, half a mile long, and broad enough to deploy aircraft. With nacelles, engines attached to the hull like outboard motors, it would propel itself at five knots and reach its station anywhere in the ocean under its own power. The pykerete had a slow melting factor which could be redressed by Habbakuk's self-contained refrigeration system.

Mountbatten was intrigued: he passed papers on to Bernal, who was at first amused, and then enthralled. The functional, engineering and strategical possibilities of the leviathan were attractive to him, and as a crystallographer he was stimulated. As far back as 1933 he had been interested in the anomalies of the physical properties of water as a liquid. With R. H. Fowler, the physicist son-in-law of Lord Rutherford, he had produced a paper for the *Journal of Chemical Physics* (vol. 1, No. 8, 1933) on 'A Theory of Water and Ionic Solution, with particular reference to Hydrogen and Hydroxyl Ions'. And now he directed himself to the molecular structure of pykerete. He brought in a young Austrian refugee, Max Perutz, in academic origin a glaciologist, who had worked with Bernal on molecular structure at the Cavendish Laboratory, Cambridge. Perutz was to get the Nobel Prize for Chemistry in 1962 for his achievements in molecular biology.

Bernal's report to Mountbatten was favourable and reinforced the latter's enthusiasm enough for him to go to Churchill. Mountbatten regaled a post-war public dinner with a lively description of what happened at Chequers:

> I was told that the Prime Minister was in his bath. I said, 'Good, that's exactly where I want him to be.' I nipped upstairs and called out to him, 'I have a block of new material I want to put in your bath.' After that he suggested that I should take it to the Quebec Conference.

In the hot bath the outer film of ice had melted but the exposed wood pulp had kept the remainder from thawing.

The next demonstration by the irrepressible Mountbatten was even more dramatic. At the Quadrant Conference in Quebec eight chiefs of staff, four British and four American, were present when Mountbatten had two lumps, one of ice and one of pykerete, wheeled in on a dumbwaiter, together with a big meat-chopper. He asked for a volunteer to hack the blocks. General Arnold (US) accepted the challenge. He used the chopper and split the ice easily. He assailed the pykerete and jarred his arm. The block was barely chipped. This was not spectacular enough for Mountbatten. He drew his revolver, fired and shattered the ice. He then fired at the pykerete cube and the bullet ricocheted, fortunately without VIP casualties.

It was left to Bernal to demonstrate less flamboyantly to President Roosevelt at the behest of Churchill. A waiter produced a jug of boiling water and two silver punchbowls. Bernal put a cube of ice in one punchbowl and poured on boiling water. The ice melted immediately. He repeated the performance with pykerete, and after several minutes it still had not melted.

British instructions had already been issued that Habbakuk was to be pursued 'with urgency and secrecy', but eminent scientists including Zuckerman were much less sanguine than Bernal. Cherwell dismissed it with 'Strongly disagree' and Charles Goodeve, Controller of Research at the Admiralty, heartily concurred with Cherwell. Pyke had made himself thoroughly unpopular with the Americans and anything in which he had a hand was suspect. So Bernal (although another 'Commie' in the US glossary) was left 'to carry the baby'.

By now Churchill was enthusiastic and Roosevelt was well disposed; but Admiral King, Chief of US Naval Operations, was positively hostile. His staff prepared papers, examining the pros and cons. Bernal had to appear before the Combined Chiefs of Staff and their planners. He argued the case for Habbakuk eloquently but was so unimpressed by the brief the Americans were putting up against it that he turned devil's advocate and told them the questions that they should have been asking – had they been better scientifically informed. His *on the other hands* were so effective that Churchill sent Mountbatten a sour message: 'The next time you come to a Combined Chief of Staffs Conference, you must not bring your Scientific Advisers with you.' (Churchill here was anticipating the remark of Senator Edmund Muskie at a US Senate inquiry where the scientific witnesses were balancing the arguments: 'Next time send us a one-armed scientist.')

Strategically, the Habbakuk project was put on the back burner, where it continued to simmer, with Perutz still on it full-time and with Bernal remaining on the Joint Anglo-American-Canadian Board, chaired by Dean Mackenzie of the National Research Council of Canada, who had rejected Pyke as a member. The scientific, technical and constructional arguments remained valid but, as Mountbatten, always loyal to the idea, explained, its priority was reduced by circumstances. Although the basic raw materials, sea water and wood pulp, were abundantly available in Canada, steel for the skeleton was in short supply. Meanwhile, the Portuguese, as friendly neutrals, had provided bases in the Azores, which made *Habbakuk*'s usefulness as an anti-U-boat station in the Mid-Atlantic less meaningful. The fitting of long-range petrol tanks to fighters had brought Normandy within striking distance for battle support from British land bases; and although it would have needed seventy or eighty torpedoes to demolish Habbakuk, the Lords of the Admiralty decided to spread the risk over more conventional carriers. (Pyke committed suicide in 1948.)

The break-up of the Bernal–Zuckerman partnership had happened before all this. In January 1943, seven months before the Quebec Quadrant conference, they had been sent by Combined Ops HQ to North Africa to learn the lessons of Rommel's successful desert retreat in spite of continued air attacks by allied air forces. They were given a list of problems to be examined on the ground, such as the effects of different kinds of bombs on different types of targets. They had collected a great deal of material and useful personnel for their research and had established their operational research headquarters in Tripoli when an embarrassed Bernal broke the news to Zuckerman: he was going to Canada immediately for a meeting about *Habbakuk*, leaving Zuckerman to carry on the North African mission. Zuckerman desperately tried to dissuade him, telling him that Habbakuk was 'nonsense'. Bernal was unhappy but obstinate. Next day, Zuckerman saw him off at the airport. Each went their separate ways – Bernal to succour Pyke and Zuckerman to join the warlords.

Bernal wrote his own verdict on his war career in *World without War* (Routledge and Kegan Paul, 1961):

> The only time I could get my ideas translated in any way into action in the real world was in the service of war. And, although it was a war which I felt then, and still feel had to be won, its destructive character clouded and spoiled for me the pleasure of being an effective human being.

Postscript

Ritchie Calder died in 1982 before completing his account of Bernal's war work. Bernal was sent across to the Normandy beaches on D-Day Plus 1, to inspect the actual problems involved in the landings that he had been planning for so long; he has left a lively account of his experiences (see Chapter 9). In October 1944, he flew to Colombo to join Mountbatten, then in charge of the Eastern Front as Supreme Allied Commander South East Asia (SACSEA). In November Bernal visited the Arakan front, then flew back to Colombo and home to Britain in December. Sir John Kendrew, later president of St John's College, Oxford, was then attached to Mountbatten's staff, and has sent us the following account of Bernal's visit.

He wrote (to Francis Aprahamian, 7 June 1982):

I have looked back at my old journal for the year 1944 and I enclose a copy of all the passages referring to Bernal's visit to Ceylon. Oddly enough, the most interesting day, 9th November, is only mentioned briefly in the journal and my memories of it are much more vivid. The problem under study in the bombing trials concerned methods of bombing troops in the jungle, because the RAF having run out of the proper fragmentation bombs was using naval depth charges instead. These made a much louder bang than fragmentation bombs and therefore the Army thought them tremendously effective; in fact they were more or less completely useless (having only a very thin casing). Mountbatten was subjected to enormous pressure from the Army to continue using depth charges even after new supplies of fragmentation bombs arrived, so he asked Bernal and myself to arrange trials in the jungle. For this purpose I contacted the official rat catcher of Kandy, who deposited a sack containing about 100 rats on my desk one morning. These we put into cages and suspended them in the jungle at varying distances from a depth charge which was then blown off. Even the rats very close to the depth charge were totally unharmed, as predicted. An incident I recall was that before one of the bombs was detonated Bernal pulled his slide rule out of his pocket and calculated that he and I were at a safe distance; there followed the most enormous explosion and we rapidly dived into the nearest ditch with rocks and earth flying overhead. Bernal emerged brushing earth out of his hair and looking very puzzled, saying that he thought he must have got the decimal point in the wrong place. Another memory of that day is that the depth charges were moved to the proper place by elephants because in the jungle there were no tracks suitable for wheeled vehicles; somewhere I have a photograph I took of Bernal with an elephant.

Sir John's journal entries are as follows:

18 October 1944 – . . . to dinner with Taylor to meet Bernal. Joubert also turned up after dinner, so we had quite a gathering.

19 October 1944 – All afternoon one long meeting at SACSEA with Bernal.

20 October 1944 – Yet another Bernal meeting, followed by lunch at the Queen's with him and Taylor and Hussey.

24 October 1944 – A busy day with Bernal – he spent the morning with me: then we had an official lunch party with the Deputy, A.O.A., Vasse, Taylor etc. – and thereafter a great meeting in the War Room. To add

to the confusion Lindsay, the forerunner of Naval Operational Research, turned up.

9 November 1944 – A day in the jungle, observing trials of bombs for jungle clearances. Bernal, Menzies and Whitworth Jones were there. A good outing, 35 miles from K.

24 November 1944 – The day full of Menzies and Waddington. We lunched with Taylor and Bernal and talked infinite shop.

25 November 1944 – Wad. and I dined with Bernal and Gregory Bateson the anthropologist.

29 November 1944 – Bernal came to lunch at SACSEA.

1 December 1944 – It's Bernal's last day in Kandy and I hoped to see him for drinks, but it didn't come off.

8

Memories of Desmond Bernal

Earl Mountbatten of Burma

As Chief of Combined Operations I was charged by the Prime Minister, Mr Churchill, with the development of the special equipment and techniques for an opposed landing across the Channel on a scale unknown anywhere before. I began by getting as many officers of the three fighting services as I could to work on this task, but I soon came to the view that I also wanted to have the best available nonservice men, preferably scientists, who had not been trained in conventional Service Staff Colleges, and who would have open minds to entirely new problems.

I discussed this idea with Sir Henry Tizard, who was then the most prominent scientist in government employ, and said that I wanted to get two first-rate scientists who would be put on to the operational analysis of devices and equipment and techniques for the landing that was to come. He was much struck with this departure from the conventional use of scientists in defence establishments, and gave me the names of Desmond Bernal and Solly Zuckerman. I wrote to both; I remember Bernal replying that he would accept if Solly Zuckerman did. Zuckerman did accept, so I got them together. I should add that Mr Leo Amery, Secretary of State for India, asked me to take on my staff another independent thinker called Geoffrey Pyke. Pyke had no scientific qualifications but in many ways was a genius. Bernal helped him, as I shall mention later on.

The particular problems that I assigned to Bernal related to the initial landing across the beaches. He worked with my Director of Experiments and Staff Requirements, Captain T. A. Hussey, RN, and made significant contributions to techniques for dealing with underwater obstacles and defences on the beaches, with the assessment of the gradients of beaches, and the consistency of their bearing surface, and with runnels and sandbanks.

Soon after the arrival of the scientists, I told my staff that they were not to be asked just to answer questions posed by the uniformed staff but to be fully brought in to the framing of the questions for whose solutions their help was wanted. Curiously this went badly at first with the uniformed staff. But then fate played into my hands.

A young naval officer went to Bernal and asked him what he thought the chances were of making a very small, light and portable echo sounder to measure very small depths accurately. Bernal asked, 'Why?' The officer replied that the matter was too secret to explain, and it was up to Bernal to answer the question. To this Bernal replied, 'No,' and that I had given an instruction that the scientists were to participate in the formulation of the problems to be investigated. He therefore wished to know more.

Rather reluctantly the young officer then told him that what was wanted was a way of finding out, without the Germans knowing what we were doing, how to measure beach gradients, the runnels, as well as the consistency of the kind of beaches we might assault. His own idea was to put a light echo sounder on a board and push it in at night from a submarine with a swimmer who would try to obtain this information.

Bernal's reply was, 'You've asked the wrong question, you should have said "How do we measure the beach gradients, and runnels, without the Germans knowing?"' His own answer was that photo-reconnaissance photography should be used, taking vertical photographs of the desired beaches at various stages of the tide and directions of the wind, and that the coverage should extend beyond the desired beaches so as to disguise from the Germans what we were up to.

This incident has been quoted before but I mention it here because it shows how strongly Bernal resisted the attempt by uniformed staff to put him into a position of merely answering questions that they were putting. He insisted in participating in the formulation of problems. This was a very great pioneer service which he performed.

The other major project on which he was employed was called Habbakuk. This was a brainchild of Geoffrey Pyke, and consisted in producing enormous unsinkable aircraft carriers built of a mixture of 5 per cent wood pulp and sea water frozen into a form of ice which became known as pykerete. It was far stronger than ordinary ice and when it melted it produced its own insulating coat. Each ship was to have a displacement of about a million tons with a 3,000-foot runway and a speed of barely 5 knots. The concept was really vast and I turned it over to Bernal for its technical evaluation. He became the principal technical man on the committee that was set up to look into this scheme.

There were two main reasons for these great iceberg aircraft carriers. One was that they could provide fighter airfields against the north-west coast of France in case it was finally decided to invade the Continent from that direction, which aircraft based on English airfields could not cover. In fact the German defences to the west were far weaker for the very reason that we could not cover it with our shore-based fighters. The second reason was to establish airfields in mid-Atlantic from which aircraft could operate to help in the anti-U-boat war.

Now to discuss what Bernal did. He was the man who supervised all the scientific aspects of the investigation into these colossal aircraft carriers or mobile floating airfields. He enquired into the properties of ice, and more particularly into those of this curious stuff we called pykerete, and he worked closely with the naval constructors. It was Bernal who persuaded me to take steps to get Perutz, a brilliant refugee scientist, released from internment to work on this project – with excellent results. It was Perutz who in particular studied the properties of pykerete, and who supervised the practical experiments of making pykerete in cold storage plants. Finally a section of one of these giant carriers was built in Newfoundland, where the pykerete was frozen by nature on a large scale.

The Habbakuk project was going along nicely, but there was a lot of opposition. At the Quebec Conference in August 1943 the American Chiefs of Staff were persuaded by Admiral King to ask for an appreci-ation of the pros and cons of the project. Admiral King, the Chief of US Naval Operations, was opposed to it. The US planners started to prepare their appreciation with inadequate knowledge and understand-ing and were somewhat lost. Bernal, who had become so expert in proclaiming the points in favour of Habbakuk, now, with his Jesuitical mind, volunteered to produce all the points against it as well. The criticisms of the project were so powerful that they turned the scales against it and I had a telegram from the Prime Minister, Mr Churchill, saying, 'The next time you come to a Combined Chiefs of Staff Conference, you must not bring your Scientific Advisers with you.'

In fact all that happened was that the priority was taken off Habba-kuk, though the work continued under Perutz. There were four reasons why its priority fell. First, these great floating airfields or aircrafts carriers had to have enormous steel skeletons, round which the pyker-ete was built, and the demand was so great that steel couldn't really be afforded for this purpose. Second, the Portuguese allowed us to use their airfields in the Azores so that the need to have a staging post and floating airfield in mid-Atlantic to help in the U-boat warfare disappeared. Third, the addition of long-range tanks made it possible to operate our fighters across wider stretches of the Channel, and reduced the necessity of having a floating airfield off the French coast. Fourth, although Habbakuk itself was supposed to withstand between seventy and eighty torpedoes before disintegrating, in the end it was believed to be cheaper and simpler and quicker to construct larger numbers of conventional carriers, using about the same amount of steel. Even if some were sunk there would still be enough left to carry on.

Desmond Bernal also helped with the artificial harbour, Mulberry. In particular he looked into the question of the suppression of waves by introducing barriers. The first idea he examined was the Russian scheme for putting perforated air pipes along the bottom through which compressed air could be pumped. This causes an aerated barrier of water, which being compressable would not transmit wave motion. I turned down this idea on the grounds that it required a great air compressor to be moored off an enemy beach, which could easily be knocked out by a single well-aimed bomb.

The next idea we looked at was what were called Lilos. These consisted of inflated canvas bags moored as a breakwater so that they would yield to the pressure of the waves and not transmit them – the air being not so much compressed as altering its shape in the bags.

Finally, of course, we turned to the simplest of all projects but perhaps the most expensive, which was to sink ships and concrete caissons. This idea was first put forward by the late Admiral John Hughes-Hallet at the Rattle Conference which I called at Largs in the autumn of 1942 to discuss the plans of the Normandy invasion. I called in Bernal, who quickly proved to be a powerful exponent.

I brought Bernal on board the Queen Mary for our trip over to the Quebec Conference – called Quadrant – in August 1943. I called him in to give a demonstration to the Prime Minister and the Chiefs of Staff of the idea of an artificial harbour providing enough shelter for disembarking. We went into one of the big bathrooms where Bernal had made little paper ships. First of all the paper ships were put at one end of the bath and an officer making waves showed that they would soon be sunk. Then Bernal used an inflatable swimming collar which he stretched across the bath. The collar absorbed the movement of the waves and the little paper ships survived. There was no doubt that Bernal's demonstration, together with his extremely able exposition of the subject, helped tip the scale with the Chiefs of Staff and made Mr Churchill more enthusiastic than ever.

Bernal, of course, was used on a number of minor projects in the Headquarters with great effect. His brilliant mind always either demolished the need for some proposed project or developed it into something more practical.

Desmond Bernal was one of the most engaging personalities I have ever known. I became really fond of him, and enjoyed my discussions and arguments immensely. He had a very clear analytical brain; he was tireless and outspoken. But perhaps his most pleasant quality was his generosity. He never minded slaving away on other people's ideas, helping to decide what could or could not be done, without himself being the originator of any of the major ideas on which he actually worked. This may be the reason why his great contribution to the war effort has not been properly appreciated, but those of us who really knew what he did have an unbounded admiration for his contribution to our winning the war.

9

D-Day Diaries

J. D. Bernal

[The following account was written by J. D. Bernal in the early fifties, at the request of the Russian journalist Boris Polevoi, and was to have been published in a series entitled 'Real Men in the Western World'. It is probably based on notes written at the time or shortly after the Normandy landings.]

D-Day Minus 1

Casual, almost unnoticed goodbyes. Everything very quiet and ordinary driving down to London. At the morning meeting, nothing unusual. I asked the Captain for the weather report. It looked bad. Can it be put off another day – or is the whole show off? I decided to do an ordinary day's work. Then the call came from the Commodore: 'Ready to start in five minutes.' We got into the car: just got to the end of the drive when I remembered I had forgotten the maps and I went back to fetch them.

It was an ugly kind of morning. As we came down through Hindhead and Petersfield, the traffic began to thicken up: more and more convoys; then the parking strips, now empty, that had been so full of tanks and guns two days before. We drove into the HQ and there separated and I went to see P. and discuss with him what things had to be looked out for; talking also to the navigators and naval construction people and, most of all, to the intelligence officers. There did not really seem very much that could be done now: even fresh information that I had got over the wire was of no use: it could not be got across to the force commanders in time for them to do anything about it.

We decided to go for a walk. The HQ was a maze of tents and huts so carefully concealed that walking through it was like walking through any well-tended big gentleman's place – wood and meadows and artificial lake. We walked out and there were the Downs and we looked out on the sea where the whole expedition lay just about to start. The air was dim and there were heavy clouds and little bits of rain. It was an immense assembly of ships but against the cloud and the sea it looked very small. We lay on the grass and talked, estimating prospects of success or failure on the different beaches. P. thought it was somewhat silly of me to go: I felt like that myself. Perhaps it was that I could not completely justify it because I would not really influence the things that were happening there and only learn things for other times that might never happen: but at the same time I felt that I had to be there, that I should never forgive myself if it did happen and I had not been.

We strolled back slowly to HQ and now things began to look more lively. I went into the gunnery room and studied the fire plan on the maps. It was time for dinner. We were having drinks in the wardroom. I was introduced to the two meteorological officers, in a way the two most responsible people in the whole show. It was on their reports that Go/Stop orders had been given. They would not drink. We talked about various meteorological things – the methods of measuring the operational accuracy of forecasts by means of weighted averages of different factors: wind, cloud heights, et cetera. Already the whole expedition was on the sea. The wireless man came in. He was most astonished. 'They are well within range and they have not spotted us yet.' Everything very quiet and peaceful and on edge. I prepared all my heads of enquiries and as there was nothing doing I went to bed at midnight.

D-Day

It was difficult to sleep, conscious all the time of what was going on across the sea. 'And gentlemen in England, now a-bed.' Wide awake at 4 a.m. In the dormitory P. and I called to each other. He was going to see if there was anything worth noticing and coming back to bring me out. He did not come back. At 4.30 I got up myself. I had gone to bed a civilian: I now carefully dressed in the very doubtful disguise of a naval officer and had to remember that I could not even walk across from the dormitory to the main building without putting my hat on! I went up to the gunnery room. Absolutely nothing yet. Probably the first waves had been detected, but they had not been fired on. The planes had gone but the weather was bad and they would probably make a very poor showing although there were more of them than we expected. But the disadvantage of the weather was balanced by what seemed to be genuine surprise. Then H-Hour came. They should be landing now and yet we knew nothing. Something was delaying things. Naval guns were not opening fire.

Then the reports began to come in: good, bad, indifferent – but 50 per cent good, which was more than was expected. The coastal batteries were firing very feebly: news began to trickle in about the landings; but very little news and long waits between them. Some messages were in code and it would take time to decode them. There seemed so little doing that there seemed a good chance to get breakfast – a hearty and thoroughly naval three-course breakfast. After breakfast there was a little more news: on three of the beaches at least the landings had been successful: foothold had been gained. From one beach, no position: from another no news at all. This did not look so good as this was a most difficult beach. Altogether we all felt very breathless about it. It seemed to be going so much better than we had hoped. The ships were unmolested, there was no air counterattack, the enemy was not present in force.

Again the reports narrowed down to a trickle. P. and I went out to walk up the side of the lake, discussing the difficulties that we had as scientists in working with such a professional organization as the Navy. We passed through green fields to the ruin of an old abbey and a strange old country house with an abandoned, weed-grown garden. It was time for lunch – the last meal in England.

The Commodore appeared: I saluted – my first salute! We got into the yard and drove off to the port. We were too early – there was half an hour to spare. We looked round the gunnery ranges on the drill ground. It was all extremely peaceful. Then we went into the little MTB [motor torpedo boat] alongside. The commodore had known the skipper's father.

There was no room for us all on the bridge. I wedged myself into various positions, sometimes hanging over backwards with my toes on the step, peering over the shoulders of the others, and sometimes sitting at the foot of the step looking astern. The crew climbed into the gun positions or lay down and crouched under some cover on the deck. It was going to be a very rough passage. The skipper handed out the Hyoscine tablets that I so well remembered S. testing two years ago. We pulled out of the harbour past many familiar buildings and then out to sea, at first in shelter of the land. Behind us sailed four patrol boats also at high speed like four birds with white flashing wings in the water. The boat was alive. It swung from side to side and leaped up over the waves to come crashing down, sending up sheets of foam.

It was no use trying to keep dry but our progress was so exhilarating that it did not matter. We began to pass the convoys of the second wave: hundreds of LCTs [landing craft tanks] and all kinds of other craft wallowing along in the waves while we shot past. The wind rose as we got towards the centre of the Channel and the seas were high but we seemed somehow to fly over them. Sometimes for a few minutes together every wave would come over in soaking spray; then we would have respite for a minute or two. Then the sky began to clear: first in little patches and then all over. The sun came out and with the wind, the waves and the sun the whole journey became altogether a joyous one. The crew were all young, I suppose the oldest was twenty-two; we were all excited but we could not talk except to point, between the roar of the engine and the crash of the waves. We began to pass returning convoys and to look for signs of action: a few were on tow, others showed holes made by mines – but their return at all meant success. This was not going to be a second Dieppe. How well I remember those battered and scarred LCTs – the few that got home.

The sky was now all blue except that a pile of clouds lay on the southern horizon; there were clouds over France. Then I saw a very thin grey line – France – that I had not seen for four years. The shapes began to appear: all around cruisers, battleships, merchant ships and LSTs (landing ships tanks) and hundreds of other ships. We could see the coast clearly now, the coast already burnt into my memory from having pored over so many maps and photographs. I knew every bit of it. The distant steeples of Langrune and Luc: the noble steeple of Bernière rising above its woods: the water tower above Courseulles: the hill of Mont Fleury and the lighthouse of Vire. The village of Rivière with the children's sanatorium: and then, beyond it in the next dip – Arromanches. Arromanches, where I had been ten years before. They were fighting still in Arromanches: you could see occasional little flashes. The lower part of the coast was covered with smoke from fires. The big ships were booming out – but a desultory bombardment. Shells seemed to be falling beyond Cap Montvieux: the sky clouded again and the sea stayed rough.

We drew near the HQ ship. Our own had not yet appeared and we decided to go on board. As the MTB swung up on the crest of a wave, we jumped for the rope ladder and scrambled on. Somehow this was more of a break even than leaving England.

Big ships are all the same, war or no war. Once inside, the atmosphere, the lights and the warmth made me realize immediately how cold and wet I was; but soon I was stripped of my coat and standing by the bar drinking excellent but generally unobtainable drinks and meeting officers I had known as a civilian and who gave instant and surprised stares at seeing me thus disguised. The Commodore had found a cabin. Then we sat down to dinner – a very sumptuous meal. We heard the latest news – how the beaches had been secured and the enemy driven from the hills but yet only a little way along the coast, and no news whatever from the Americans. It was 9 o'clock. People said 'The King': he was speaking on the radio. I moved across the bar: there was only one chair and the ship's cat was lying on it: I sat on the edge. The strange, halting speech went on: it was difficult to realize that we were near the middle of the twentieth century. The anthem struck up – rather casually, the officers stood to attention.

I went back to the cabin and worked on my notes. Suddenly the hooter went. I put on my helmet and came out on deck. It was overcast now and beginning to get dark. There was a familiar hum of planes in the sky and crackling from every ship came flak, those beautiful streams of red fire: fountains rising and falling, swaying and twisting in the sky. 'They can't resist firing,' an officer said. 'It looks so nice. If we did not issue them with tracers, they probably would not fire at all. I suppose we had better let them have it; it doesn't do the enemy any harm but it makes us feel good.' The bombs began to fall – few and far off on the shore. Then another hum of planes. I found to my surprise that my teeth were chattering. I was not rationally scared but somehow the isolation of being in a ship was worse than any raid on land: 'to mew me in a ship is to enthral me in a prison that were like to fall' (John Donne).

The fire parties passed me: that was familiar. Then everything went quiet again. I went back to the cabin and found at the door an engineer officer just back from the beach, who thought it was his cabin. I would be the last person to dispute it with him so I went on deck again in time for another warning. When that was over I went back to pick up my things and to talk awhile to the officer about what he had seen. Then I very quietly slipped into the Commodore's cabin and lay down as I was on a bench at the side. It was very quiet now and the ship was too big to move much in the swell, but I found it difficult to sleep.

D-Day Plus 1

I was up again at four and looked out. It was the grey before dawn. It seemed somehow quieter but the wind was more to the north on the beaches and there was a heavyish swell. I walked up and down the decks; at 5.30 the Commodore joined me.

I told him if we wanted to go on the beaches we should have to start at 5.45 or the tide would be upon us. He enquired for a boat from the officer on watch. I picked up all my things and was just turning to move them from the cabin when the crash came. We're hit, I thought, and very close at that. Smelling the smoke and feeling sudden exhilaration at actual rather than imagined danger, I stepped out on deck.

We had been hit forward about thirty feet from where I was. In a few seconds flames were rising to the height of the mast. The wind was throwing them through the companionway and back along towards me. I worked it out: it was a bad conjunction. The ship lay head to the wind and the fire would be driven along and have most chance to spread. The sea was rough and the chance of getting out the boats, which were almost certainly too few, would not be too great. I was probably the only man on board without a Mae West (life belt) – I had procured an excellent one but had left it behind at HQ. I remembered studying the statistics of the bombing of ships, how one in three caught fire, and remembered the burnt-out hulks in Tripoli harbour. But all this was not frightening, it was simply irrevocable fact.

The fire parties started running up with the hose. An officer passed me carrying a small hand extinguisher. There was another roar of planes and the whole anchorage was alive with flak: I could see them high up in the sky. They were Spitfires without doubt, but bombs had been dropped and everyone was on edge. Any plane in sight would have been fired on.

Then, in practically no time, the flames began to die down a little. I moved forward. At the edge of the deck I could see the body of one of the AA gunners lying up against a rail, clearly dead, already pale and distorted; it was less than ten minutes since the bomb had fallen.

The flames were now nearly out. I met the Commodore. 'This incident seems to be over,' he said. 'I think we should go on our way.' He hailed a boat and we went down the violently swaying ladder into it. The night before I had seen a whole divisional staff, from the general to a signalman corporal, lowered rather unceremoniously with ropes into their boat, but I had a naval uniform and naval men are supposed to be able to manage such things without assistance. I chose the right moment as the boat came up and jumped, and was lucky to find my feet. It was an odd American craft, less like a boat than anything I have ever seen or felt: it rolled, pitched and roared. We had the alternative of lying in the hold and seeing nothing or sticking our heads out and receiving a bucketful of water in the face. The Commodore, who wore his full uniform, preserved it dry until almost the last, when an unexpectedly large wave came right over and soaked him to the skin.

We made our way laboriously towards the shore. It was very quiet; all the firing had died down. The ships hung at anchor. A few boats were moving about. I could see the shore more clearly: the village of Trevières, the dunes, the great marsh of Heurtot; on the shore, piled-up stranded LCTs. The coxswain had orders not to land: he came up very reluctantly. While we were still in deep water we struck and jammed – it may have been an obstacle, but more likely the top of some drowned craft or vehicle. It took us some time to get clear: we could see around us other wrecks. We saw that unless we were prepared to swim for it we would not get ashore. We hailed DUKWs but they were too far away or outward bound.

There was nothing for it but to turn back. Now we were moving in the teeth of the wind, very slowly and much beaten by the waves. We got depressed. In weather like this it would be practically impossible to land supplies or reinforcements for the assault troops. After two hours we got back to the despatch boat. It was more difficult to get on board than ever: both vessels were pitching: there was nothing for it but to judge one's time and jump. Both of us were feeling pretty sick, wet and cold but we revived in a little time, enough for coffee and sausages. After that there was nothing to do. The tide was in. There would be little to see on the beaches and there would be no chance of getting there because it was now too rough to get any boat up and clear of the yacht.

We were a mixed company: a general with a colonel in attendance, an American major and a senior commodore. My commodore's spirits began to rise again. He told stories, and before we knew where we were we were playing cold poker for drinks while the captain obligingly looked the other way. Underneath we all knew that the fate of the whole expedition depended on the weather and God seemed definitely against us. Was this a historic vengeance for 1588 (*flavit et dissipaverunt*)? I remember that the Met experts had forecast 'a complicated series of depressions such as had not been seen in June within living memory'. The weather would continue bad until noon Wednesday, after which it might change but could not be relied on.

And then, almost miraculously, the change came. The clouds flew away, the wind dropped, the sea quieted almost at once. The strong, bright sun shone over all the ships and the woods and the fields of the coast. The whole anchorage seemed to shake itself and stir into activity. LSTs came on to the beach. Merchant vessels moved up. The sea became alive with small craft. The Commodore left on the MTB to make his report. I joined in with the American major for another shot at the shore.

This time we made it, hailing a DUKW that rolled us right along to the exit: we jumped out of it on to French soil at one of the precise spots that I had studied so often and whose history and geology were more familiar to me than any other place on earth. The major seemed somewhat bewildered. I suggested walking with him part of the way to the corps HQ where he was due. No challenge from the military police.

It was an incredible scene: the confused row of stranded craft, the queue of vehicles coming off the beach, some broken-down or bombed, derelict tanks here or there. The various beach organizations were setting up their posts with their standards; the military police directed the traffic; bodies were laid out in rows under blankets; a soldier nearby had already collected his first souvenir – a pair of sabots. We walked off the beach up a dusty road: there had been heavy bombing just inland, shelling as well. The broken houses reminded me of Plymouth. We walked up the hill.

Then we saw our first civilians, blue-bloused peasants, quiet and almost surly, none of the enthusiasm of the Arabs or even the Italians in Tripoli. A woman passed by and smiled but ever so slightly. Further up, the houses were not so knocked about, their doors were wreathed in roses but the roses were dusty – and dust, road dust and bomb dust, gathered everywhere. I saw a middle-aged man and his wife passing along the street: he looked an intellectual: I think he was the village schoolmaster.

I said, 'Bonjour, monsieur.' We talked for a few minutes.

His wife said, 'I have waited five years for this.'

He said, 'You gave us an impression of force that the Germans never gave us. They have gone, and in such a hurry too; they left everything behind, their food, their pigs and even their poultry and rabbits. Forty of them there were yesterday and they have all gone now. The last days were hard; there were many bombs on the village; ten people killed.'

I said, 'I hope this will mean the end of the war to you.' They were polite but seemed a little doubtful. Later I addressed some nuns very politely but was received with frozen and taciturn virtue.

The American major got a lift. I went on alone through another village and then at the crossroads, sitting round the war memorial, I saw some different people. For a moment I did not take it in and then I saw they were prisoners sitting and taking it easy, with a couple of guards with Tommy guns sitting beside them. They looked a mixed crowd in their grey-green uniforms: some strange faces from Eastern Europe or Asia, all looking fresh; they could only have been fighting a few hours and were just bewildered. I went to look for a battery and followed a path that seemed free from mines, only to find that it led to wire and another minefield. I walked through a field of fresh corn and looked at some craters through it and came back to another lane where some soldiers asked me whether it was safe from mines. I asked for the battery and the sergeant offered to show me. I knew he was taking me to the wrong battery but I did not want to seem rude so I went with him. He wanted news of what was going on: they all wanted news. It surprised me in a way to find soldiers who were actually interested in a war in which they were fighting.

I examined the battery and took photographs as a stream of infantry came up a path from the beaches. Now things were coming over more and more rapidly. It was getting very hot: I wished I was not wearing so many clothes but I dared not take anything off for fear of lowering the dignity of the officer caste, though some of the men were stripped. I walked down another way to the beach and met an officer from our HQ who had come in with the assault wave.

We stood in the road and chatted. There were some mines and a bit of wire: very like English 1940 defences, slightly less formidable in fact; at the other side of the ditch a dead cow lay, one stiff leg pointing upwards ludicrously. There was a sudden shot and a crackle of tiles: probably a sniper but we could not be sure; no one paid much attention. Afterwards I learned more about the snipers. We compared experiences and another officer who was by took copious notes. He was trying to get an impression of the whole thing – I hope I see them some day. We could not go on because they were blowing up beach mines to clear the exit. Everyone was taking cover – the usual bangs I have heard so often on trials – and I left the officer and went off to the beach again.

There were the obstacles, the timber ramps, the hedgehogs: all the things I had peered at so often through magnifying glass on air photographs and plotted. I went down to look at them: the obstacle-clearing gang were hard at work. The willing ones were shinning up the posts and defusing the mines and shells. They complained that it was dangerous work: they did not mind defusing, they said, but it was unfair to make blast waves whilst they were at work. I took their photos not feeling too happy about it myself, as I was effectively just as close to the shells as they were. However, I did not think the shells were sensitive enough to matter. From time to time there would be a somewhat feeble whistle that made us take cover, but usually we only took cover after the explosion so as to keep clear of the bits and pieces falling from the sky.

The traffic was moving in spasmodically. I met the commander in charge whom I had met the night before and the lieutenant who ran the party. They had been working thirty-six hours with hardly a break and there was still an enormous amount to do. There were far too few of them for the job. They told me the story of the assault. The lieutenant was a tough young man who had got wounded in the shoulder in the assault but had had it patched up and gone on working. They were all very pleasant and keen and made a number of suggestions for improving the methods of getting better equipment. I promised I would do what I could. If we ever have this kind of thing again we will have to treat obstacles more seriously: so many landing craft lying about sunk by the mines or holed by the obstacles. While we talked, the stream of traffic started again: one of the men saw something in the ground, shouted to the tank to stop and pulled a mine from under the sand just in front of it. The top of the beach was full of derelict mines and a few booby traps which they obligingly showed me – otherwise I would almost certainly have got caught in them as they were concealed. More obstacles were being blown up and rather heavy pieces of steel were being thrown into the air; we took cover behind stranded craft. The men were drinking soup out of tins: they offered me some energy tablets which were very good, and asked for news eagerly but not anxiously. My news was now stale but I gave it for what it was worth.

Everything was quiet again: no gunfire or planes. I went to look at a German strongpoint. The gun which had done such execution in the assault now lay broken in front of it and was being used for hanging out clothes to dry. I remembered 1939 and the hanging out of washing on the Siegfried Line! The fortifications were poor: I had never believed much in the Atlantic Wall but I felt very irritated at the stupid and slipshod construction of this work and felt like sending a note to the Todt organization telling them to sack the people responsible. It had been taken by the infantry after causing a fair number of casualties. Very few Germans were killed but there were a good many prisoners.

The afternoon wore on. I moved to the part of the beach where I had predicted peat and clay – it was all too true. There they were just where the eighteenth-century abbey had stood. There they were as they had been laid down in the Neolithic Period on top of the clay of the last ice age and in it, like flies in amber, was stuck every kind of vehicle except the jeep: tanks, lorries and even DUKWs. Some were being pulled out, but others were getting bogged again. I thought – it is always the same way: I may be right, I may even know that I am right, but I am never sufficiently ruthless and effective to force other people to believe that I am right and to act accordingly. All this was so unnecessary: it all could have been avoided if people had not thought that my objections were just theoretical and statistical and that they were practical people and need pay no attention to them.

I went back to the rendezvous with the American major but he was not there and I strolled down the beach, now very long and flat at low tide. I passed more files of German prisoners embarking. I met the commander at the water's edge. It was getting late: nothing much was moving on the beach. It did not seem as though we would get off. Then, half a mile away we saw a DUKW: we ran through pools of water in between stranded vessels with the tide racing in over the flats. The DUKW was just going into the water, but it halted as we hailed it and took us aboard.

It took us to a control ship, from which after an almost interminable wait we got a small craft back to the despatch boat. It was a perfect, quiet evening with promise of better weather to come. I felt strangely happy: for all the confusion and the inefficiency, I felt here at last something was doing; and everyone had been so cheerful and kind and giving a feeling that they were really trying to do something – not just their duty, but something they wanted to do.

Back at the despatch boat there was an officer who had been that day the whole round of the front line: he told us what the various formations were doing and how the beginning of the counterattack was developing north of Caen. It was still a very small war: in that one summer day, walking and hitchhiking on jeeps, he had covered the whole front. An hour later he went back to England to report. For that time, the five of us in the wardroom knew more about what was going on on the second day of the second front than anyone in the world. It had been a long day, full of change and contrast, idleness and activity, wind and sun. Tonight was the real chance for the Luftwaffe to wipe out the concentration on the beaches and the ships in the anchorage but although the flak had already started, I could not attend to it. I was asleep in my bunk.

D-Day Plus 2

Nothing had happened. At 5 a.m. I was awakened by the steward with a cup of tea! After another excellent breakfast, we went off in the lieutenant's boat at 5.45 a.m. on a somewhat ceremonious round of delivering notes for the elder commodore. Arriving at the first control ship, we enquired politely, 'I hope you had no trouble in the night?' to receive the answer, 'Yes, a spot of trouble, four killed and twenty wounded.' It was a beautifully peaceful morning: the risen sun was just behind clouds, the water, at first glassy smooth, was just beginning to ripple in the morning breeze: the whole anchorage was asleep: not a ship moving and the beach deserted.

We sailed along to another control ship opposite Bernières. There the Colonel and I were dropped and waited in vain for a DUKW: the DUKWs were not out yet. After a while we spied a lumbering LCM (landing craft motor) with a heavy lorry in front. We hailed it and got on board. Now we were off with a rush to Île de Bernières: those rocks, the charting of which had caused such a flurry a fortnight before. We moved slowly in. The quiet of the morning was suddenly broken by staccato shooting and flashes of fire and the burning of houses and a church in the village. A minor battle was in progress. We heard later the story, which may or may not be true, that parachutists had reinforced the snipers in the village, particularly in the great Norman church tower which dominated the whole country. They would be difficult to get rid of short of blowing the whole place down: already there was a yawning hole near the base of the steeple.

We moved in cautiously. A number of ships were stranded on the rocks – I could see them now, vast, ill-defined, patches of seaweed. A man in the bows was taking soundings with a pole – 6 feet, 5, 4; 6, 4, 3 feet – and all at once we were aground. The boat swung round and managed to back out. I said to the coxswain, 'I think I know something about these rocks: you will never get in at this stage of the tide; better wait for it to rise a foot or two and then I think I can show you the way in.' We drew further out and waited. I was impatient. All my chances of seeing Bernières beach at low tide were vanishing. I would never be able to verify the height of the rocks or the position of the submerged forest.

After twenty minutes we started again. This time I gave a course. 'Steer for the church, but then, to allow for the wind, you had better bear a little to the right.' He looked at me very strangely. I tried to remember which was port and which was starboard but gave it up, deciding to stick to my landlubber's language in spite of my uniform. We edged forward slowly but I had no chart and there were no maps of any kind; I could only guess the position rather vaguely and so a few minutes later we ran on the rocks again! This time there was a grating sound and when the coxswain tried to reverse, one of the propellors was caught. We drifted clear again and tried to edge forward but it got more and more difficult. The wind was freshening and catching in the large hood of the lorry and kept swinging the boat away from its course, and there was a strong tide running as well.

It took us the best part of two hours to get in that mile: but at last we landed on a firm sand beach and the lorry, with us in it, rolled out. We came to the exit: the battle of Bernières seemed to be over but we were told the town was full of snipers. I badly wanted to see the church but I could plead no military reason and had to admit that I would feel a fool if I were shot for the love of Norman architecture. So we stuck to the beach defences. More strongpoints carefully camouflaged but very poorly constructed. Some seemed to be just as they were left with all the German gear lying about – even a bottle of wine undrunk. I imagined people feared booby traps and I kept my hands off as well. There were the breaches that had been made in the sea wall to pass the tanks through and the bridges they had used to get over it, more obstacles, more stranded craft. We talked to the men on the beach. They told us about the firing in the town and how eight snipers had been found to be French women in German uniforms and they had been promptly shot. I wondered how true the story was. If it was true, was it affection or politics that made them do it?

We went on down the beach to the west: behind, in the water meadows between the church and the dunes, the tank machines were clearing mines; they looking for all the world like hay-tedders, until they belied this by suddenly becoming enveloped in a pall of black smoke from an exploded mine. We passed another strongpoint constructed of logs and sand rather on the Jap style. By now it was nearly noon. The tide was up and the beach was in full activity. Enormous pontoon rafts, every inch loaded with vehicles and men, were coming in and discharging. Every kind of small craft, from LTCs to DUKWs, were bringing their quota. Men just ashore were looking around them; the older hands were digging in against expected air raids, or hanging out their clothes to dry, or cooking meals. Sitting against a telegraph post of a little railway, a midshipman was writing his first letter home. Behind, the clearance of mines went on, punctuated by explosions and smoke.

We approached Courseulles, a fairish town with a little port and a great reputation for oysters. There were more strongpoints, some better constructed in the Todt style. Near the port mouth there was a most elaborate system of cover ways and strongpoints protected by an anti-tank ditch. The cover ways led up to what must have been a head-quarters. Most of the houses were punctured by hundreds of shell holes, not destroyed like our blitzed houses. One gaping wall opened into the officers' mess. The Germans had made themselves at home: they had put a red-brick stove against the fireplace and installed a bar: on it stood in incongruous contrast an empty bottle of champagne and a full glass of milk. On the other side there was a piano, strings torn to pieces by splinters. On the wall, covered with splinter marks, were Nazi inscriptions: an eagle grasping a swastika-covered world in the middle, on one side 'Unser Glaube ist der Sieg'.

The little port was very quiet. The restaurant still advertised oysters and shellfish. The oyster beds were there, still intact between high wire fences. I forgot the month when I advised some of our people to try them. In the basin, barges were already unloading. It reminded me of Ireland: everything was so peaceful. A little further along, the Colonel ran into some old friends whom he had not seen since he went out east. There was a padre and some other officers: they talked about the people they knew. The Colonel was explaining why things had not been so good in the jungle – the troops went into action directly listening to the shouts and the jeers of the Japs. You had to tell the men, 'This is not like proper war: it is dangerous and frightening.'

In the background four soldiers were leaning against the sides of a shell hole in an old garden wall: through the wall you could see others digging some trenches and laying out their camp. 'I would not go up the road, if I were you,' they said. 'The wood is still full of snipers.'

We wanted to get back to the beach, so we crossed the river again and came into the marshes full of bomb craters. In a ditch a tank lay with its crew beside it. We talked to the officer. 'We had a tough time getting through,' he said. 'There were a lot of defences in the dunes which we could not see and mortar fire from the hills behind. When we did get through we found the floods, or rather the first tank found them when it disappeared in the blown-up culvert across the stream. The local people were specially helpful. They showed us the sluice to let the water out. Then we bulldozed some rubble over the tank and went on ahead. I got stuck in the ditch because I could not see it under the water.'

We came through the dunes on to the beach again. Now it was only a narrow strip a few feet wide, and consequently it seemed fuller of men and vehicles than ever. We looked for a while for the beach master and rather fancied going out to sea in his amphibious jeep, but we reflected that we were only sightseers and he had a job to do: we waited until we could board an outgoing craft that took us to the nearest control ship. There I met a naval officer from HQ whom I knew. 'It will be a long time,' he said, 'before we can get any craft to take us to our ship.' Meanwhile he invited us to a meal in the wardroom.

We sat down to enormous plates piled high with stew and peas. They wanted news of the land fighting which we gave them for what it was worth. They told us something of what happened at sea. While we were talking an explosion, much larger than that of intermittent gunfire, shook the vessel. They looked out: some distance away a landing craft had been blown up by a mine. We came up on deck and looked at the shipping going by, LSTs and merchant ships mostly now. He got us a boat to take us some miles away to another HQ ship, not ours this time. There we found to our mortification that we had just missed the bus and would have to wait two and half hours before there was anything going our way.

I went down to the wardroom and tried to write up my notes but it was warm and stuffy and difficult to keep awake. I came up on deck again in time for another air-raid warning. The cruiser nearby was firing more often now at the shore, and so were the other warships. Behind the hills there suddenly rose an enormous column of black smoke, like the genie in the story, gradually rising and filling the sky. We wondered whose oil dump had been set on fire. The shelling seemed to me to be directed fairly close to the shore to our left. It did not sound too good: the Germans had been miles away from there yesterday. Afterwards we learned that it was a tank thrust that had almost reached the coast before it was driven back.

It was evening again before we got back to the despatch boat. I had to go back. I did not want to: I had got attached to the place and its strange inconsequent life, its near-peacefulness and distant dangers. I was taken back by the same torpedo boat that had taken me over. The weather was quieter, dull and raining at times.

I stood on the deck as long as it was light and watched the procession of craft coming in: but it was thinner now than yesterday and at times we were almost alone in the sea, with only smoke showing on the horizon. It was wet and dark. I went down below to the chart room where a boy of twenty was plotting our course in the lanes through the minefields. Although the weather was calmer, it was very difficult to stand here: there was a succession of sudden lifts and jarring crashes as the ship hit the sea again. There was an old commander there: we started talking. I talked, perhaps too much, mostly of the war in the early days and the various scientific developments. The young man was interested and keen. It was the last picture . . . a brightly lit room, no more than a box, with its charts spread out on the table and all round the crash of waves and the roar of engines. I fell asleep standing. Then it seemed to be quieter. When I looked out we were just steaming past the Forts, and in a few minutes we were at the landing stage.

10

The Peacemonger

Ivor Montagu

The scientific work I do can be done by others, possibly soon, possibly not for some years; but unless the political work is done there will be no science at all.

Paul Langevin (1934)

Apprenticeship

It is often said somewhat condescendingly of someone who has indubitably distinguished himself in one area that he has wasted his talents by dispersing himself in directions that were not, indeed should not, have been his primary concern. This is especially so when the victim being weighed and found wanting has concerned himself with matters of life and death, happiness and prosperity, the future lot of humankind. It seems that these are matters restricted to a narrow circle; the scientist especially belongs below the salt.

There was never any chance that Desmond Bernal could have fitted such a servile role: not because of ambition but because of conscience. A colleague, Dr Alan Mackay, has described how 'His science, politics, culture and personal affairs were all mixed up together and very closely related to each other in a whole, where each part illuminated and contributed to the rest.' He was born that sort of person. The times and circumstances around him did the rest.

Bernal's interest lay in everything. For him the rift between C. P. Snow's two cultures did not exist. No place, no period was alien to him, no aspect of the universe, no pattern of human behaviour or relationship in society. He pursued facts. He wanted to experience, to find out. He wanted to know all that available resources would disclose, and expand those resources where he could. But the essential was not simply the gratification of curiosity but an ethical pursuit.

I have noticed this to occur where a scientist has a religious background: most powerfully, perhaps, in those who have rejected the background, or at least set it aside in its supernatural aspects, rather than in those who conform indifferently to the orthodoxy that happens to surround them. Bernal, of course, never indulged in the self-deception of knowledge worship, imagining that everything is known about anything, but accepted that everything is knowable. He believed that the more that could be discovered and understood the more humankind could benefit. Compassion, morality and his awareness of historical development attracted him to socialism. Philosophic materialism and Engels's explanation of the dialectic led him to Marxism. His ethical compulsion made discovery insufficient: knowledge must be shared, and sharing must lead to action. Engels's dictum 'Freedom is the knowledge of necessity' was perceived by him as a flash illuminating the path to practical change. The twenties and thirties completed Bernal's training in sociology. His finals tutors were Baldwin, MacDonald and Chamberlain; his study courses were the 1926 General Strike and the world economic crisis of the thirties; and his research professors were Mussolini, Hitler and Locarno.

I saw Desmond briefly at Cambridge in the early twenties. Up from 1921 to 1924, a small group of us met, usually in the rooms of Maurice Dobb. We imagined ourselves a ginger group for the University Labour Club and busily settled the affairs of the world. Bernal turned up occasionally. He was older than most of us and reserved, a listener rather than an animator. (Years later I discovered that he had amusedly preserved the tie I had invented to protect the anonymity of our juvenile cabal – black with large red spots.) After Cambridge I took an entirely different route. Our paths ran roughly parallel in that decade but never crossed. Bernal at this time pursued his scientific career, almost commuting between labs in London and Cambridge, pioneering new techniques, emitting theses, publishing books and articles, heightening his reputation by lecturing and attending conferences across Europe, marrying, begetting, delightedly enlarging his personal acquaintanceship not only with academics here and abroad but also with practitioners in literature and the arts, the churches, the establishment and the labour movement, preaching social responsibility to scientists and scientific possibilities to politicians.

It was, however, in the thirties that world events pushed harder, deflecting the channel of his extra-mural life. In 1931 the famous second International Congress on the History of Science took place in London. The high-powered Soviet delegation caused a sensation, and the speeches of Bukharin and Hessen fortified for Bernal his conception of the indissoluble link between mankind's technological problems, social formations and the theoretical as well as practical development of science. This experience concentrated his attention on the modality as well as the history of science and led to lasting personal friendships with Soviet scientists.

In 1934 another crucial encounter occurred in Paris with three eminent French colleagues, all renowned physicists: Paul Langevin, Frédéric Joliot-Curie and Pierre Biquard. At this time Hitler had come to power, and already his regime had shown its character to all except the wilfully blind. The threat to peace was manifest. Fascist-inspired riots in France were faced by the Popular Front. Langevin, the senior of the group, sought to rally not only scientists but members of other cultural professions. Desmond was impressed by Langevin's advice that it was a scientist's duty not only as a citizen to promote action by his colleagues, but to bring his specialized knowledge to bear on technical problems to furnish practical aid that perhaps no one else could give. The words of Langevin that head this memorial essay were spoken to Bernal in conversation and Bernal constantly recalled them to the end of his life.

Within a year of Bernal's return from France an organization, For Intellectual Liberty (FIL), had come into being in Britain, working in parallel with Langevin's Comité de Vigilance des Intellectuels Antifascistes in France. In Cambridge a Scientists' Anti-War Group was carrying out a practical study of the operation of explosives, incendiary substances and poisonous gases in various circumstances and settings in order to be able to criticize current official ideas of protecting civil populations and make useful proposals for their improvement. I should not be misunderstood here as implying that these bodies he played a part in initiating were in any sense his 'creation'. Such ideas only come to anything if the soil and the moment are suitable.

It was never Bernal's way to give up doing anything he thought essential in order to do anything else. He simply squeezed the new responsibilities in somehow, neglecting nothing already undertaken. Consider the year 1937, when already FIL was at full stretch. By now his strictly scientific work had reached a richly productive and innovative stage, a progress marked by his appointment to the Chair of Physics at Birkbeck and election as an FRS. He spoke and wrote for a variety of peace organizations. He was already an active participant in both the

National Peace Council (NPC) and the International Peace Campaign (IPC), addressing – with Peter Wooster and A. F. W. Hughes – a crowded gathering of MPs in the House of Commons committee rooms on the defects in current air-raid precautions, and sitting on a Scientific Advisory Board appointed by the TUC. Yet he found a snatched moment in which to do research near the top of the Jungfrau and to write a foreword to the catalogue of a Barbara Hepworth sculpture exhibition.

1937 was also the year that Gollancz published the Left Book Club topical, *The Protection of the Public from Aerial Attack* by a group of Cambridge scientists. Bernal had participated in the programming of the research involved and helped from time to time and with an introduction, but by that time he was less often in Cambridge. The coordination of the work fell mainly on Peter (W. A.) Wooster and his wife Nora. Needham, Pirie and Waddington were among the experimenters, who also included volunteers from Manchester and the Nobel Prize winner R. L. M. Synge. The subjects covered included gasproof rooms, gas masks for children, fireproof paint, and thermite explosions. Where Bernal particularly came in was in summarizing the knowledge at conferences and public meetings. He helped disseminate the knowledge by writing for *Nature* and by arguing before committees. He was in close touch with Liddell Hart, the writer on military affairs.

No wonder that when he was late for meetings, his friend Hyman Levy enquired, 'Where is that sink of ubiquity?' and everyone knew whom he meant.

For Intellectual Liberty included only some six hundred members, the academics scattered in centres of learning, the artists, et cetera joining as individuals, though one writers' body merged as an autonomous section. Several MPs joined. Aldous Huxley was the first president. Kingsley Martin, Henry Moore, Philip Noel-Baker, Ronald Kidd, C. P. Snow, R. H. Tawney, Leonard Woolf, P. M. S. Blackett and E. M. Forster were among those who turned up occasionally at the executive (which met in London). The minute book attests that of the twenty-nine executive committee meetings that FIL held from 1936 to 1939, the sink of ubiquity missed only three. (The nearest to his score were Margaret Gardiner as secretary and Margaret Fry who took the chair.) The group promoted meetings and seminars. It collaborated with other bodies, sent delegates to conferences abroad, arranged tours for visiting speakers, and arranged asylum and often employment for suitable academic and other refugees. It organized letters signed by VIPs that were reported in the press, with protests or pleas on issues of the day:

Spain, China, Poland, Czechoslovakia, the saving of victims threatened
with fascist execution. (Deputations were received by the president of
the Liberal Party and Labour Party's current chairman and secretary
urging the two parties to collaborate against Chamberlain; that, of
course, got nowhere.)

The readiness of people, of the highest possible standing, to sign
such appeals is astonishing and provides a useful corrective to the
versions of the period that so often – and incorrectly – assert the
generality of British opinion of those days to have been on Chamberlain's side. In fact, the letters grew more and more authoritative,
beginning with a modest 7 signatories in the spring of 1936, but rising
to 100 in late 1938. Perhaps the climax was the August 1939 letter
signed by 375 staff members of British universities – including 6 heads
of colleges and 70 professors – 'strongly urging' the inclusion of
Winston Churchill in the government.

In January 1939 Bernal attended a conference at All Souls arranged
by Sir Arthur Salter and there met Sir John Anderson, Lord Privy Seal
in the Chamberlain government, responsible for civil defence. Sir John
was no fool. He realized that by this time Desmond was probably better
informed about civil defence than most of his own advisers. Soon
afterwards Desmond was appointed to the Civil Defence Research
Committee, and assigned to Subcommittee B, dealing with the study of
the physics of explosives, blast and penetration. In 1941 he did some
work for Bomber Command and was regarded as so invaluable he was
prevented from going to Moscow to lecture on protection from air
raids. Instead he celebrated the tercentenary of Comenius in Cambridge. After that he joined Mountbatten. His apprenticeship as an
anti-war monger was over.

'The world will never be the same'

Bernal emerged from his wartime experiences with many doors open
to him. His professional future was assured. He was Head of the Physics
Department at Birkbeck. He stood well with the establishment and had
accepted appointment to government advisory boards in several fields.
His views on the social responsibility of science and on the necessity for
the systematic organization of knowledge across frontiers were powerfully influencing distinguished contemporaries as well as the younger
generation. Honorary academic degrees and awards were showered on

him – from the Royal Medal of the Royal Society to the US Order of Freedom with Bronze Palms.

One might have imagined that the end of World War Two would reduce Bernal's sense of the urgent need to work for world peace. When the news of the Hiroshima explosion reached Bernal he immediately exclaimed, 'The world will never be the same.' Of course, not Bernal alone but many physicists realized this: Einstein, for instance, and many of the physicists engaged on the Manhattan Project, had for this reason begged Roosevelt not to let the bomb be used before first warning the Japanese; but it is doubtful how far the statesmen and politicians with power of decision grasped the significance of what had happened.

Before the year was out Bernal had published an article rich in prophetic detail.[1] He pointed out that the atomic bomb as so far used was 'only the first, the smallest and the least efficient of such weapons' and that 'soon bombs hundreds of times more powerful would be prepared'. He noted that such a bomb could produce Hiroshima-style devastation over an area such as the whole of London, and that 'effective defence against such bombs, in spite of certain reassuring statements, seems extremely unlikely'. A war between two countries possessing such bombs would therefore result not only in terrible slaughter of civilians and devastation of cities, but leave an aftermath of disorganization, disease and starvation far worse than that to be seen now in Europe after six years of war. 'The question that must be answered by the people of the world,' he declared, 'is whether they are going to tolerate a condition of affairs in which any inhabitant in any part of the globe is to have this threat perpetually hanging over them.'

Bernal concluded that mankind, faced with the greatest responsibility it had ever consciously encountered, could not meet it by 'lamentations about the slow development of man's moral sense in relation to his physical powers, but rather by united and organised efforts of the peoples of the world to see that these powers are not left in the hands of those who would use them for narrow and destructive ends'.

When the Charter of the United Nations Organization was carpentered together, Roosevelt had insisted on the provision of a Security Council with in-built British, Chinese, French, Soviet and US vetoes for two reasons: first, because he had lived through what had happened to the League of Nations at the hands of the US Congress and hoped, by propitiating the US Senate, to avoid a repeat performance; and, second, because he thought that the consequence of the veto would be to encourage the Great Powers to seek compromise, since without unanimity no enforcement action could claim to carry authority under the UN Charter.

The thinking was ingenious – but this is not the way it worked out. The makers of the bomb, confident that their exclusive ownership of such power made it unnecessary to compromise with anyone, adapted the Security Council to a forum not for seeking agreement, but for discrediting it. They did this first by provoking, then mocking, the inevitable sequence of *Nyets* to which their intransigence drove Molotov and, finally, by devaluing the United Nations, blaming it for the helplessness their own confrontation tactics had imposed upon it.

This, then, was the Cold War, the egg whose hatching has shaped the world of today. The first nongovernmental countermove was a Polish–French initiative that took place at Wrocław in August 1948. It resembled those antifascist activities of the thirties that sought to rally personalities. It was entitled a Conference of Intellectuals for Peace. The object was to voice fear of what might result from a conflict between the late allies, and to restore for peace the cooperation that had gained victory. The English invitees included Bernal.

At no conference before or since have I seen so exalted and so assorted a galaxy of celebrities, in so appropriate a setting. The city was a heap of ruins. Rising above the rubble, approached precariously by planks over mud, was the germ of an art gallery; blazing posters vied with brilliant ceramics. At a café table in the open, Picasso sketched Polish citizens and the Pole Feliks Topolski sketched Picasso. Within the hall, poets, sculptors, musicians, architects, scientists and theologians, historians and political journalists, priests and film directors, parliamentarians and former prime ministers jostled to speak. Bernal himself harped on a favourite theme: the obstacles a new arms race would present to humanity's chance to benefit from scientific cooperation. People strove to get through to one another. Survivors of enemy occupation and the concentration camp spoke of horrors uncomprehended to others who, by comparison, had had a quiet war. The Soviet speakers denounced Westerners without imagining that some they abhorred were respected colleagues of their listeners, and then plaintively enquired whether their speeches had not been liked. Some misunderstandings were inevitable.

In the upshot, optimists claimed there was enough common ground to warrant dialogue continuing, pessimists complained that no common recipes for instant total solutions had come into view. Some returned home to scoff, others to set up national groups. The great majority agreed on a liaison committee to ensure that what had been born survived. The second step followed six months later: the convocation of a world congress, jointly by the Wrocław Liaison Committee and the Women's International Democratic Federation, to be held in April 1949.

The new gathering was conceived as an entirely different affair from the tentative efforts against war and fascism in the thirties and the similar model in Wrocław. Gone was the elitist pattern. True, VIPs still played a prominent part. The organizing committee, headed by Frédéric Joliot-Curie, included among its vice-chairpersons not only Bernal but two Nobel prizewinners, celebrated lawyers and poets, a former president of Mexico, an ex-premier of Italy, the head of the Academia Sinica, international trade union leaders, and representatives of women's and youth organizations. Now, instead of invitations being issued on a strictly personal basis, an appeal to appoint representatives to take part was signed by 72 famous names and launched throughout the world in millions of copies. Over 2,000 full delegates from 72 countries representing, in all, organizations whose membership totalled over 600 million participated either in the Salle Pleyel at Paris or – liaising by telephone – in an 'overflow' at Prague where nearly the whole Chinese delegation and others, delayed by visa refusals, held parallel sessions. In Paris upwards of 60,000 people came to a great public meeting in the Stade Buffalo.

Why was the response so swift and so general? From 1939 to 1945 the world had had its bellyful of war. Instead of ushering in the era of cooperation for which so many people hankered, the United Nations appeared to be paralysed by distrust. It appears that the lately victorious Alliance was in danger of being broken into rival blocs, and resources were to be devoted not to reconstruction but to a new arms race. The rationale of the new movement was simple. It could be summed up in a few phrases from speakers at its outset. As Ilya Ehrenburg put it, 'War is not a natural calamity, like a tempest or an earthquake. War is man-made and man can prevent it. The vast majority in all countries want peace. There must be some way by which they can assert their will.' Or, in the words of Frédéric Joliot-Curie, 'Peace will not be delivered to us on a plate. It must be won.'

The conception was clear. The UN was an organization of governments, not a voice of peoples. If these governments reached deadlock, could the peoples bring nothing to help resolve the impasse? Let as many people as possible come together and themselves consider, argue and, where the consensus was considerable, support the aims that even ministers so generally professed. The logic seemed impeccable; the

official response completely contrary – it is strange how often governments that boast of free speech or their own forms of democracy seem
embarrassed by efforts of unofficial movements to widen discussions.
Restrictions on travel, loss of employment, disqualifications arose
immediately in the parliamentary countries, prison or worse in the less
sophisticated.

Already after Wrocław the French ecclesiastic Abbé Boulier had been
censured and penalized by his archbishop for participation. Kenni
Zilliacus, a British Labour MP who broke his party's ban on the Paris
congress, was suspended from the Labour executive on his return. Paul
Robeson, whose singing at the congress inspired everyone who
attended, was misquoted in the US press and labelled a 'traitor', with a
sequel of threats to murder him, a genuine attempt to do so at
Peekskill, and confiscation of his passport by the US State Department.
Years of professional ruin followed in what should have been the best
years of his voice. Bernal had his first encounter of this kind just before
Paris. He was about to leave for a scientific meeting in New York when
his visa was cancelled. In Britain a small peace committee had been set
up, headed by the science writer J. G. Crowther, in response to the call
from the Paris assembly. A few of its members were invited to Moscow,
where a Soviet Peace Committee was holding its first conference in
August 1949. Here, in the Hall of Columns, Bernal spoke on his
consistent theme of the responsibility of scientists for peace. In a
prophetic warning, he said this was a particular responsibility because
their work had contributed so much in the past to waging war. In the
present era of a decaying system, war was still the most profitable
investment of the monopolists.

> The direction of research is increasingly turned to war uses. It is radar, long-
> range rockets, the atomic bomb, radioactive poisons and the last perversion
> of science in bacterial warfare that now occupy the thoughts of men who
> could use their knowledge for peaceful construction. . . . In Britain more
> than 60 per cent of the current expenditure in science is for war purposes.

In the USA directly or indirectly, Bernal said, most of the current
expenditure of $1,000 million a year on science was also for war. The
restrictions of secrecy this imposed on science had sinister effects. 'Not
only is the atom a secret but the habits of bacteria and even the size of
the earth are becoming military secrets.' Soon it would spread to all
aspects of science. 'In the United States it will soon be the case that no
one who is not and has not always been an open enemy of the Soviet

Union will be allowed to teach or research in science – and Britain will immediately follow suit.'

As soon as reports filtered back to Britain, the British Association suspended Bernal from its council; it confirmed the removal a few weeks later. It is perhaps hard to see, with hindsight, what they found so remarkable in this oration. In those days, however, even such modest data were unfamiliar. Only twelve years later the retiring US President Eisenhower, in his farewell speech to the nation, issued an almost identical warning from his own point of view, by his denunciation of the military–industrial complex. That was the ambiance of the world of those days, the days in which McCarthy flourished.

In March 1950 at Stockholm, the international committee designated to prepare the successor congress to Paris launched a world campaign for signatures to an appeal against nuclear weapons. The text runs:

> We demand the absolute banning of the atomic bomb, weapon of terror and mass extermination of population.
>
> We demand the establishment of strict international control to ensure implementation of this ban.
>
> We consider that the first government to use the atomic weapon against another country whatsoever would be committing a crime against humanity and should be dealt with as a war criminal.
>
> We call on all men of goodwill throughout the world to sign this appeal.[2]

Eight months later, when the second Peace Congress opened in Warsaw in November 1950, the total number of signatures collected for the Stockholm Appeal had reached 473,154,259. Nearly three quarters of this colossal total was collected from China (222,500,000) and the USSR (115,275,000). The figures from France – 14 million; Italy – 17 million; the two Germanys – 19 million; Japan – 6,068,666; USA – 3 million and Britain 1,343,340 are also remarkable when one takes into account the particular circumstances and resources in each case. Undoubtedly the magnitude of the response was due not only to the simple wording but to the fact that in those days its content appeared the most natural and obviously acceptable thing in the world. Scarcely anyone refused to sign the appeal. A generation of indoctrination had not yet resigned the unthinking world to confrontation and 'living with' the bomb.

The very success of the campaign must have aroused the authorities' concern. The British Peace Committee had hoped to host this World Peace Congress themselves in the Sheffield City Hall and had given a press conference with Bernal as chief spokesman. Attlee, then Labour

Prime Minister, replied with an attack on the congress, denouncing the Stockholm Appeal as totally one-sided.[3] Under pressure, a government spokesman admitted in Parliament that there was no power to forbid the congress or to limit the legal actions of alien visitors. But there was power to refuse admission to any visiting alien if, as an individual, his presence was considered undesirable by the Home Secretary, and no reason need be given. Delegates setting out, especially from far away, could not know whether they would find themselves acceptable or not. In the upshot, hundreds were turned back.

In 1940 Joliot-Curie had been welcome enough in Britain. When his initiative and foresight had saved vessels full of heavy water (essential at that time for nuclear research) from falling into Nazi hands, the authorities sent a destroyer specially to fetch him. Only a short while earlier he had given lectures in Britain with no trouble at all as befitted the High Commissioner for Atomic Energy of France. Now, the Stockholm Appeal was going round the world. So, as a peril to British security, he was unceremoniously turned back. So were famous French novelists, a French admiral and a distant cousin of the royal family. Picasso got in – and drew a sketch on the wall of Bernal's flat. Among the Soviet citizens rejected were the Metropolitan Nikolai Krutitsky and the composer Shostakovich (a few months later, he was welcomed to Oxford to receive an honorary degree; was he all that dangerous?).

Amidst the chaos an invitation was received, and accepted, to transfer the congress to Warsaw. A frantic interception of delegates en route began. Some two hundred British delegates, most of whom had never possessed passports in their lives, had to be rushed through the formalities. The Czechs offered the loan of a dozen transport planes. The British Peace Committee ingeniously promised free transport and hospitality to journalists from every British national newspaper and allotted all of them to the last plane scheduled to leave, to deter the authorities from thinking better of their restraint and suddenly cancelling landing permits at Heathrow. In the end the number of participants reaching Warsaw by land, sea and air totalled 2,065 and proceedings began on time.[4]

The main business of the Warsaw congress was the establishment of a World Peace Council. The Paris–Prague congress in April 1949 had discussed what sort of organization should be set up in order to reflect and express the breadth of the feeling for peace in all countries, whatever differences there might be either in detail or in political grouping. Joliot-Curie had put it in a simple phrase: 'There is a place for everyone in the world movement for peace.' The World Peace

Council elected by the Warsaw congress numbered 250. Immediately afterwards the council elected a twenty-person Bureau to function between councils, with Joliot-Curie as president; Desmond Bernal became one of the vice-presidents.

Bernal found himself surrounded by friends and colleagues in or near his own special interests – the Joliot-Curies, Biquard, Madame Cotton, Einstein's pupil Leopold Infield, Oparin. The council brought him in touch with poets, artists, novelists, architects and singers from all parts of the world whose work he had admired but whom he had never met. Not only were there experts who could open up fascinating new fields of knowledge, like Jose de Castro, with his UN Food and Agriculture responsibility for South America, there were also legendary heroes of popular struggle and grave statesmen, whose hopes for peace and international cooperation matched his own. There was going to be a lot of work.

There were already more than 100 national peace committees at the time of the Warsaw congress. Each was entirely independent. No other structure would have been wise or practical. Each was a different size and had a different status in its country, each had different relations with its government, with its national traditions and society and with other peace organizations. Worldwide there were the three great signature campaigns: the Stockholm Appeal in 1950 for the abolition of nuclear weapons; the 1951 appeal to the five heads of state of the US, Britain, the USSR, China and France to meet and seek to reverse the Cold War; and a second appeal on nuclear weapons, drawing attention to their increasing danger, in 1955. There were no more. The second and third campaigns scored their hundreds of millions of signatures, but the impact was a declining one and each exhausted the human resources of the national peace committees.

There were the great congresses – five in ten years – 1949 Paris–Prague, 1950 Warsaw, 1952 Vienna, 1955 Helsinki, 1958 Stockholm, and later 1962 Moscow and 1965 Helsinki. Warsaw, in a formal sense the first, provided a pattern that became typical. The attendance was around 2,000, comprising national delegates based on the comparative strengths of the national peace committees, the travelling distance, the accommodation and the size of the available hall; other guests and observers came from international peace organizations invited by the preparatory committee. The plenary sessions were held in the great auditorium, and far more delegates wanted to speak than could be offered time. Commissions, studying separate questions in separate places simultaneously, enabled more people to speak and made dialogue possible. Then there were the struggles to gain consensus in the drafting committees, the paperchase to have the resolutions, reports and translations ready for delegates to take home.

What counted was not just the sessions and the set oratory but also the encounters. Friendships struck up on staircases, arguments continued in corridors, all bred confidence. The great majorities for concluding resolutions at the congress were often sneered at by a hostile press.[5] I believe the most important outcome of the congresses was not the phrases of particular resolutions but the ending of a sense of isolation, even among those who came from minorities or from small and neglected countries.

Of course it is important to know what the movement did unite on. The resolutions at Warsaw may be summarized as follows:

- the banning not only of nuclear weapons but also of bacteriological, chemical and radioactive weapons, indeed all weapons of mass destruction;
- the reduction, under control, of all armaments;
- condemnation of aggression wherever it might occur and of armed intervention in the affairs of any nation;
- peaceful solution of the Korean war by procedures in conformity with the UN Charter;
- promotion of economic and trade relations between all countries, without discrimination and to mutual advantage, strengthening the independence of all nations large and small;
- a whole array of specified international exchanges in science, tourism, knowledge of national histories and culture, the arts, literature;
- ban on all war propaganda anywhere.

With hindsight it may be hard to understand the justification for the outcry among opponents or the measures taken by the British government to prevent such a conference being held on British soil. After all, most of the propositions are constantly on the agenda of the UN. Those relating to armaments, trade and international exchanges are frequently supported by majorities there, but never seem to get anywhere. The fate of two others is especially ironic. The ban on war propaganda was proposed at the UN, but was rejected by the Western countries on the grounds that it would breach the principle of free speech. In later years the World Peace Council spent a great deal of time on the question of aggression, pressing for a definition to be adopted, arguing that without this any rule against it would be weakened. The World Peace Council set up several committees of diplomats and experts in international law on this subject, to devise foolproof drafts. When the question eventually came before the Security Council the proposal by the USSR was defeated, the West arguing that any definition might handicap the policy of a nation in some situation not foreseen in the definition.

Bernal's dedication required an enormous effort. There was not only participation in the congresses and their preparation: his own wide experience in scientific committees, many of them international, had taught him that collective leadership was essential, especially in a movement with so many diverse interests, so many real differences to reconcile. To help provide this, he must attend everything, be ready for every duty, learn to know every unfamiliar colleague and how best to work together. All the members of the bureau shared a similar loyalty. Absence from World Peace Council business, except by a few of the most distant, was rare. Nine council meetings took place in ten years, and seventeen bureau meetings. These were normally at the headquarters of the secretariat, first in Paris, then Prague, next Vienna and eventually Helsinki. All these were reasonably accessible from London but, at this time more than ever, Bernal had to try to be everywhere at once. He had to order his complex life so that travel for the peace movement did not hinder his normal and ever-increasing responsibilities in his profession. It is difficult to categorize his travels at this period as being precisely for one purpose or another. It is rare, when he was abroad, for him not to be asked to squeeze in a radio interview or give an extra lecture; he himself would be eager to visit an old friend in a laboratory to discuss the latest research.

His work for the World Peace Council did not seem to affect the respect in which he was held by his scientific colleagues or the calls on him by peace organizations in Britain. The World Peace Council and its British supporting body, the British Peace Committee, of which Bernal was also a vice-president, were both blacklisted by the Labour Party. However the proscriptions were not watertight and he was in great demand for conferences and articles. How he found time to write his major book on peace, *World without War* (1958), is hard to understand.

The Mantle of Elijah, 1958–63

The pressures on those who seek to follow the Langevin formula, combining a whole-time commitment to scientific work with no lesser devotion to peace, can become intolerable and sooner or later the price must be paid. Irène Joliot-Curie died in 1955 of leukaemia. Maurice Goldsmith tells the story in his 1956 book on Frédéric Joliot-Curie and quotes Frédéric; 'Irène died from what we call our occupational disease. We are more careful nowadays but in the 30s . . .' he paused. 'I'm not finding this easy.'

Fred himself became ill in 1958. He took a holiday, fishing by the sea. Suddenly he felt bad. A train brought him swiftly back to Paris. The engine driver knew whom he carried, and when station bureaucrats would not allow the ambulance into the station, he refused to let the engine reach the buffers. The bureaucrats gave way but precious moments were lost. Joliot died after surgery. They say the delay may have caused his death.

Frédéric Joliot-Curie was a heroic figure and, as a Resistance leader, a hero of France. As a scientist and Nobel prizewinner he was accorded a lying-in-state, as his wife Irène had been two years earlier. Each time, small-minded bureaucrats tried to limit those who should take part in the ceremonies. In the case of Irène, Frédéric put his foot down and won. Probably it was de Gaulle himself who intervened when her husband died. I remember standing two at a time as we members of the World Peace Bureau, taking our turns of duty with Cabinet members, acted as guards of honour at both funerals. (Those of us who were at the time 'défendu de pénétrer en France' were given special permission to enter for a limited, forty-eight-hour period.) People filled the streets, following the coffin to the chosen place of burial. They pressed into the cemetery, not far from the Joliot-Curie home, until it was overflowing.

There were speeches by Desmond Bernal for the World Peace Movement, by a comrade for the Communist Party, of which Fred was an executive member, and by Biquard, his intimate scientific associate. I have never heard Desmond speak so movingly. Though he spoke in French he rose to an eloquence I have never heard before or since. Afterwards everyone in the area of the cemetery queued up to grasp the hand of the chief mourner, Joliot's eldest son. How his wrist withstood the ordeal is hard to imagine. As we passed the bureau, members filtered off to Frédéric's house where lunch had been prepared. Almost as soon as we were seated, without hypocritical pretence, we fell into discussion of what everyone had thought of but none had mentioned: the next step in the future of the World Peace Council.

Of course, the council itself would carry out any elections. We were so scattered and could meet so seldom that some exchange of views was necessary even though not all the bureau were present. It did not take long before everybody's opinion fixed on Bernal. To Desmond himself the suggestion obviously came as a complete surprise and was as clearly unwelcome. First, his modesty meant he had never seen himself as a suitable candidate. Second, because he had been so close to Joliot, he had probably avoided visualizing a time when he would no longer be there. Moreover, it was one thing to cram together two bellyfuls of work, peace and science, quite another to top them with a load of extra responsibilities. Whatever sympathy I felt as a friend, I did not see how I could come to his rescue. He had inadvertently dug a pit for himself. How can one make a speech to thousands declaring work for peace the noblest task of humankind, which no obstacle should deter, speaking with total conviction, and a few minutes later argue with friends and colleagues that circumstances make it impossible for oneself to act on that principle?

Those who sat together at lunch knew the movement, and they knew Bernal's qualities through and through. They knew his scientific reputation, his all-round knowledge, his global understanding of peace, his readiness to speak and write, his firmness and clarity in the chair, his tirelessness in those long drafting sessions that lasted into the small hours. They realized that confidence between him and Joliot had been total. This would be no new broom but a loyal disciple who would continue the traditions they were all devoted to. All that Desmond could usefully do was to leave things open and to argue that he could not possibly presume to Joliot's place. He could not, certainly could not, be president in rank but, if he must, he would be in function. He agreed to become chairman of the presidential committee. Collective leadership should be enshrined in the official structure.

It is interesting to compare Joliot-Curie and Bernal. Their differences were skin-deep, and what they had in common was more fundamental. Fred was slight, athletic and handsome. Desmond was more generously built and slightly lumbering. Joliot's eminence in his scientific work was world-renowned, Desmond's specialities were better known to his fellow scientists than to the general public. Yet Desmond was probably the more prolific polemicist and publicist, and his crusade on the organization of science made him better known among the new generation of scientists. Both men were Marxists. Joliot stood high on the French Communist Party; Bernal, though he worked closely with Communists and often wrote for Communist Party journals, did not actually hold a party card, at least after 1939. The war work of both men had been valued highly by those in Allied Command positions during the 1939–45 war. Both had the personal qualities essential for international work for peace: faith in its possibilities, utter contempt for racism, warm feeling and calm conduct, integrity and fairness, a total absence of egotism, and respect for others.

Within a few weeks the full formal meeting of the bureau and the next council meeting had confirmed Bernal and the new Presidential Committee. From the first he was welcomed by the whole movement. A certain naïveté made him friends easily. Nehru was an old friend. With Khrushchev he was soon on amiable terms. He met Ho Chi Minh and, in 1954, Mao Dzedong. The Chinese leader was curious about Bernal and later had *World without War* translated in order to read it himself. He did not like it at all, and thought it utopian, though his opinion never seemed to affect the attitude to Bernal of Chinese delegates.[6]

Bernal was physically a stronger man than Joliot but the burden he had assumed was increasing. The character of the peace movement was changing and becoming more truly international. Logistical problems and the fact that the World Peace Council started in Europe meant that at first our attendance and attention concentrated too much on European problems. Soon the developing countries themselves discovered us. Bitter experience taught countries, especially those that were still colonies or dependencies, that it was difficult to get into the UN, and almost as hard to gain a hearing there, let alone sympathetic action. Not so with our council meetings and congresses. New countries joined until we counted more delegations than even the UN at that time. They often sent their most able spokesmen, for they learned to value a hearing before active listeners, spread almost worldwide, more than no hearing in New York. Certainly the movement had never ignored the right of peoples to independence and self-determination.[7] From the beginning we had the sense to realize that the peace movement must give priority to extinguishing flames before continuing fire drill or testing new apparatus. We could not afford to neglect a single conflict, however small. We had to seek a just and peaceful settlement, not only to curtail suffering but to try to prevent its extension and escalation. Korea, North Africa, the Congo, Vietnam: in all these our discussions played a useful part.

The decisive word on this issue came in a debate in Vienna in 1952. The meeting was held at a time of conflict between the British and Egyptians. An Egyptian editor and a Latin American urged us to demand the withdrawal of all occupying British troops. An English pacifist clergyman argued, sincerely, that this was an internal colonial problem and outside our scope.

With tears in his eyes the Egyptian, standing on a chair, urged, 'We do not hate the British. We do not want to shoot their soldiers, but please tell us how else we can get them to leave our country?' The clergyman (a very decent man) replied, 'Well, I can see how the independence problem in the case of fighting is bound to concern us. But I propose that the question of independence or self-determination cannot rightly concern us unless there is shooting going on.' A Pakistani delegate jumped up. 'But that is ridiculous,' he expostulated, 'if you say that the World Council of Peace cannot concern itself with independence problems unless we start shooting!' Henceforward no one disputed that such problems were causes that required our attention so that we might contribute to a solution *before* shooting broke out.

The geographical expansion of the World Peace Council required meeting places in more than one area. A meeting about Africa, Latin America or Asia would be better held nearer those most knowledgeable and most affected than in, say, Stockholm. A modicum of guests from elsewhere should be capable of providing sufficient alternative experience to relate the problems to a general background. Local conferences had occurred quite early in the history of the council, but now they were essential; in addition, locally organized and conducted area organizations came into being (for example, the Afro-Asian Solidarity conference with headquarters in Cairo).

Some of these developments had begun during Joliot-Curie's time, but Bernal's years of peripatetic habit, his desire to fulfil every duty, and the increasingly international nature of all these tasks imposed upon him an energy expenditure beyond what anyone could support. He hated to admit that he could not do A tomorrow because he was doing B today; to him that would have appeared to be an immodest claim to be essential at B and an impolite hint that A was less valuable. In such matters he could not say no.

The travelling was superhuman. A list of Desmond's main commitments in 1962, when he was sixty-one years old, gives some idea of how he gave of himself:

January   Paris lecture on 'The Theory of Liquids'.

January–February   Lecture tour to Chile, Argentina and Brazil. Lectures, laboratory inspections and a summer school, 'The Image of Man'.

March   Berlin, two lectures and the opening of a new building for the Institut für Strukturforschung.

April–May   Seven lectures and conferences in the New York area, including a series at Yale and a conference at the American Academy of Sciences. Topics included molecular structure, the origin of life, the order of disorder, crystal synthesis and the solid state, and fossils and meteorites. Also private talks with US peace leaders.

May   World Peace Council preparatory committee, Stockholm.

June   Bakerian lecture, Royal Society. Lectures on 'The World Without the Bomb' and scientific topics in Accra. Further lecture series in New York.

July   Meeting in Munich to celebrate fiftieth anniversary of the discovery of X-ray diffraction.

September   WFSW symposium on higher scientific and technical education, and World Peace Council general assembly, Moscow. Pugwash meeting, London. Private conference on disarmament, London.

October   Special World Peace Council meeting on Cuban missile crisis, Finland. Scientific lectures, Amsterdam and Groenigen.

November   International colloquium on negotiated settlement of the problems of Germany, Brussels.

Then came 1963: ups and downs and the shadow. A conference was held in January in Oxford to set up a European Federation against Nuclear Arms and the World Peace Council was invited to send observers. A group was selected, including a poet and a bishop. When the conference assembled, the invitation was cancelled, on the initiative of a US representative. I had never seen Bernal so furious at the unpardonable insult to guests who had come from as far away as China.

In February he spoke as a World Federation of Scientific Workers delegate at a UN conference in Geneva. He was invited to a conference of solidarity with Cuba in Rio de Janeiro, but he was refused a visa, just like the old days! His health was beginning to cause problems and in May and June 1963 he went into hospital for tests and a rest. He attended the Royal Society discussion on 'New Materials' in June and was elected a member of the RS's International Affairs Committee. He was also elected president of the International Union of Crystallography. In June 1963 he again visited the USA. On the day of his return he had his first stroke.

'But I don't like gardening'

The first stroke was a warning. It was not disabling. Desmond had to rest in hospital and listen to the doctors. They told him the story of a father and son, both dons. The father, like Bernal, had a stroke and was told to rest and take up gardening. But the father was impatient. He ignored the warning. A second stroke followed. End of don. Years after, it was the turn of the son. He too had a stroke, obeyed medical advice and lived till past eighty. 'But,' complained Bernal, 'I don't like gardening.'

It was not in Desmond to move in second gear if he could crawl in top. As soon as he could get about again, everything – within the limits of his rusting machine – went on as before: Birkbeck, research, writing articles, working on books, even receiving protest delegations. In a short time he was even pushed in a wheelchair to a first-floor cocktail party launching a Penguin edition of *Science in History*.

He had to stop going abroad but needed to keep contact with the Helsinki office. The British authorities were no help. He and I, both having known the then Home Secretary at Cambridge, wrote explaining the position and asking if the French World Peace Council secretary could come over to discuss business, promising he would do nothing but meet Bernal at his home or London office. The negative answer came upon glorious parchment wrapped in pink tape. A veritable Cabinet decision. Thereafter I always visited Desmond before I went abroad, to learn and expound his ideas, and called on him as soon as I got back to relay what had happened.

He went on holiday to St Ives with Margot Heinemann and their daughter, Jane. I visited him there. Resting? Before lunch he had me walking up the hilly point, puffing to keep pace, desperately trying not to miss what he was saying about signs of possible life in hundreds-of-millions-years-old Canadian meteorites. At last Desmond was prevailed upon to return home to sleep after lunch. Using the opportunity I got into bathing shorts and sneaked across the sand for a swim. Suddenly I heard a shout behind and there was Desmond similarly clad, insecurely testing his footing at the water's edge.

Lectures in Britain yes, but meetings abroad he had to miss and this loss of immediate contact was a source of weakness to all World Peace

Council affairs. What was to be done? Bernal was the sort of man who would rather die at his post than forsake a duty that others thought he could usefully attempt to do. His colleagues could not bear to face reality. Some had hesitated at his appointment to the presidency, but there were none who could bring themselves to end it now. Where indeed could they find a sucessor to match the first two personalities in quality and qualifications? At last, in 1965, Desmond made up his mind. He knew that the movement must face his going. He would make one last effort. Attended by a doctor, a long-time friend, and staying in a small house amid trees and flowers on the edge of Helsinki, he would preside over his last congress.

Scattered memories come to mind as I think of Desmond.

On a plane, flying from Warsaw to Paris on a day of blazing sunshine, Bernal standing up and peering out of windows and, when asked what he was seeking, pointing out along the route the shadows and the darker colours that disclosed the whereabouts of Neolithic settlements.

In Moscow, in the Park of Culture and Rest, we encountered a strange and primitive mechanism which he immediately recognized: a long pole with a weight on the short end and a chair on the other, so fixed on a swivel in a split post that a handle could wind it through 180 degrees in a vertical plane. Instantly seating and strapping himself into the chair, Bernal bade us turn the handle, and immediately he rose yards into the air in a complete semi-circle so that finally the pole stopped with him upside down and his hair sweeping inches from the ground. I have never forgotten the intensity of his gaze as he concentrated on observing the unusual horizon and on storing his sensations in this unfamiliar position.

In Berlin, an archbishop – whose name and nationality I will not identify – beguiled those responsible for the list of speakers to squeeze him in by showing them his speech – only three pages. Once on the rostrum the archbishop thrust his hand into the pocket of his robe and brought out a thick sheaf, stunning the interpreters by embarking on the ample monologue all in Latin. In the body of the hall, his hand cupped to his ear, Bernal sat listening to it avidly without blinking an eyelash, perhaps the only one in the great audience who could make a word of sense out of it.

Now, in Helsinki, the Congress of Peace, National Independence and General Disarmament, the last scene. Pablo Neruda reading aloud the poem of tribute he had specially composed a few moments before, his poet's voice· caressing the Spanish syllables: 'Bernal . . . Bernal . . .'

as the thousands present rose cheering and Bernal, himself presiding, this time indeed blinking – and unable to rise.

Before he left Helsinki, Desmond resigned the chairmanship of the presiding committee. Two months later, in September 1965, came the second stroke, the disabling one. Now he could not rise or turn in his bed without help. He had lost the use of his right hand. Only those close to him could decipher his efforts to communicate. Yet still the mind imprisoned in the carcass was clear and unaffected. Still he worked. He reviewed new books. Francis Aprahamian helped with proofs of his own last book and late scientific papers. He enjoyed visitors and was not deserted. I learned to lace peace movement reports with anecdotes that enabled him to make an easy reply, smile when in agreement, or even try to cap them with a few left-handed scrawls. He did not die until 1971. Even then he made his body go on working. By his request it was turned to medical use.

In Britain two separate memorial meetings took place: a posh one, with tributes from eminent scientists; another, smaller, more intimate, with peace movement friends from home and abroad.

His scientific papers are housed in the University Library at Cambridge. His books and papers on peace are held in the Bernal Peace Library at Marx House, on Clerkenwell Green. Birkbeck each year gives a lecture in his name; the Peace Library gives (and usually prints) peace-orientated lectures. Various national peace committees abroad have established Bernal scholarships in his memory, open particularly to Third World applicants. This would have given him particular pleasure. The pretences that a nuclear war could be long, limited, or victorious, that nuclear arms could be in any context a defence, have become an absurdity, and that more are needed for deterrence a lie. Civil defence is a mockery that doctors refuse to take part in. I think of Desmond, always ready to listen to any peace advocates, discuss with any peace group, speak or write on peace whenever invited, join organizations he felt had the same aim when this was acceptable. I think of him on the first Aldermarston march, and on every successive Easter march until walking for him was over and he could walk no more.

No life could be happier than that of one who, like Bernal, was ever stretched to the limit and who strove to contribute to mankind in all that he could do. His life goes on in what others do today.

234 J. D. BERNAL

Notes

1. The article appeared in the newssheet of the Women's International League for Peace and Freedom, November 1945.
2. The original text, drafted by Joliot himself, was in French. Several different translations into English have been made but the above is the finalized version. Even so, 'men' is a slip that could not recur nowadays.
3. I have never been able to understand this ploy. Perhaps Attlee did not see the Stockholm Appeal and the text was misrepresented to him.
4. About one quarter of the participants were women.
5. At the Warsaw congress, voting on the two main resolutions was 1,655 votes for, 3 against and 2 abstentions. Of the 329 guests and observers present, 76 requested permission to add their own votes.
6. *World without War* is essentially an examination of the increased possibilities for human welfare once material and scientific resources are no longer wasted on war. Mao thought it liable to spread illusions and weaken the struggle for peace. He considered the first necessity was to understand the causes of war and the need to eliminate them. Sinologists can perhaps tell us that the Chinese tradition of scientific literature is exclusively realist and Mao's judgement is therefore valid for China. At least for Britain, the utopian approach has always been popularly understood, from More to Morris, and operates in a contrary way, fortifying the belief that improvement is attainable.
7. There is a passage to this effect in the Warsaw address to the UN, but no special commission study or resolution.

11

Bernal at Birkbeck

Eric Hobsbawm

From 1938 until his retirement in 1968, Bernal was Professor of Physics and later of Crystallography at Birkbeck College in the University of London.[1] It would be absurd to confine even the professional life of such a man to the academic post he held, especially as Bernal's connection with his college in the war years was inevitably intermittent or marginal. Nevertheless, the thirty years' association with Birkbeck was fundamental both to Bernal and, not less so, to the college, of whose staff he was for all this time much the most distinguished and eventually the most widely known member. Birkbeck provided Bernal's permanent base as a professional academic, from the time he arrived there until he was no longer able to work. For a while he even lived above his laboratory in the college. This contribution on Bernal at Birkbeck needs no further justification.

The college was in some ways uniquely suited to a person of his interests and convictions. It had originally been founded in 1823 as the London Mechanics' Institution, on a suggestion by the editors of the *Mechanics' Magazine* in London, J. C. Robertson and Thomas Hodgkin, the latter best known as a labour economist and Ricardian socialist who won the praise of Karl Marx. The actual founders of the college, whose foundation stone still bears their names, were a less inflammatory type of radical, largely belonging to the Benthamite school and ranging from Brougham, James Mill and George Grote to that crucial figure in London artisan politics Francis Place. These sponsors showed a distinct lack of enthusiasm for the heterodox economics that the London mechanics, as militant as they were skilled, tended to favour by inviting such persons as Robert Owen to address them. Conversely, leading London employers who supported the venture were rather more interested in the technical education of their operatives in a period of major technological and scientific transformation of industry. The teaching of the social sciences soon fell victim to these disputes – they were not reintroduced into the college until after Bernal's retirement – the mechanics tended to abandon the college, and it settled down for the best part of a century as a night school of self-improving clerks and other young persons of intellectual interests from socially modest family backgrounds. From 1863, such part-time students were able to study for the degrees of the University of London as 'external' candidates, and they did so in large numbers.

However, besides the practice of evening study for persons earning their living during the day, the college retained two important parts of its original heritage. From the start, and indeed long before the teaching staff won this right, students were associated with its government in a democratic manner, and the college also remained attractive to people of a radical mind, heirs of the artisan radicals and Chartists of its early years. It was one of these, Sidney Webb, the great Fabian, a former Birkbeck student, who eventually secured the full integration of his old college into the reformed and reorganized University of London. It was admitted 'as a School of the University in the Faculties of Arts and Science for evening and part-time students' from 1920 and received its royal charter as such in 1926. Its undergraduate students, though obliged, in the words of the charter, to be 'persons who are engaged in earning their livelihood', studied (and continue to study) for exactly the same examinations as full-time internal students of the university.

At the time Bernal joined it, Birkbeck as a full university institution was still relatively young. Indeed it still continued to prepare students not only for degrees but also for the Intermediate examination which qualified for university entry and was usually taken by school leavers at the end of a complete secondary education. It was an institution that enjoyed one unparalleled asset: a body of students whose very presence demonstrated determination to study at the cost of exceptional sacrifices of time and energy. It was the quality of this student body that attracted to the college staff of high calibre and often of democratic sympathies, and there is little doubt that it also helped to attract Bernal. No other body of university students in London could compare with them except the evening students, who at that time were also taught in the social sciences at the London School of Economics (LSE, another institution in which Sidney Webb had a hand).

In other respects the assets of the college were still modest. It operated in a cramped Dickensian corner of the City of London off Fetter Lane in conditions that few university institutions would have regarded as acceptable, even by the spartan standards of scientific laboratories among the late Victorians. It had little money, even by the modest standards of the time, though by herculean efforts it managed, fortunately, to raise enough to begin what was to be its new building in the centre of the Bloomsbury university precinct before 1940. Its teaching staff, though reinforced by young academics seeking posts in the sparse university job market of the inter-war years, was still partly in the awkward frontier zone between night school and university college, although it did contain a number of able and original persons not necessarily of the orthodox academic type: the mathematician Paul Dienes, who had once reorganized Budapest University for the soviet republic of Bela Kún, Stephen Potter, the inventor of 'lifemanship', and 'Professor' C. E. M. Joad, better known outside philosophy faculties than within them.

'We were conscious', reports a student of the 1930 intake, 'that the college had only just made the grade.... Broadly speaking, there wasn't a great deal of distinction about the place.... And so you reach a stage where the teaching was effective, but there's no great vigour in the place.' Fortunately science, and particularly physics, had already begun to take off, in consequence partly of the award of an FRS to a Birkbeck chemist but mainly of the arrival, earlier in the 1930s, of Patrick Blackett, the later Nobel laureate and President of the Royal Society, as Professor of Physics. 'The fillip that the college and the student body got from the knowledge that he was a "bright young man" right bang up to date with Physics, who'd accepted an appointment at Birkbeck, struck us all as magnificent: the college was made by attracting a young man like that.' Blackett left after four years to accept the chair at Manchester. 'And then we thought, oh hell, we've just seen this wonderful development and we can't possibly be as lucky again. And then the news broke that Desmond [Bernal] had been appointed and back we went to our sense of satisfaction, we felt we were all right again.'[2]

I have quoted A. J. Caraffi, Clerk to the College from 1952 till his retirement in 1979, at length, in order to show both the sort of institution into which Bernal came and the enormous significance, at this stage of its development, of the arrival of scientists of the very first order such as Blackett and Bernal. Naturally one should not exaggerate the impact of single individuals, however distinguished, on a college. Birkbeck, unlike some of the provincial university colleges then still under the 'external' umbrella of the University of London, had established its credentials as a university institution by the time World War

Two broke out. While its students and tradition attracted some, and its location in London probably most potential faculty members, it could hardly have recruited a staff of the quality it did in the 1930s – and this almost certainly applies also to Bernal – without academic credentials.[3] Nevertheless, the arrival of two such obvious scientific luminaries as Blackett and Bernal clearly established the status of the institution in the eyes of the students, the staff and the university world, as nothing else could have done.

The ambiguous and complex relations between Bernal's outside reputation and his position in Birkbeck raise a question that may as well be discussed at the outset, all the more so since Bernal's reputation was political as well as scientific. What was his place and role at Birkbeck during the thirty years he worked there? He was in the first instance a departmental head, a director of research, and a teacher, and he made his impact on the college primarily as such. Though his membership of the important Professoriate Committee necessarily involved him in general college policy, he was not much drawn to general college administration, and, except for a relatively brief spell as a Governor of the college in the 1950s, he took relatively little part in it. In the event this made him vulnerable. His later struggles on behalf of his discipline were to be so frustrating and unhappy partly for this reason. He had neither time nor much taste for college politics, though he was to be forced into them.

Neither was he notably active in the public and social life of the college, his own department always excepted. Unlike a figure such as, say, Harold Laski in the LSE, he was not a familiar presence to all college members, irrespective of departments, though probably all could recognize that vast head on a stumpy body, that bushy mop of hair, that characteristic slightly – and as time went on increasingly – unbalanced walk, like (his own phrase) Edward Lear's Pobble. He simply did not have time for very much activity outside his department and his superhuman schedule of scientific, lecturing and political engagements, both in Britain and across the globe. Thus – to take an example at random – in the academic year 1961–62 he made visits and delivered lectures to the universities of Concepción and Santiago in Chile, and Rio de Janeiro in Brazil, to the Humboldt University in Berlin, the University of Munich, Yale (a series of postgraduate lectures on 'Molecular Structure, Biochemical Function and Evolution'), the Ghana Academy of Science, and the Gordon Research Conference on Physical Metallurgy in New Hampshire; and to the Physical Society of France, the British Association, and institutions and associations in

Glasgow, Newcastle and Manchester; he also gave a number of lectures 'to various scientific and student societies' (not forgetting the Birkbeck College Natural History Society) on 'the topics of meteorites, origin of life, structure of liquids, scientific documentation, and relations of science and industry',[4] as well as the Royal Society Bakerian Lecture. His engagements and travels on behalf of the world peace movement must be added to this extraordinary load.

He was certainly by far the most overcommitted member of Birkbeck College. Every morning, when not absent on one of his numerous journeys, he would pass the door of his admirable and indispensable secretary, Anita Rimmel, and invariably greet her with the entirely true statement, 'Glad you are here – there's a lot to do today.' It is hardly surprising that, while loyally taking part in such activities as he was expected or asked to, his energies at college tended to be concentrated on his department. Nevertheless, his colleagues saw a good deal of him, especially after the college moved into Bloomsbury where, for a time, most of them were at best a couple of hundred steps distant from each other, and physically a good deal closer in the tiny senior common room and staff dining room. Few of them can have failed to hear or overhear those remarkable stories about the more surreal aspects of his wartime activities, of which he had a large and well-polished repertoire. Bernal had little in the way of social chitchat, at all events of the usual senior-commonroom kind, but he was both a riveting and – when he wanted – a very funny as well as an encyclopaedic talker.

So, despite the difficulties, he fitted well into a college to whose students and traditions he was deeply attached, as were most of his colleagues. Seen from the outside, he and it belonged together. When Rosalind Franklin planned to transfer to Birkbeck, she wrote to a friend:

> Whatever one may have against the man, he's brilliant, and I should think an inspiring person to work under. And Birkbeck College is, I suppose, more alive than other London colleges. It has only part-time evening students and consequently they really want to learn and to work. And they seem to collect a large proportion of foreigners on the staff, which is a good sign.[5]

Bernal and Birkbeck made a good marriage. Whatever Bernal's politics looked like from the outside, within the college they served to reinforce the college tradition: anti-elitist, democratic, ready to receive foreigners and to treat women as equals. Franklin's friend and biographer Anne Sayre rightly argues that her treatment at Birkbeck, which disarmed a notably prickly champion of women's rights and made her years there

almost certainly the happiest of her scientific career, was due to the
'Communist notions' of the equality of male and female workers with
which Bernal had imbued his department.[6] But even outside the Physics
Department, Birkbeck was not an institution that encouraged male
chauvinism.

As for the college itself, most of its members admired him, were
proud of the presence of so distinguished a scientist, and found it
impossible not to like him as a man. Most of them felt what the Clerk
of the College described to him on his retirement, with transparent
sincerity, as 'the sense of pride and privilege to have served alongside a
man of genius'.[7] Most of his colleagues were younger than he, but the
fact that the college made virtually a new start after World War Two
diminished the distance between the prewar survivors – and Bernal
only just dated back to before the war – and the expanded and
expanding postwar staff. While the politics of the Birkbeck staff covered
a very wide range, it is highly probable that most of the teachers, and
almost certain that the great bulk of the students, saw themselves as
occupying positions to the left of centre, though only a small minority
shared Bernal's views. Even in the worst periods of the Cold War of the
late 1940s and early 1950s, the atmosphere at Birkbeck was a great deal
more relaxed and tolerant than in some other academic environments
of the time. The fact that Bernal was a well-known Marxist and
identified with the USSR created no real problems, at all events in
public. Birkbeck had, after all, a long tradition of political radicalism
and heterodoxy.

Behind the scenes there were undoubtedly some who feared and
detested his politics, but to the best of my knowledge this never
emerged into the open. Such differences may nevertheless have contributed to Bernal's later difficulties in college politics. On the other
hand the more general struggles over the future of the college, which
also embittered his life in the 1960s, reinforced the personal as well as
the public bonds between him and the majority of the college staff,
who fought for the maintenance of Birkbeck's traditional character.
Within this majority Bernal was both prominent and influential. In
short, Birkbeck accepted, enjoyed, admired and loved Bernal the way
he was. For him in turn, Birkbeck was not only his home and professional base but also an integral part of his life as a man of the left as
well as a scientist.

Between the Birkbeck of 1938 to which Bernal was appointed and
the Birkbeck of 1945 to which he returned there lies the rupture of the
war. The college barely survived it, though it successfully resisted the

decision actually to close it down – it could hardly be evacuated like all other parts of the University of London – and during the years of bombing it stubbornly maintained its services to working students by daylight on Saturdays and Sundays. It was unable to resist the bombs of 1940, 1941 and 1944. For practical purposes the college was reborn in 1945 when evening courses were resumed and Bernal 'really began his work as Professor of Physics'.[8] As readers of Chapter 7 of this volume will have realized, his own wartime activities gave him hardly any chance to think much about his small and depleted department, struggling along among the ruins. While at Princes Risborough he was obliged briefly to return as formal chief of it to settle a dispute over the dismissal of the head steward by the acting departmental head, but his move to Combined Operations forced him to break even formal connection with Birkbeck for the duration.

Nevertheless there was some continuity between the prewar and postwar college, if only through the other physicists who dated back to 1939, partly survivors of an older Birkbeck like H. R. Nettleton, partly relics of the Blackett era like R. E. Siday, and a then recently arrived MSc student C. H. Carlisle, later Bernal's successor in the Chair of Crystallography. Thus Blackett's own interest in cosmic rays had left behind a team, dormant during the war, that pursued these phenomena in locations ranging from Holborn underground station and the roof of Senate House to a pony stable in a colliery and the Jungfraujoch. The subject was remote from Bernal's interests – or rather it was near the margin of his astonishing range, but would hardly have been on his own research programme.

His own interests at the time are indicated by the titles of his lecture to the Royal Institution in 1939 on 'The Structure of Proteins' and of his inaugural lecture in the same year – never apparently written up – on 'The Structure of Solids as a Link between Physics and Chemistry'. Since Bernal, waylaid by some preliminaries, had no time to arrange the slides prepared for him by Dr Isadore Fankuchen, who worked on Bernal's then-favourite tobacco mosaic virus, his brilliant improvisations around the unpredictable pictures would have indeed needed a good deal of restructuring for formal publication. He had already begun to mobilize research grants, though the £450 he received in 1938–39 from the Department of Scientific and Industrial Research for 'X-ray and optical investigation of the molecular structure of crystalline and liquid crystalline protein' would hardly impress modern workers. Still, in 1938 it represented the equivalent of many a lecturer's annual salary.

Little more need be said about this brief prelude to Bernal's continuous career at Birkbeck from 1945 to 1968.

European university professors, especially natural scientists in charge
of laboratories with sizeable staffs, complex machinery and the expenditure of substantial funds, long tended naturally to drop into the role
of sea captains, military commanders or business chieftains: sovereign
under God or a remote authority. Even in Britain, which in this respect
gave them less personal power than the *Herren Professoren* of Germany
or the academic barons of Latin countries, and in the oligarchies of
Oxford and Cambridge little more than the status of first among
equals, it was often enough their right and duty to tell subordinates
what to do, though preferably with paternalist affability and gentlemanly courtesy. This role suited neither Bernal's personality nor his
politics.

As a person, he was a man who liked to inform but not to command.
What gave him authority was the sheer sense of intellectual superiority
and knowledge which he could not help giving to most of us, and
certainly to most of his juniors. His ideal as a scientist was to be a
citizen of a republic of equals, though most colleagues and collaborators could not help seeing him as 'more equal' than they were. In
orthodox military terms, he was very far from obvious officer material,
unlike, say, a man like Blackett. His politics also committed him to an
ideal of a self-governing democracy, in which all members of his
department, from students, office staff and technicians to himself,
shared in the common task; it also committed him to a firm belief in
trade unionism. Moreover, as an experimental scientist he knew that
no sharp line separates workshop and laboratory, technician and
scientist, head and hand. Indeed contemporary Marxist views about
science and its history insisted on the unity of all these. The unity of
theory and practice at the laboratory bench was essential not only for
the advancement of science but also for the students' success in
learning.

Bernal's first major impact on his department was therefore as a
democratic reformer. On his fiftieth birthday the oldest teacher in the
department summed up his achievements between 1945 and 1951 in
words that do not require paraphrasing:

Bernal commenced quietly at the right end by tackling the deplorable
conditions of our laboratory assistants, on whose services the efficient work-

ing of the overcrowded laboratories is dependent; soon we came to have happy, energetic student-assistants, given full facilities for study and participation in College social life. Simultaneously he turned his attention to raising the status of the posts of Steward and Lecture-Demonstrator and planned for the creation of a workshop – now efficiently manned and equipped by a trained engineer and assistants.

Bernal now turned to the teaching laboratories which possessed one part-time demonstrator! As a temporary expedient, successful and promising students were appointed to help at the 'practicals' and were replaced gradually, as research developed, by post-graduate workers. Thus has come into being – what after all our pre-war efforts, seems to me almost miraculous – the stimulating company of some dozen demonstrators whose learning and research experience has been of inestimable help to the evening student working under difficult conditions.

Then came the fruitful concept of staff–student meetings, each class normally electing two representatives to voice the suggestions of their constituents. This free and frank exchange of views has become a vital characteristic of the Physics Department.[9]

One may add that it provided the model for the staff–student consultations that eventually became general throughout the college.

Quite apart from such reforms – for which Bernal must take personal credit, since the department had to be virtually recreated on (and literally in) the ruins of the prewar Birkbeck – his style of leadership encouraged both devolution and sociability. As for the first, let me quote the tribute of one member:

> There are many kinds of Professors, ranging from the extremely bossy and interfering, to those so incompetent as physicists that they just bury themselves in administration. No-one could accuse you of either extreme! We like particularly your device of leaving us alone to get on with the job and yet popping in sufficiently frequently to see how we are getting on, so that we don't feel neglected. . . . We hope that future generations of Professors will model themselves on your example.[10]

As for the second, another tribute recorded, '. . . he has brought us together at staff meetings, at colloquia, at Christmas parties and summer outings, and enlivened us with talks on his travels and his optimism.'[11] I do not think he simply saw this sociability as part of his professorial obligations, which he took very seriously – down to his lectures for undergraduates, which were much appreciated though Bernal was by no means an ideal lecturer. (He was too fast, too dense, and apt not to finish his sentences since he lost interest in them once the main idea had been expressed. As one student complained, 'We can either listen to you or take notes, but not both.') He gave parties because he also liked having a good time, just as he lectured because he was interested in what he was saying. He was, after all, not only a scientific prodigy and a revolutionary, but also an intellectual bohemian of the emancipated twenties and thirties, a party-giver and a party-goer. (His complete lack of any sense of music – as distinct from all the other arts – had the advantage of encouraging conversation.)

His was a happy department, not least because it was headed by a man who did not mind being sent up by his juniors, and who in any case took ideas and causes more seriously than himself. When it was proposed to name the rooms in the new Torrington Square research laboratories after various eminent figures in the history of science – a pompous project hardly in keeping with the rickety spaces waiting for their names – the nuclear physics group, located elsewhere, decided to name their laboratory after Oblomov: and Oblomov Room was to remain its quasi-official name.

All this may appear trivial to readers who were not Bernal's colleagues or students or otherwise associated with Birkbeck. Yet it is not irrelevant to the portrait of either the scientist or the man – or of the miserable disputes that soured his later years at the college. These arose primarily out of the growth and branching-out of the Physics Department which in 1945 'as regards research, was simply moribund'.[12]

Bernal's major scientific aims were based on 'his vision of a biological science based on known molecular structures' or the carrying of physics and chemistry into biology.[13] He had indeed long planned an institute for protein structure, and he carried this idea with him from Cambridge to London and attempted to realize it after the disruptions of the war. As it happened, the Birkbeck contribution to the major achievement of molecular biology in the immediate postwar years, the discovery of the genetic code, was marginal, though his position in the field was such as to attract both Crick and Rosalind Franklin to his laboratory, and workers at Birkbeck – Furberg, inspired by Carlisle, and Franklin – were associated with it.[14] Bernal himself later realized that he had failed to recognize or seize the opportunity to develop this work. Nevertheless, the organic work of his crystallography team pursued 'an underlying plan' whose main objective, in his own words, was 'to follow up the structure of globular proteins which I first started in 1934 as well as that of crystalline viruses which I started in 1936'.[15] And while Birkbeck was merely one of several physics departments in London and clearly inferior in size and resources to the major establishments of the university, it was, thanks to him, the largest and most important centre of crystallography in London, though work in the field went on elsewhere (with Lonsdale at University College, with Randall and Wilkins at King's and with Ubbelohde at Imperial College). Moreover, it was the only one to undertake regular teaching for a master's degree.

It was Bernal's ambition to establish such a centre for teaching and research that was to precipitate the conflicts in which he became involved in Birkbeck. From the outset he faced two difficulties which were administrative and political (in the academic sense) rather than scientific. In the first place, such an institute or research unit cut across the traditional dividing lines of the sciences – physics, chemistry, biology – insofar as it had to involve the collaboration of workers from different disciplines. In the long run it would inevitably encroach upon areas regarded as their own by other disciplines, or kept by them as preserves for future work. Since the initiative in these projects came, in London at least, from the physicists (Bernal in Birkbeck, Randall – himself inspired by Bernal – at King's College), it implied a subordination of other disciplines of the then enormous prestige and self-confidence of physics, which was likely to arouse jealousies and resistance. In the second place, the federal structure of the University of London made it difficult, if not impossible, to set up viable central institutions independent of the colleges. *De facto* those that were successful largely engaged in activities that did not fall into the range of any college and eventually turned into something like colleges themselves (for example, the School of Oriental and African Studies). Those that overlapped with college teaching and research found difficulty in developing beyond the stage of being service centres (for example, specialized libraries and places for intercollegiate meetings). This difficulty was exemplified by the Institute of Historical Research, which failed to achieve the high ambitions of its founders. In London the choice lay, in practice, between research institutes entirely independent of the university (such as those set up by the Medical Research Council) or research units virtually operating on the basis of a single college.

Here, however, Bernal's position was weak, since his base was a small, relatively poor and to some extent marginal institution, not comparable in size, area, prestige and past traditions of research with University College, King's College or Imperial College. In any university with severely limited resources, proposals to extend one department unfortunately, but inevitably, suggest to other departments that their slice of today's and tomorrow's cake will be reduced. In a small and physically cramped college like Birkbeck, these rivalries made themselves felt more rapidly and more desperately than elsewhere.

In fact, Bernal's own success exacerbated them. He returned to a college in ruins, and indeed initially in 1945 he could not work there at all but had to use the Royal Institution, where he collected together the elements of his team – C. H. Carlisle (later his successor as Professor of Crystallography at Birkbeck), Werner Ehrenberg (formerly of Prague, later Professor of Physics at Birkbeck), Helen Megaw (later Director of Studies at Girton), A. D. Booth (the founder of what was to become the Department of Computer Science) and not least Anita Rimmel (Bernal's invaluable secretary and personal manager), subsequently joined by J. Jeffery (later Professor of Crystallography). Essentially this biomolecular work was funded for five years by the Nuffield Foundation, though funds also came for other purposes from the Department of Scientific and Industrial Research via the Building Research Station.

Since this accommodation was only temporary the college had to find room, and managed to acquire from the university two ruins in Torrington Square which were to remain Bernal's research base, and indeed his home, since space was also found for a flat for him at the top of these buildings. 'It was', as has been said, 'Bernal's fate always to work in old buildings,' or more exactly in scientific slums.[16] The buildings were made approximately habitable (including a fairly spartan structure on the space occupied by a static water tank during the war), and came to occupy a small place in the history of the arts as well as a more substantial one in that of the sciences, since Picasso later improvised a drawing on the wall of Bernal's sitting room during a visit to attend a Peace Congress in Britain.[17] The buildings were occupied from 1947 and officially opened by Sir Lawrence Bragg in 1948.

In spite of the material difficulties, Bernal's department at Birkbeck flourished and diversified. He might still be far from the institute of his dreams, but he had a team of teachers and researchers, now largely integrated into the department, and the students and researchers who came to him from all parts of the globe almost certainly formed the most cosmopolitan group in the college, even excluding the visitors who briefly alighted in Malet Street and Torrington Square to pay their respects to Bernal in his capacity as a personality of the international left, and especially of the peace movement, to which he understandably devoted more time and effort than even a man of his energy could afford. In many ways the middle 1950s marked the high point of his career at the college. Though the first signs of physical trouble appeared in 1951 when – as he recalled in retrospect – he began to find mountain climbing exhausting, he seemed at the height of his powers. He brought new people onto the academic staff, notably D. J. Bohm, a theorist of great gifts who had been forced to leave the USA during the McCarthy period. He attracted workers of brilliance, notably Rosalind Franklin and Aaron Klug, who soon formed a powerful team which took charge of research on viruses and promised the college an important place in the spectacularly advancing field of molecular biology. He returned to research himself, and explored an old interest of his – one of the many – the structure of liquids. Those who talked with him or his co-workers in those years remember both the sense of satisfaction they felt, the air of excitement, as well as the admiration of all who came into contact with him, of the rapidity of his understanding, the brilliance of his comments, the stimulus of his suggestions, his own joy at other people's discoveries.

Yet the promise of these years was not to be quite fulfilled. There were certainly other reasons – for instance, Rosalind Franklin's death in 1958. But above all, uncertainty about the future of Bernal's department, and especially of crystallography, stood in the way. The departure of Klug and his group for Cambridge where, in 1962, he joined Bernal's research student and admirer Perutz, underlined these uncertainties.

They existed because the fundamental fact about Bernal's domain at Birkbeck was that it was necessarily and inevitably engaged on expansion, which meant competition for scarce resources – especially from 1952 when six of the crystallographers previously supported by outside grants were taken on the payroll of the college as teachers. Not merely did his department grow; it developed activities that themselves required resources, and were increasingly difficult to contain within a single department. This was notably the case with crystallography, Bernal's main interest, and computing, which he was one of the first to try to use in connection with crystallographic and other physical calculations. In fact both the subsequent departments of Computer Science and of Crystallography were offshoots of Physics. Bernal himself planned the separation of Crystallography from Physics from the early 1950s. It was this that led to the conflicts that helped to shorten his life. These must be briefly elucidated.

Crystallography did not fit into the established pattern of the sciences. Administratively it hardly existed as a separate field, except for the department of Bernal's research student Dorothy Hodgkin at Oxford. Of the three dozen departments in which X-ray crystallography was pursued, about one third belonged to Physics, one third to

Chemistry, and the remainder to miscellaneous disciplines. Birkbeck crystallography which, as we have seen, had the major responsibility for work and especially teaching in this field in London, shared this ambiguity: its position rested entirely on the personal interest of the Professor of Physics; the largest London unit in the field, unlike University College, did not actually possess a post specified as crystallographic.[18] The expansion and formalization of what was inevitably an expensive subject were therefore likely to raise difficulties within the college. Moreover, as was to become increasingly evident, while the Chemistry Department had no strong views on the subject, there was some feeling in biological departments that crystallography might with advantage be drawn into a wider department of 'life sciences', thus reinforcing the position of the biological disciplines within the college. Some other departments were also bothered by the potential diminution of their share in college resources, not only financially but in terms of space. Filling pint pots with quarts was a way of life in Birkbeck. The new building opened in Bloomsbury in 1951 had already proved too small to hold all the expanded activities of the postwar college, as was the extension of that building opened in 1965–66, which had to find room for the crystallographers since the Torrington Square accommodation was not permanent.

While interdepartmental resentments were to be expected, it is at first sight surprising that the college as such created difficulties for what was probably its most prestigious department (headed by what was certainly its most distinguished academic figure), whose expansion would certainly increase the college's own standing. Indeed, as we have seen, in the immediate postwar years (under the mastership of the zoologist W. Gordon Jackson, 1943–50) Bernal had little to complain about. The difficulties began and later increased under the mastership of John (later Sir John) Lockwood who presided over the college from 1952 to his unexpected death in 1965.

They were certainly in part due to the pressures of finance and space. It would be unfair to underestimate the real difficulties of those whose job was to allocate resources in a small college from demands that, looked at in a purely college perspective, might appear disproportionate: in 1967 Physics and Crystallography together occupied over 40 per cent of the space of all science departments and about 20 per cent of all college space: twice as much as the next-largest of the other nineteen departments (Chemistry).[19] That Crystallography fulfilled general university functions that other Birkbeck departments did not would not necessarily reconcile other claimants within the college.

Perhaps the difficulties were also in some ways diplomatic. To divide Bernal's empire (if the term is not offensive to the memory of so committed an anti-imperialist) might well have been seen as giving him two departmental voices in college affairs – a crystallographic and a physical one – for the loyalty of Bernal's people to him, and their shared interests, were well known. To what extent personal feeling, hostility to his politics, or difficulties of communication arising out of his multiple activities outside college had a role, we can only speculate. They certainly played some part. The fact remains that under the mastership of John Lockwood the college showed a distinct lack of sympathy for Bernal's projects.

Whatever the earlier uncertainties and frictions, the matter became acute in the early 1960s. Up to that time the expansion of crystallography had proceeded (from 1952 with official recognition and resources provided by the university, including a better building in Torrington Place), and the eventual separation of the two departments appeared to be accepted. It had been formally proposed by Bernal in 1956[20] and included in the college's submission to the university for the quinquennium 1962–67 – which would be Bernal's last complete quinquennium before his retirement. Moreover, whilst previously Bernal had merely asked for the setting-up of separate subdepartments or departments of crystallography and computer science, he now asked 'to resign the chair of Physics at the end of the session 1961–62 if it could be arranged that I transfer my title of Professor to one of Crystallography for my last six years of service in the College'.[21] What he certainly had in mind was to safeguard the continued existence of crystallography at the college after his retirement, and his proposals made a decision on this matter urgent.

The conflicts about crystallography in the early 1960s coincided with and were complicated by a much wider conflict within the Birkbeck staff about the college's future. Broadly speaking, the Master proposed a complete change in the character of the college, which was to shift increasingly to the teaching of full-time students, that is, of students at the undergraduate level, mainly school leavers, and to move out of its cramped but central site onto some more ample site on or beyond the outskirts of London, on which major building and expansion could take place. For this, resources looked like being available in the light of the government's then enthusiasm for financing university expansion. The difficulties of some departments in attracting sufficient part-time undergraduate students provided some arguments for such a course, and decisions to allow some departments to accept full-time school leavers were taken in 1963. In fact, the new course proposed the liquidation of Birkbeck College as it had existed since its foundation, and while it had some support within the college, it met with even more opposition. Not surprisingly, hostility to the new plans was passionate. A state of virtual civil war within the Birkbeck staff existed for some years, and did not come to an end until the death of Sir John Lockwood. Following this, a powerful outside Academic Advisory Committee on Birkbeck College under Sir Eric Ashby was set up which eventually came down on the side of a traditional Birkbeck of part-time students on its central site. Not unexpectedly, Bernal found himself on the side of the defenders of the traditional character of the college against the Master and his supporters. Insofar as those who took a jaundiced view of Bernal's departmental projects anyway were also engaged on the opposite side in the battle about the entire future of the college, the conflict about crystallography took place in an exceptionally acrid and embittered atmosphere.

The details of college wars are apt to be of no great interest to outsiders, and indeed they are relevant to readers of this volume chiefly because hostilities reached their peak at the time of Bernal's first tragic stroke, and undoubtedly contributed to it. I recall the Academic Board on 2 July 1963, which Bernal attended straight off an all-night journey from New York, delayed for several hours of waiting in a heatwave, in an aircraft without air conditioning. He looked like death, and I remember noticing that his speech already seemed slurred. Indeed, halfway through that board, which he attended only because any failure to fulfil his college duties might have been used against him, he asked a colleague to take over from him. That evening he had his stroke.

The essential facts are that in 1962 it became impossible to oppose Bernal's plans by mere delaying tactics, since a professorial title was conferred on his colleague W. Ehrenberg, a charming, crippled, and witty Central European experimental physicist. It had been claimed that financial restrictions made it impossible to establish a chair for a new crystallography department, but since two professors were now available and paid for in Physics, the transfer of one of them to head the new department would not cost anything during the current quinquennium. The question of the future of crystallography had therefore to be faced head on.

Bernal's proposal to create a department, put forward at a special meeting of the Professoriate Committee on 13 June 1962 was therefore strongly opposed, as he recalled, with 'objections which I had failed to anticipate and whose grounds at the time I did not fully understand'.[22]

Even the lame compromise to which he reluctantly agreed was thrown out by the Governors of the college, who asked the Academic Board to reconsider the entire matter, which it did by passing the buck to a special committee of science professors on the future of crystallography. Its report was in turn to go back to the Professoriate and the Academic Board and thence to the Governors.[23] As usual in academic affairs, a compromise was cobbled together. On 8 November 1962 the Professoriate recommended that a Chair of Crystallography was to be established – evidently for Bernal – but until 1967 at least it should continue in the Department of Physics. It was a victory of sorts for Bernal, but won at enormous expense of time, effort and worry. Moreover, Bernal knew that his opponents, headed by Lockwood, were in no sense reconciled to defeat.

Paradoxically, Bernal's stroke helped him to achieve what he had so long battled for. It made it impossible for him to carry on as head of the Physics Department and he therefore resigned from the direction of the department which Ehrenberg took over as acting head.[24] It is symptomatic of the rancorous atmosphere of that period that Bernal was thereupon refused the right to take part in various committees of the Professoriate on the grounds that he was no longer head of any department. He complained about this to the Academic Board[25] – as always in measured and indeed notably understated terms – and immediately raised the question of the separation of the two departments again:

> It is claimed, however, that nothing has occurred since then [8 November 1962] to put this decision [not to separate the departments] into question. My point here is that something has happened, namely my own illness.[26]

He was, he argued, both capable of and willing to run the Crystallography Department, but not both departments; and Ehrenberg in turn could not represent the crystallography interests as well as the physics ones. He was still seeking to ensure the future of crystallography after he went.

Over the resistance of the Master the Professoriate accepted his arguments and Crystallography was finely emancipated in 1964 – at the peak of the college civil war. Perhaps at this point the opposition of the majority of the college staff to Lockwood's plans provided Bernal with effective allies. His concern for the future of crystallography alone could not do so. It could create enemies, without necessarily providing allies who were willing to fight with sufficient commitment for a man they admired and a case they might approve while, as nonexperts, not quite understanding why it seemed so important to him. Now the cause of Bernal was the cause of all adversaries of Lockwood. Perhaps also his colleagues were moved by pity for Bernal's physical ruin, by a revulsion against the idea that his wishes should be refused in such circumstances, by a bad conscience about the contribution such a refusal had made to his personal tragedy. Certainly the logic of his argument was now accepted. Crystallography at Birkbeck thus survived Bernal.

The whole miserable episode is an example of the tangled and trivial civil wars with which anyone with experience of colleges and universities will be familiar. None of them do credit to their institutions, but few of them end in tragedy. It is difficult for anyone who was involved in these disputes at the time, or who surveys the record impartially, to look back on them with anything except a sense of shame. They forced a scientist of extraordinary gifts to pursue his and his colleagues' work under constant threat of strangulation, in constant uncertainty about its very survival. They certainly contributed to Bernal's death. How much, it is impossible even to guess, for Bernal's health was already undermined, and he had also in those years to bear the increasing strain of his work for world peace, which involved a heavy load of travel, and strenuous, difficult, and endless meetings and negotiations, often into the early hours of the morning. He was doing too much. Even the devotion of Anita Rimmel, of his literary research assistant Francis Aprahamian, and of all his department, could not sufficiently reduce his multiple load. Nevertheless, there can be no doubt that the Birkbeck troubles shortened his life.

Bernal officially retired from his new Chair of Crystallography in 1968, but in practice his pyrrhic victory concluded his effective career at the college. The year 1965, which saw both his own second, and much more serious, stroke and the sudden death of Sir John Lockwood, virtually brought it to an end. By dint of the heroic efforts of Anita Rimmel, who now also acted as the interpreter of his increasingly incomprehensible speech, he continued in action. However, his physical decline accelerated. As time went on his extraordinary brain, imprisoned in an almost totally inoperative body, came to be virtually cut off from communication with the outside world. For a man such as he there could be no greater tragedy.

The college, its old vocation reconfirmed, and headed by a new Master, recalled his life at a memorial meeting on 24 January 1972. Readings from his works were interspersed with tributes from friends and admirers of his genius: C. P. Snow, J. C. Kendrew, Solly Zuckerman and P. M. S. Blackett: three lords, two Nobel laureates, two OMs. It was characteristic of Bernal's career that men and women who were never tired of expressing their admiration for his remarkable gifts, and of acknowledging their profound debts to him, gained more public and scientific honours than he. Passages were read from a lecture on 'Science and the Humanities', given long ago at Birkbeck, from *World without War*, from *The Social Function of Science* and from his *The Origin of Life*, written in conditions of almost superhuman difficulty towards the end of his life, and published in 1967.

In the life of any academic institution a professor is only an episode. As a person Bernal survives his death as part of its history – of those thirty years in Birkbeck College which, as the college told him on his retirement, would doubtless be seen 'as the peak of achievement in the Faculty of Science' – and in the memories of his contemporaries, whose ranks will inevitably thin. His contributions to shaping the college survive, notably in the form of the Crystallography Department which he did so much to establish and to perpetuate, at so great a cost to himself. Conversely, the life of a man whose activities were so various and wide as Bernal's cannot be written primarily in terms of any college, even one with which he was so long associated. They reached far beyond it. Nevertheless, 'affection for Birkbeck and the promotion of its ideals'[27] played a very important part in Bernal's life. He saw himself as part of that tradition in which the London Mechanics' Institution had been founded at the Crown and Anchor Tavern in December 1823, for a constituency shortly thereafter defined as that of the working classes – 'who work and do not employ journeymen'. Throughout his career it remained, as it still does, a college for adults working during the day, at any rate during their undergraduate years. To teach and collaborate with such people, and in such a tradition, was not, for Bernal, a fortuitous by-product of accepting a chair at Birkbeck rather than some other institution. It was an integral part of a life devoted to science and humanity.

Notes

1. In addition to recording my own memories, I have consulted a number of people who were much more closely involved with Bernal than I. I am particularly indebted to Anita Rimmel, Harry Carlisle (in person and through his 'Serving My Time in Crystallography at Birkbeck: Some Memories Spanning Forty Years, 1938–1978', unpublished manuscript, 1978), John Jennings, Alan Mackay, Peter Trent and A. J. Caraffi. Some of them, as well as Brenda Swann, have kindly made documents available to me.

2. Transcript of recording made by A. J. Caraffi, 29 September 1972, pp. 2–3.

3. The present writer, who applied for academic posts a few months after leaving the army in 1946 – when institutions still had to rely on their pre-1939 reputation – was advised against applying to certain colleges by his Cambridge research supervisor, on the grounds that 'it will not do you much good to have been there'. Birkbeck was approved.

4. Birkbeck College (University of London): Report on the One Hundred and Thirty-Ninth Session, 1961–1962, pp. 38–40.

5. Anne Sayre, *Rosalind Franklin and DNA*, New York, 1975, p. 138.

6. Ibid., p. 174.

7. From the Clerk to Professor J. D. Bernal FRS, 25 July 1968.

8. H. R. Nettleton's Foreword to 'To Professor J. D. Bernal from his Staff', 10 May 1951, p. 3. This booklet of thirty-two typescript pages was presented to Bernal 'on the occasion of his Jubilee and as the eve approaches of our general exodus from Breams Buildings'. It contains a frontispiece showing the geographical provenance of the members of his department. They came from Austria, Burma, Canada, Chile, China, Czechoslovakia, Egypt, France, Germany, Greece, Holland, Hyderabad, Ireland, Malaya, Norway, Pakistan, Spain and the USA.

9. Ibid, pp. 1–2.

10. Ibid, p. 17 (E. P. George).

11. Ibid, p. 3.

12. Ibid, p. 1.

13. Robert Olby, *The Path to the Double Helix*, London, 1974, p. 254. For Bernal's scientific work, see also Dorothy Hodgkin's memoir of his life in *Biographical Memoirs of Fellows of the Royal Society*, Vol. 26, December 1980, pp. 17–84.

14. Crick went to Cambridge, and Franklin, initially not attracted by the technique of work on single crystals in which Birkbeck specialized, did not join the college until 1952, after an unhappy spell at King's.

15. J. D. Bernal, 'The Department of Crystallography', *Lodestone* (Journal of Birkbeck Students Union) 55, No. 3, Summer 1965, p. 41.

16. Olby, p. 261.

17. This section of the wall was transferred to the Institute of Contemporary Arts after Bernal's retirement.

18. J. D. Bernal to Master, Birkbeck College, 3 January 1956.

19. *The Future of Birkbeck College: Report of the Academic Advisory Committee on Birkbeck College*, May 1967: Tables between pp. 38 and 39.

20. See note 18.

21. J. D. Bernal to Master, 17 November 1959.

22. J. D. Bernal, 'Case for the Maintenance of the College Policy for a Department of Crystallography' (undated), p. 1.

23. To simplify the complexities of academic constitutionalism, the Governors were the ultimate governing body of the college; the Academic Board, consisting of members of the academic staff, could 'consider and make representations to the Governors' on 'all academic matters'; and the Professoriate Committee, consisting of the heads of departments of professorial rank, represented the group of departmental chieftains. The Governors contain and contained a minority of representatives of teachers, students and former students.

24. See J. D. Bernal, 'Note for the Professoriate Committee, 11 December 1963, on the application by Professor Bernal for the separation of the Crystallography Laboratory and the Department of Physics', 11 December 1963.

25. Letter to Academic Board, 3 December 1963.

26. Note of 11 December 1963, p. 2.

27. Clerk to Bernal, 25 July 1968.

12

Science in History

Peter Mason

Leonardo da Vinci painted his masterpiece *The Last Supper* right across a wall in the church of Santa Maria delle Grazie in Milan. Although his consciously scientific treatment of perspective and movement created an overpowering effect, he was accused of letting his science stand in the way of his art. His response was to affirm his belief that painting itself was really a form of science. Four and a half centuries later, Desmond Bernal wrote his remarkable book *Science in History* – also a masterpiece – and was accused of letting his history and his politics stand in the way of his science.[1] His response was to affirm his belief that science, history and politics are inseparable.

The analogy goes deeper. Leonardo's painting soon started to crack and decay because of a basic flaw in his technique, and it has required repeated restoration ever since. Bernal was able to revise his own work, three years after its publication in 1954, but he could not manage to go on patching up the gaps and cracks that were appearing in it for much longer. When a third edition came out in 1965 he explained in its preface that recent advances in science, especially the applications of electronics to computing and automation, were affecting world economics and politics to an extent that defied immediate analysis. A complete recasting would be required. In the meantime he had rewritten the section on 'Science in Our Time', deleting some of the historical treatment of the social sciences to make room for the new material.

Far from regarding this as an interim measure, though, Bernal made the astonishing statement that 'it is highly unlikely that another edition of this book on the basis of the present one will ever be produced'. Did he really feel that the sheer speed of technological change had swept his analysis back into the melting pot? Could it be that a man who had successfully charted the course of a hundred branches of science over five millennia should now be unable to put into perspective the developments of just one more decade? Neither of these propositions seems plausible. The difficulties that *Science in History* was running into indicated a deeper-seated malaise, some basic flaws in the optimistic, scientistic model that had inspired its creation.

In a lavishly illustrated version of the third edition, published in 1965, Bernal revealed his disappointment at the way science was being misused. He wrote a new preface, very different in tone from his earlier ebullience: 'The great adventure of science seems, very sadly, to lead to such an end as negates all its original promise through the ages.' The power of science had increased enormously in both capitalist and socialist countries yet, paradoxically, the grievous gap between the developed and the underdeveloped worlds was still widening. Something was really wrong.

Bernal faced up boldly to the shifting basis of his world-view, describing his position by the word *provisionalism*: 'It is more than scepticism,' he said. 'It is a conviction that whatever we think now, people in a very short time will think differently and better.' As nearly three decades have passed it is now up to us to think differently and, if possible, better. Three areas where, it seems to me, we could most usefully revise Bernal's historically based view of science are (1) the nature of science, (2) craftsmanship and elitism and (3) the future of science.

The Nature of Science

This is not simply a matter of definition. Hubert Dingle once took Bernal to task for not defining precisely what he meant by science and actually listed ten different usages of the word in *The Social Function of Science*. Bernal retorted indignantly that a single definition would be futile since throughout its history science has been so linked with other social activities that any attempted definition could express at best only one of the aspects it has had at some period of its growth. He evoked Einstein in support of this view:

> Science as something existing and complete is the most objective thing known to Man. But science in the making, science as an end to be pursued, is as subjective and psychologically conditioned as any other branch of human endeavour – so much so, that the question 'What is the purpose and meaning of Science?' receives quite different answers at different times and from different sorts of people.[2]

This dynamic idea of science in the making gave the coherence to Bernal's historical view of science right up to the twentieth century. But then the road he had travelled so confidently from the Stone Ages up to the great discoveries of the nineteenth century suddenly branched out in all directions as those discoveries found application throughout society: to lighting, transport and communication; to entertainment and health; to production and war. He made a heroic attempt to map the labyrinthine paths through the jungle of modern science, using graphical and tabular methods to which I shall return later, but this did not help him with his agonizing paradox. In 1964 he wrote in his contribution to *The Science of Science*, 'We have the potentiality of the age of abundance and leisure, but the actuality of a divided world with greater poverty, stupidity, and cruelty than it has ever known.'[3]

Bernal's view of the issue was perhaps obscured by his deep-rooted faith in the virtue of science itself (however defined!). Snow remarked on Bernal's need for a faith, and Pirie even pulled his leg about it during their protracted but friendly debate about the origin of life, observing that faith seems to be an occupational hazard for physicists. He recalled Gowland Hopkins's comment on the craving for certainty that had led so many physicists either into mysticism or into the church and similar organizations.

Fortified by his faith in science, Bernal returned to the fray in *The Science of Science* as the champion of the powers of science – in the right hands. The transition from the divided world of 1964 to the age of abundance and leisure would be hazardous, he granted, but he could still 'feel confident that the ultimate pattern will, so to speak, impose itself the moment its logic is fully appreciated'. He chose three main areas of scientific advance to illustrate his thesis: the provision of energy in unlimited quantities (with the aid of nuclear power); the widespread use of computers; and the fundamental understanding of biological processes. All three areas have turned out to be of enormous significance, but not simply as exemplars of the wonderful powers of modern science. Energy has not become as cheap and unlimited as Bernal envisaged, and the efforts to provide it have changed the political economy of the world as well as contributing to the social problem of inflation. Computers are having an increasing effect upon society, eliminating menial tasks and providing more time for leisure, though too often in the form of unemployment. Biological advances have spawned agribusiness, which is doing its bit towards widening the gap between rich and poor; sociobiology, with its proclivity to support racism, and genetic engineering, bringing the ability to create and patent new forms of life, have introduced a new breed of scientists who divide their time between the laboratory and the Stock Exchange.

It is easier from this distance to see that science was not only increasing its powers: it was also changing its nature. Quite apart from the obvious social and political dimensions it was becoming less dependent upon fundamental discoveries (particularly in the physical sciences) and far less dependent upon the contributions of the Great Scientists.

The way in which the fundamentals of physical science were losing their significance can be illustrated by a specific example: the measurement of time. In *Science in History* Bernal starts off with the early Egyptian calendar, which only required an accuracy of a few days in order to prepare for the flooding of the Nile, and goes on to the mechanical clocks of medieval Europe, when social life required timing to better than one hour. The social function of the clock in this society was to replace the watchman who rang the chimes at hourly intervals; the resulting device, still known as the watch, was the prototype of modern automatic machinery. Bernal also refers to Needham's description of the hydro-mechanical clock of I-Hsing in AD 725, accurate to about 2 minutes in a day, and the steady improvements culminating in Su Sung's famous tower clock of AD 1090, accurate to about one minute in a day. His interest was not fully aroused, though, until the seventeenth century, when the measurement of time became crucial for navigation and, therefore, for the national security of the colonizing powers, Holland, France and England. Accuracies within a few seconds a day were now required to be maintained over several weeks' sailing in the Atlantic. Galileo had discovered an ideal controller in the pendulum, and Huygens went on to design a timekeeper, now rejoicing in the classical title of 'chronometer', of adequate accuracy on land but of no use at sea. Bernal relished the story of the self-taught Yorkshire carpenter John Harrison, who built a chronometer accurate enough to meet the British Admiralty's criterion of keeping time within 3 seconds a day over a six-week voyage. He qualified for the award of £20,000, but then discovered that it was harder to extract the money from their lordships than it had been to build the chronometer.

The accuracy had now been improved to just a few tenths of a second per day, a hundred times better than the great Su Sung clock six centuries earlier, and ten times better than the best clocks of even thirty years before. Could scientific advance keep up this astonishing and accelerating rate of progress? It could and it did. The advance has continued, via quartz crystal clocks and 'atomic' clocks, to accuracies of one millionth of a second per day. But what, really, does such an accuracy mean?

Even in this age of jet travel, radio communication and fast living, it is difficult to imagine an ordinary person needing a clock that keeps time to better than, say, one second in a year. That is about two milliseconds a day, an accuracy that was reached in around 1920. Beyond that, a timekeeper is perhaps better thought of as a scientific instrument rather than a clock. The atomic clock can do such possibly useful things as measuring variations in the rate of rotation of the earth, but even for a scientific instrument quite preposterous accuracies would be achieved before long if the present rate of progress were to be maintained. A conservative extrapolation indicates a daily error of about one ten million millionth of a second well before AD 2100, that is, within 120 years. This is about as long as it takes an atom in a crystal to execute a single vibration, an accuracy that is hard to imagine being significant for anyone, atomic scientist or otherwise. So in this particular instance it seems that sufficient accuracy for ordinary purposes was reached half a century ago, whilst another hundred years at the present rate of progress would lead to accuracies surpassing anything that could be of any significance to anybody.

Here, then, is one branch of science whose nature has changed profoundly. Further fundamental discoveries in time measurement have now become unimportant, in contrast to both the seventeenth and nineteenth centuries when they were crucial. This is a clear example of that saturation of basic physical discoveries which had not yet become so evident when Bernal was, as he put it, searching the past for clues to the future. I shall look briefly at the related question of elitism before coming to examine how Bernal's views might now be modified with the benefit of hindsight.

Craftsmanship and Elitism

In 1938, while Bernal was writing *The Social Function of Science*, Brecht was writing *The Life of Galileo*. One of the many things in which these two imaginative writers concurred was a historical view of science in which the 'great man' theory was rejected and the role of the craftspeople was recognized. 1938 was also the year when Bernal moved from Cambridge to Birkbeck College in London, set up over a century before as the London Mechanics' Institution with the object of teaching physics, chemistry and economics to working men.

Brecht posed the general question in his poem 'A Worker Looks at History':

> Who built the seven gates of Thebes?
> The books are filled with names of kings.
> Was it kings who hauled the craggy blocks of stone?

ending with the bitter observation that, even in the legendary Atlantis,

> The night the sea rushed in,
> The drowning men still bellowed for their slaves.[4]

Bernal, in his treatment of early history, followed Farrington in emphasizing the connections between scientific advance and the existence of an exploited class. The Ancient Greeks, for instance, virtually stood on the threshold of an industrial revolution, but failed to cross it. Having an ample supply of slaves, they lacked the motivation to develop labour-saving machines.

Brecht illuminated the relation of craftspeople to science in a passage where Galileo is trying to persuade the young Medicean prince to look through his telescope. He has already insisted that, for the benefit of the lens grinder, the discussion should be in the vernacular rather than in Latin, and he starts to lose his temper with the philosopher and the mathematician who are part of the prince's sycophantic entourage:

> Your Highness! My work in the Great Arsenal of Venice brought me in daily contact with draughtsmen, builders and instrument-makers. These people taught me many a new way of doing things. The question is whether these gentlemen here want to be found out as fools by men who might not have had the advantages of a classical education but who are not afraid to use their eyes. I tell you that our dockyards are stirring with that same high curiosity which was the true glory of Ancient Greece.

(To which the philosopher drily replies, 'I have no doubt that Mr Galileo's theories will arouse the enthusiasm of the dockyards.')[5]

Bernal too was conscious of the problems of class distinction in science, both ancient and modern, and he discussed the disadvantaged positions of women and technicians, though perhaps from the point of view of the interests of science as much as of the individuals themselves. He seemed to feel that, if women were to be fully involved, science would be essentially the same but moving at twice the pace. He made no exploration of the striking imbalance in the name index of *Science in History*, which runs to nearly a thousand people of whom only twelve are female.

It must have come as a shock to Bernal in 1957 when, at one of his greatest moments, during a conference in Moscow held by the USSR Academy of Sciences with the sole purpose of discussing his *Science in History*, he was taken to task for overstressing craftsmanship and technology and undervaluing science. And this from none other than Kölman, one of the Soviet scientists who had so impressed Bernal, and his radical associates at the 1931 History of Science Congress in London, with his materialistic interpretation of scientific progress as a response to social needs. Bernal replied that, if he had been guilty of excess, he was only redressing a gross error in previous histories of science, and asserted a credo that could as well have come from Galileo: 'I have repeatedly been impressed, every time I have studied in detail the developments of science itself, with the enormous importance of the working craftsmen.'

A second paradox now presented itself with the clash between elitism and democracy. Here was science, proliferating so rapidly that its leading exponents could scarcely keep up with a small fraction of it, yet cut off from its roots if the workers, the craftsmen and the people at large were not consciously involved in it. And Bernal's historical studies as well as his Cambridge background conspired to confirm in him the idea of the great scientific breakthrough, the revolutionary jump – usually associated with some illustrious figure of science – which for three centuries had been the distinguishing feature of the impact of science on the well-being of humanity. He sought to counter the isolation of the scientist by reiterating the importance of direct contact with the needs of the people, the contact that was vital to the Ionians in the golden days of Greek science, or to Galileo prowling round the Venetian arsenal. Looking forward to the rational utopia, the new Atlantis that he always believed must come, he gave a ringing affirmation of his faith in science:

> Scientists will recognise their weaknesses, a lack of contact, not so much with the seats of power as with the people who can be the real beneficiaries of science. When that contact is renewed and improved we can hope to have a world where science ceases to be a threat to mankind and becomes a guarantee for a better future.

He was not unaware of the limitations his background and position as an eminent scientist imposed upon him. He wryly recalled sitting at dinner next to a Soviet historian who said to him, 'Oh, well, you are a scientist, you only deal with the present. Now I am a historian and I deal with the future.' Let us take the hint and approach the future of science by enquiring not what marvels science may produce to meet human needs, but rather what are the human needs that most urgently call out to be met.

The Future of Science

Among the most successful features of *Science in History* are the tables Bernal uses to summarize his analysis of developments in science and society. He locates one such table at the end of each major period of advance: from the Stone Age to the Iron Age; the classical Greek era; feudal science; the scientific revolution from the Renaissance to New-ton; the capitalist era of the Industrial Revolution; then, showing the strain of specialist proliferation, he is forced into presenting separate tables for the physical and the biological sciences in the twentieth century. Lastly comes a grand summary, a chart tracing the interactions of science and technique right through from the Old Stone Age up to 1965. This great synopsis terminates in eight broad scientific lines of development leading into the future, an eightfold way that is a logical projection of the more benevolent aspects of the scientific enterprise as it was developing in 1965.

If it is assumed that science is intrinsically of human benefit then the problem is indeed simply the technical one of ensuring that these developments proceed with the greatest efficiency attainable. But it is just this assumption, Bernal's central article of faith, that has been coming under suspicion – at first from the anti-science movement, but now from an increasing number of people who are confronted by the coupling of science to nuclear warfare, the correlation between high technology and unemployment, and, Bernal's saddest paradox, the widening gap between the richer and the poorer nations.

Could the essential fallacy perhaps lie in the scientistic dogma itself, the faith in the intrinsic virtue of science that happily projects its great social successes of three centuries indefinitely forward into the future? When Faraday was asked what was the use of the electromagnetic phenomena that he had just been demonstrating, he gave his unfortu-nately immortal reply, 'Of what use is a newborn babe?' He little knew how that phrase would be beaten almost to death in efforts to justify fascinating and personally rewarding but otherwise useless work for generations to come. But this powerful delusion cannot be disposed of simply by quoting individual examples, however convincing they may be. It is essential to gain a wider, Bernalian perspective of the problem, and this can only be achieved by approaching it from the other end. Instead of trying to guess what clever new things science may have in store, we should start from the end point of science: the essential and the discernible human needs.

In Table 12.1, the needs are the reasonable primary requirements of an affluent twentieth-century person; they stop short of secondary and largely contrived 'needs' such as electronic toothbrushes programmed to release a succession of interesting flavours while playing any selected tune. I also omit the needs of those scientists who are insatiably curious to find out more about the workings of the universe. Those are genuine enough but they are irrelevant to the major present needs of humanity. In the very long term, of course, there might be a resurgence of a general interest in problems on the borderlands of metaphysics – fundamental particles, infinities, et cetera – but for the near future these will remain the esoteric interests of a tiny group, a minority art form.

Table 12.1 shows how successful that power of science has been in overcoming the physical constraints upon the flowering of our affluent individual's personality and talents. But this impression must at once be qualified by the observation that the majority of the world's population are far below the requisite level of affluence. The productive and distributive forces have grown rapidly but not rapidly enough to make even the basic benefits of science universal. If we try to assess the benefits that modern science has already brought to the world we are faced with a population increase from about three-quarters of a billion in the mid-seventeenth century to around five billion today. On a conservative estimate one fifth of the population today subsist at a level inferior to the affluent eighteenth-century standard of Table 12.1. This depressed class of one billion people is thus numerically larger than the world population two centuries ago, and its members cannot be expected to see science as a fairy godmother. Our scientific world has failed even to keep pace with the inflation of population.

The first priority for future science should clearly be to help in redressing this balance. The last two breathtaking centuries of physical science have given us the ability to meet all reasonable physical needs (as well as many times overkill) and the point now is to use that ability. This will require social, economic, and political skills as much as scientific. If this sounds like an echo of the complacent nineteenth century, the time when the young Max Planck was advised by von Jolly, professor of physics at the University of Munich, to study philology or music because there was nothing left in physics to be discovered, the

Table 12.1   Science and Human Needs: The effects of 200 years of science
(including technology)

NEED	1780 level	1800	1850	1900	1950	1980 level
FOOD Ample supply, wide variety, hygienic water supply	MEDIUM		agricultural chemistry refrigeration	vitamins nutrition		HIGH
SHELTER Warm, comfortable, labour-saving house. Clean, safe, well-lit cities	LOW	gas lighting arc lighting	electric lighting reinforced concrete	neon signs plastics fluorescent lighting		HIGH
CLOTHING Comfortable, durable, variety of colours, materials and styles	MEDIUM		synthetic dyes	rayon	nylon terylene synthetic leather	HIGH
HEALTH Medical services, expectation of healthy life beyond 60 years	LOW		contraception	antiseptics immunization X-rays radioactivity	synthetic drugs antibiotics genetic medicine	MEDIUM
LEISURE Radio, recorded speech and music, photography, cinema, television, games	LOW	photography limelight		cinema gramophone	radio television computers automation	HIGH
TRANSPORT Swift, comfortable travel by land, sea and air	LOW	railways steamship		bicycle motor car aeroplane	synthetic rubber jet flight	HIGH
COMMUNICATIONS Rapid worldwide communication by sound, by sight or in writing	LOW	electric battery	telegraph	electronics telephone	radar transistor	HIGH

N.B. The levels shown are for an affluent individual in an advanced society.

resemblance is only superficial. The essential difference is that today there is next to nothing left to discover, *in physical science*, that will have any real significance in meeting ordinary human needs.

It is necessary to specify the physical sciences because there are surely many important basic discoveries still to come in biology, for instance in the areas of genetics, medicine, and the brain. From the viewpoint of significance to humanity the life sciences still have a long way to go. The Bernal chart of the future will contain a strong element of biology, though it will be increasingly dominated by the science and sociology of production, especially by the applications of electronics and computers.

Suppose that some entirely new physical phenomenon were to emerge, looking as miraculous today as X-rays seemed in 1896. What major unfulfilled human need could it meet, and why is there no hint of it in the imaginative literature of our time? After all, Johannes Kepler and Cyrano de Bergerac were describing voyages to the moon centuries before Verne and Wells, all long before the somewhat anticlimactic landing of Apollo XI in 1969. Atomic bomb explosions were described with chilling realism by H. G. Wells and Olaf Stapledon (amongst others) years before that searing flash lit the sky over Alamogordo. Submarines, aeroplanes, motor cars, genetic engineering, robots, satellites, television and most of the other technical props of our civilization formed part of the future worlds imagined by these writers either before the basic discoveries had been made or, at the least, before they had been applied. Yet when we ask what the modern writers of imagination are envisaging for the near future the answer is, essentially, nothing. Nothing, that is, in the nature of any radical yet remotely plausible innovation. There are fantasies set on other planets, often including time travel, but even in fantasy these seem unlikely to have much effect on the lives of ordinary people. A modern science-fiction writer such as Ursula Le Guin may use an other-world setting but the purpose is still to explore social and psychological problems of this world.

All of this emphasizes the role of science as a social tool for the satisfaction of human needs. That is the wisdom painfully acquired by Brecht's Galileo and delivered during his closing speech when he is nearly blind and under house arrest at his villa in Arcetri:

> I maintain that the only purpose of science is to ease the hardship of human existence. If scientists . . . are content to amass knowledge for the sake of knowledge, then science can become crippled, and your new machines will represent nothing but new means of oppression.[6]

Earlier in the play the young Galileo had argued the case for pure science, asking how we could design the machinery for pumping the water to the fields if we could not study the greatest machinery that lies before our eyes – the machinery of the stars. Today, though, we know enough to design any desired form of irrigation. The problems for the future lie more in such questions as who is to make and sell the pumps? Who is to decide where they go, and who is to use them? Miniature cultural revolutions, such as the Lucas Aerospace workers' attempts to ensure that their work was socially useful, suggest that the crucial decisions are not going to be left indefinitely with the technocrats and accountants who have so dismally failed to keep up with the humane powers of science over the last two centuries.

I have looked at these three aspects of modern science in Bernal's spirit of provisionalism. The ideas that basic physical science is effectively complete, that the age of the great men of science is yielding to the renewed importance of the scientific workers, and that the proper use of science in the future is above all a political question, these are ideas that will be put to the test over the coming decades. In the 1930s Bernal's linking of science with politics aroused great hostility, but it has now become commonplace, even among his political opponents. In 1981, for example, Edward Teller wrote an editorial urging physicists to respond to the leadership of President Reagan's administration by applying their technical ingenuity to problems of national defence.[7] He claimed that astronomy had a widespread appeal, and if it were used to attract young people into physics 'the short-range result may well be to preserve peace and save freedom'.

Teller himself is a prototype of the modern scientist in a capitalist society. He is a living refutation of Bernal's dictum 'In simple words, science implies socialism.' It has now become clear that science does not necessarily imply any particular one of the various forms of allegedly capitalist and socialist societies that now exist.

It could be argued that the individualistic and competitive nature of modern science has made it a better match to a capitalist society which has similar values than to a socialist society, thus explaining the greater productive efficiency of science in the capitalist countries. But science itself is changing, and the values of future sciences seem more likely to be in tune with Bernal's dictum after all. This was certainly the view of Albert Einstein who was saddened to see technological progress often resulting in unemployment rather than easing the burden of work for all. He saw unlimited competition as a huge waste of human potential, crippling the social conscience of the individual:

This crippling of individuals I consider the worst evil of capitalism. Our whole educational system suffers from this evil. An exaggerated competitive attitude is inculcated into the student who is trained to worship acquisitive success as a preparation for his future career.

I am convinced that there is only *one* way to eliminate these grave evils, namely through the establishment of a socialist economy, accompanied by an educational system which would be oriented towards social goals.[8]

Bernal's study of science in history is a monumental contribution to solving these wider problems. It has become an essential foundation both for understanding the changing role of science in society and for working towards that higher civilization in which his faith never wavered.

Notes

1. J. D. Bernal, *Science in History*, London C. A. Watts; 1st edition 1954, 3rd edition 1965, illustrated 3rd edition 1969 (also published in Pelican Books).
2. Ibid., p. 5.
3. J. D. Bernal, in M. Goldsmith and A. Mackay, eds., *The Science of Science*, revised edition, Harmondsworth, Penguin, 1964, p. 288.
4. In B. Brecht, *Selected Poems*, translated by H. R. Hays, Grove Press, NY, 1947, p. 109.
5. B. Brecht, *Collected Plays*, vol. 5, edited by R. Manheim, and John Willett (eds.), Vintage Books, NY, 1972, p. 422.
6. B. Brecht, *Plays*, vol. I (*The Life of Galileo*, translated by Desmond I. Vesey), Methuen, London, 1960, p. 330.
7. E. Teller, in *Physics Today*, February 1981, p. 136.
8. A. Einstein, *Monthly Review* (USA), 1949.

13

Building Tomorrow

Chris Whittaker

When he was at Cambridge, Bernal was renowned for the wide range of his reading. Church architecture in Romania in the sixth century was the subject of one book he was seen to be carrying.[1] Thirty years later, Francis Aprahamian remembered him chatting knowledgeably on the subject when they visited that country together. But at the time, he may have thought that he was extending his horizons too far. On 9 May 1921, he wrote in his notebook:

> ... the last day of my twentieth year. My character is childish and vacillating, all has been subordinated to my mind and it has nearly choked itself with useless knowledge. Two, three years ago, I could think widely, fearlessly, happily; now, full of the cares of life, of the ills of my body, thinking is becoming a task and learning an impossibility. In a fortnight comes the Tripos finding me already stale, unprepared, much forgotten, much never learnt, yet without ambitions for a first and knowing well the blow a second will mean to me.[2]

He did get only a second. But by the time that he wrote his first book, in 1929, he had certainly got back his youthful buoyancy. Probably by then he had also had his first discussions with progressive architects and artists of the time.

The World, the Flesh and the Devil, subtitled *An Inquiry into the Future of the Three Enemies of the Rational Soul*, naturally showed an intimacy with the more abstract architectures of the crystallographer. He used these in the first chapter of his 'essay in prediction' to explore yet more distant horizons, those of space travel. He took half a dozen pages to speculate on how a space station for 20,000 to 30,000 people might be constructed:

> ... the essential positive activity of the globe or colony would be in the development, growth and reproduction of the globe. A globe which was

merely a satisfactory way of continuing life indefinitely would barely be more than a reproduction of terrestrial conditions in a more restricted sphere. But the necessity of preserving the outer shell would prevent a continuous alteration of structure, and development would have to proceed either by the crustacean-like development in which a new and better globe could be put together inside the larger one, which could be subsequently broken open and re-absorbed; or, as in the molluscs, by the building out of the new sections in a spiral form; or, more probably, by keeping the even simpler form of behaviour of the protozoa by the building of a new globe outside the original globe, but in contact with it until it should be in a position to set up an independent existence.[3]

These imaginings have since been realized for the first few, less than seventy years after he wrote. Whether they should be extended for the tens of thousands whom Bernal foresaw in outer space is one of the most important (and potentially energy-consuming) questions to be asked today.

Certainly Bernal was foreseeing the molecular and genetic engineering of today when in the same chapter he wrote of:

... when materials can be produced which are not merely modifications of what nature has given us in the way of stones, metals, woods and fibres, but are made to specifications of a molecular architecture.

... a world of fabric materials, light and elastic, strong only for the purposes for which they are being used, a world which will imitate the balanced perfection of a living body.[4]

The tented tensile structures of the engineer Frei Otto and his followers have brought us to the threshold of the Millennium Dome; but at the scale of the town, we have yet to begin.

The World, the Flesh and the Devil set a pattern for all Bernal's writings: the bringing together of far-in-the-future possibilities on the one hand, and on the other the discussion of those obstacles to taking the first attainable steps in their direction. It is a book of exciting predictions in some of the fields of his own future work, a book to which 'I have a great attachment,' he said, thirty years after writing it.[5] It did not, however, look closely at the built environment; or at politics, except for warning that civilization might be brought to a crisis by 'its failure to arrange secondary social adjustments'.[6]

The next book in which he had a hand was more forthright and down-to-earth: down-to-earth to the point of burying a whole ruling class. *Britain without Capitalists* was written by an anonymous group of economists and scientists.[7] Bernal contributed a long chapter on building, taking in building materials, housing, town planning and industrial location, with detailed criticism of the architectural profession. To read this chapter for the first time today gives rise to a hollow feeling. Although a horizon of opportunity was opened up, little of it has been reached half a century later. The network of relationships between the professions, clients, industry, suppliers and bureaucracy is as tangled as ever.

In the 1940s and 1950s, new, alternative steps were taken, particularly by client-led consortia; this chapter was an opening effort to formulate and clarify them. Bernal indicts, with a wealth of quotation and reference, the appalling housing conditions of the day (four of the old inner London boroughs with over 19,000 families at more than two to a room); there is 'soul-killing ugliness' in the old streets;[8] classrooms are 'miserable, bare, bleak and uninspiring'. 'The ugliness of the majority of buildings being erected today is due to the failure of the architectural profession.'[9]

Technical opportunities at that time were thwarted by legal restrictions and by vested interests. 'Standardisation of methods and materials is prevented by the multiplicity of firms and the smallness of individual orders. Research and experimentation are virtually nonexistent; and where research organisations exist, they scarcely touch the problem of costs at all.' The lack of planning of building programmes caused considerable waste. Fluctuations and uncertainty led to unemployment of building trades operatives. Neither the operations themselves nor the manufacture of components or materials could be a balanced process with a steady load throughout the industry.

Social-cost auditing was an unknown discipline in the thirties; but he attacked the lack of it. It would be necessary to employ sociologists in planning teams at the earliest stage in order to avoid 'stunting a generation and ruining the health of millions'. Bernal saw local control as the proper way to guide the industry to meet social needs: 'it will be desirable to have conferences with tenants' committees.' This was a very long way ahead to be pointing. It is only in the last few years that democracy has, sporadically, got this far.

Bernal did not presume his way into detailed matters of design – 'the question of architectural aesthetics is much too controversial a one to be discussed here' – but he was clear that 'there exist endless opportunities for cost reduction in the manufacture of building fittings, not implying cheap design but simple straightforward good design, studied in relation to the manufacturing process and not on some antiquated handwrought model'.[10] This was something that Walter Segal was already achieving, perhaps not very far away, and sustaining throughout his life.

The wearing effects of ugliness were repeatedly mentioned. In school design, starting with examples being built at the time when *Britain without Capitalists* was being written, one can see a continuing achievement throughout the postwar decades. I feel sure that Bernal must have been in conversation many times during the mid-thirties with the pioneer designers. The 'Building' chapter refers to the Bauhaus at Dessau and to Walter Gropius, one of its founders, who was by now designing in Britain as well as contributing to *Circle*, the journal that presented the Modern Movement to a British audience. Bernal's promotion of simple, straightforward good design is in a direct line from the teaching of Gropius at the Bauhaus that standardization could make available excellence for all (as it did, for the first time in this country, when Utility furniture was produced after World War Two).

Standardization has had a chequered reputation, especially in housing. Andrew Saint, who deals fully with the common ground between scientists and architects from the thirties on, in his excellent book *Towards a Social Architecture: The Role of School Building in Postwar England* (1982), criticizes the utopian science of H. G. Wells and Bernal for romanticizing industrial production and ideas of prefabrication. At the time, however, for Bernal himself, utopianism was a delusion. He writes of the town planner working in Moscow:

> ... these tremendous possibilities do not mean however the existing city must not be reckoned with. On the contrary, it must be very carefully studied and transformed. The town-planner would be no Utopian schemer working out ideal cities in his head, but a realist who had the power and the possibilities of changing the life of the people.[11]

Again, a few pages later, his careful qualification at the end of this next quotation is noteworthy:

> Only when the designer is unhampered by the limits of proprietary brands, price rings, monopolies and out-of-date byelaws dictated by vested interests, will he be able to do justice to the tremendous services that will be required of him. Even the wildest flights of a Corbusier could become realities in so far as they correspond to the needs and desires of the people.[12]

That is, the users of the buildings and the citizens within the public spaces needed to be involved. Utopianism was as much a danger for Bernal as for his later critics. But the difference is that he was creatively aware of the origins of where we stand, of the possibilities at the smallest and at the wider levels; and he was ready to set out a

programme on many fronts for taking the first steps towards a new
world.

This was his vision at the end of the 'Building' chapter:

> Instead, in a Britain without capitalists – bright clean towns, wide streets and
> safe crossings, no smoke. The houses, spacious and beautiful, set in a belt of
> trees, gardens and playgrounds outside every door. Schools within easy reach;
> hospitals no longer scarce, deficient, crippled by dependence on sparing
> private charity; no longer cramped smoky factories, hurling soot and fumes
> on to the workers living around. The housewife's task lightened and her
> time freer. No longer, with properly planned towns and industries, an
> arduous task to get to and from work. All the possibilities of a full and
> developing life found in the towns to be had also in the country; the
> agricultural worker possessed of similar opportunities to those enjoyed by
> the city capitalists. Such is the contribution which a building industry,
> reorganised, modernised, freed from vested interests and unplanned mono-
> polistic profit-making, has to make towards the future of this country.[13]

It is a vision worth repolishing today: some of the smoke has abated
but we have acid rain and exhaust fumes instead. Factories – many of
them – are no longer cramped and smoky because they have ceased to
exist. It is still an arduous task to get to and from work, normally
hazardous on the roads and once in a while lethal by suburban train or
underground. But worst of all, few of us, whether planners, poets or –
might it be? – crystallographers, have the time or genius to put it all
together and make an argument that will spur the rest of us on.

How did it come about that Bernal got to the centre of urban needs
and to the changes that had to be made? There will have been
approaches from those who wanted to establish bridgeheads between
scientists and architects; but sometimes nothing so formal was necessary. Berthold Lubetkin remembered:

> For a while in 1935 Bernal and I both lived in the same house in Gordon
> Square. It was not easy to conduct sustained conversation since the four-year-
> old son of the landlord was constantly being psychoanalysed and obviously
> disapproved vocally of this experience. However, on occasions, Desmond
> Bernal visited me on the way down from his den upstairs. Glancing at
> architectural periodicals on my table, he expressed concern about the
> prevailing dualistic approach to architecture. The emphasis on planning,
> fulfilling most of the utilitarian functions such as circulation and services,
> attracted his attention and cautious approval. But he was also aware that
> design requires in addition the imprint of human decisions. Reducing
> everything to the pure technical aspect contained dangers for the future. In
> addition to efficiency and fitness for purpose, architecture must convey a
> promise, an appeal, or a warning.

The nineteenth-century architectural muddle and eclecticism were the result of a dualism between the purely measurable and material characteristics of buildings and the 'spiritual quality' which then needed to be incorporated into them to render the stark structures more acceptable. But this could not help the degrading quality of architecture. 'Such dualism, polarity of reason and emotion, individual and society, inevitably split the integrated unity of calculable aspects of reality from the mysteries of aesthetic appreciation as a separate entity.' He took, as an analogy, dualism in philosophy. The attempt to banish emotions or to limit their scope and the desire to reduce reality to pure integration of mechanical particles could not explain the history-laden influence of custom, practices, preferences, which sustain our choice. Finally, instead of abolishing such vague spiritual values, it became necessary to reintroduce them, having endowed them with a separate existence. The attempt to exorcise such values from nature and reality simply opened the back door to a host of allegedly banished immaterial experiences.

He saw the main danger of architecture at the time in an expansion of formalism which the functionalists claimed to have rooted out once and forever. By denying that direct sense perception is but a precondition of knowledge, we are in danger of confusing sensations with a sensationalism that creates an architecture of appearances – which is but an appearance of architecture. In following this road, architecture might easily degenerate into a meaningless stylistic quarrel following current fashions.

On several other occasions when I spoke to D. B., he was at pains not to emphasize too strongly the purely functional side of architecture. He was very pleased when I reminded him about Mayakovsky's example – talking to trade union representatives, he stated that in his speech, he is producing an electric current of some 30 microwatts, but if he read, on request, his poem to the Sun, it would lift the roof above.[14]

Lubetkin was working, only a short walk away, on projects with an important social purpose and in a style that gave them his strong personal stamp. Further, his atelier was the centre of gravity of the Modern Architecture Research group (MARS), the English offshoot of CIAM (the Congrès Internationaux d'Architecture Moderne). Late in 1937, MARS held its first exhibition in London, showing its plan for the city. The skeleton for the plan was a number of north–south high-density units, about half a million people in each, linking to a metropolitan spine along the Thames, these units separated from each other by large parks. The plan had a strong impact on the architectural and town planning professions before and after World War Two and can still strike an echo in their consciousness.

Today, the land to the north of the King's Cross and St Pancras termini is being fought over by competing development and community interests. It is perhaps the last place in inner London where decayed old uses might have been replaced by the healthy structure that the MARS plan demonstrated.

In the same year as the MARS exhibition, an article by Bernal on 'Art and the Scientist' was published in *Circle: International Survey of Constructive Art.*[15] The editors were Ben Nicholson, Naum Gabo and Leslie Martin, a young architect who was later to become Chief Architect of the London County Council. The group involved in publishing *Circle* helped enable large numbers of European modernist artists and architects to find both refuge and creative opportunities in England. In the years before the war, London was the centre for that movement.

Jeremy Lewison, author of the catalogue of the 1993 Ben Nicholson exhibition at the Tate Gallery, evokes their aims:

> Constructivist artists regarded their art as restoring the values of society. Order within modernist architecture and geometrical art implied social and moral order at a time of growing disorder and fascism in Europe. The new transparency in architecture permitted by the combination of reinforced concrete and glass ... created a sense of openness symbolic of truthfulness and morality.

Both the modernists' designs and what they were campaigning for were part of the same struggle.

Bernal's article showed that it was modern engineering that had planted the seeds from which constructivist art had sprung, and that these artworks in turn led to mathematical study and insights. He referred to the work of both Barbara Hepworth and Ben Nicholson ('curiously reminiscent of the equally formal stonework of the Tiahuanaco culture'), and finished by calling for artists and scientists to end their isolation both from one another and from the most vital part of the life of their time: 'There are no ready-made solutions but if the goal can be seen the way to it will be found.'

Barbara Hepworth, who had been introduced to Bernal by Margaret Gardiner, had a show at the Lefevre late in 1937 and Bernal contributed a foreword to the catalogue. This was 'not an aesthetic criticism'; instead, it discussed 'the relation of an extremely refined and pure art form to the sciences with which it has special affinities'. Bernal pointed to the kinship with the Neolithic of the second millennium, although 'it is never Hepworth's intention to recreate the neolithic'. The work calls for some form of social utilization:

Megalithic art was not aesthetic in intention, it represented the centre of a ritual which must have been so important in its time as to absorb the greater part of the free energies of its creators. If such art is to be of use now it needs to find the same public setting. Its geometrical character does in fact bring it immediately in relation with the developments of modern architecture. Its forms require to be combined integrally with those of buildings to which they would give a completeness that is at present lacking. So far this integration has not taken place, but one step to it might be the use of such pieces as shown here in modern domestic architecture and it is to be hoped that the present exhibition will lead to an appreciation of this possibility.

Publicly, Bernal steered clear of aesthetic criticism, although he was closely in touch with Hepworth and other artists. Privately, he was ready to make suggestions. Hepworth had written to him thus:

When you criticized my carving, you were quite right, sculpturally, spatially and constructively; but enough about myself. I really should be so interested to know where you think your idea of constructive art differs from ours. The biggest criticism that is levelled against it is that it is limiting. But it seems to me the freest possible thing – unlimited in its scope; your idea of regular irregularities seems a proof of this. You always seem to me to be searching for, discovering and applying basic laws and principles (whenever I have met you); I know no one more fitted to make those laws clear to other people.[16]

By this time, Bernal was the protagonist for the modern art movement and for architecture within that movement; for architecture as the outward face of the building industry and for the bringing of architecture into the scope of scientific knowledge and rigour. Earlier in 1937 he had spoken at the Royal Institute of British Architects (RIBA). E. J. Carter was the institute's librarian at that time. He remembered:

It was largely through the initiative of Justine Blanco-White [the wife of C. H. Waddington] that Bernal was asked to read a paper on architecture and science at an informal meeting at the RIBA. The audience was not large and was mostly young. The paper, somewhat enlarged, was published in the *RIBA Journal* of 26 June 1937. At that time science in any form of debate, theory or practice had little part in the affairs of the institute or in the practice of architecture, although the RIBA had its formal association with the Building Research Station, and Sir Raymond Unwin, the president of the RIBA, was chairman of the Building Research Board and eager to support any development of scientific understanding and practice in the profession. As many of the younger members were beginning to understand, the profession was wide open for just the stimulus that Bernal, more than anyone else, could give. It was also significant that the impetus came most vigorously from the left.

In the last paragraph of his paper Bernal summed up his general intention:

> What I have tried to show in this brief survey is that architecture and science are not two exclusive disciplines, that neither can fully flourish unless it retains a living contact with the other. In the formal, the functional and the structural aspects of architecture, science can point the way to new processes, new materials and new arrangements. It is for architecture to use these and to combine them into a new tradition. In its turn, science stands to gain by the widening of its field of enquiry and by the appearance of new problems to solve. Finally, since both architecture and science depend for the fulfilment of their latent possibilities on the development of a state of society compatible with that realisation, their interests are jointly involved in securing it.

Bernal had already sketched what these latent possibilities were:

> The modern large building needs to swim delicately on the vibrations of the earth. . . . It might ultimately be possible to make most of the living parts of houses completely out of air, so that people in them could enjoy all the advantages of being in the open air, without any of the inconveniences. . . . In future houses must admit of far greater possibility of alteration with the seasons and with the whims of their inhabitants. The possibilities of doing this easily and cheaply are latent in the new materials and processes that science could give to architecture.[17]

He ended by pointing out that we shall never get good buildings unless we can make good towns to put them in. Towns in 1937 already needed to be either underground or so heavily armoured against aerial attack as 'virtually to be underground' and so 'it will not be sufficient to have good architects and good scientists. It will also be necessary jealously to preserve and extend peace and liberty.'

Bernal himself had been intimately involved since September 1935 in studying and exposing the threat that town dwellers face from the likelihood of war. In 1934, the Cambridge Scientists' Anti-War Group had been formed (see Chapter 7). After a Home Office circular on air raids was sent to local authorities, the group had carried out a series of tests of their own, which showed that the government's reassurances on gas masks, on dealing with incendiaries, and on gasproofing rooms were seriously open to question – while, furthermore, having to gas-proof a room in one's house or flat would make it not normally habitable and so put a further 7 million people into overcrowded conditions.

The group's findings were published by the Left Book Club in 1937. Some of the experiments described sound now a bit *Dad's Army*ish – 'the thermite was arranged in such a position that water could be poured over it from a height; a bucket being balanced on the overhanging branch of a tree'[18] – but since the official tests for gasproofing had been performed on discarded telephone boxes, they were enough.

Bobby Carter's recollection, quoted above, continued,

> Bernal never let an opportunity pass to enlarge his knowledge of the air-raid precautions scene. I remember being with him on Hampstead Heath in 1937, I think, when the LCC had organized a grand late-night festival of fireworks and music – Handel's *Water Music* and the *Music for the Royal Fireworks*. When the fete was over and the tens of thousands of people were leaving, Bernal stationed himself at a small bridge over a stream on one of the exit routes to time and analyse the social and technical problems of the passage of crowds through narrow openings to air-raid shelters.

In November 1937, the *New Statesman and Nation* brought out a supplement on air-raid precautions in which Bernal tore apart a government bill on the subject. 'Once the Government, by their own propaganda, have roused the country to a realisation of the air danger, the public will demand more adequate protection and demand that it should not be paid for by the people who can least afford it.'[19] Bernal's estimates of costs, leading to a gross figure greatly above the government's, were derived from an estimate of ten shillings per household to give blast and splinter protection.

Lubetkin, Freddie Skinner and others from the MARS group, in practice together under the name Tecton, produced a book published by the Architectural Press in early 1939, *Planned ARP*. Bernal reviewed it in the *New Statesman and Nation* of 25 March 1939:

> It is a sad commentary on our present rulers that the first realistic discussion of a problem of such vital importance as air-raid precautions should not come from the Government departments concerned, considering all the facilities and information that are available to them. But it is at the same time happy arguing for Democracy that the task the Government has failed to do is being done by public-spirited citizens and municipalities. 'Planned ARP' marks a second stage in the battle for effective protection for the civil population. Professor Haldane made it impossible for the Government to continue with a policy of providing only brown paper refuge rooms. In a book which is a model of popular presentation, Messrs Tecton now show with clarity and precision the character of the shelters needed to give any required degree of protection. They have been enabled to do so because they could bring together the resources of a modern and intelligent architectural firm with those of a far-seeing borough council.

The result is that 'Planned ARP' embodies in itself a most theoretical account of the principles of ARP and the exact and practical application of those principles to a typical urban area. The Finsbury plan cannot be ignored even by the Home Office and now that it is available to a large public it will be difficult for any local authority to present to its citizens anything that falls far short of the standard it sets.

The great value of the Finsbury scheme is its thoroughly scientific basis. Starting with a reasonable forecast of the character of modern air warfare and a knowledge of the area, secured by admirable and detailed surveys, it sets out logically to solve the problem of how to give the maximum protection to the citizens at the minimum cost and in the shortest time. To solve this problem a new principle has been introduced, that of the danger co-efficient. If they had done nothing else, the authors of 'Planned ARP' would have made here a vital contribution to all air defence problems.

While he was testing and exposing the soundness of official preparations for war with its dreadful impact on citizens at home, Bernal, as his RIBA address showed, was at the same time urging a deeper coming together of scientists and architects. In January 1938, for instance, Bernal outlined a scheme for 'Scientific Research for Britain' in the journal *Nineteenth Century*, in which he proposed, *inter alia*, research bodies for transport, consumers, cookery and domestic engineering, and research associations for civil engineering, pottery, glass and woodworking – many of these foretelling, for example, the Consumers' Association, founded in 1957. Again, in May 1939 he moved the vote of thanks to Serge Chermayeff, a brilliant executant of the Modern Movement, after he gave a paper at the Architectural Association in Bedford Square. Bernal argued persuasively that:

> architectural students enjoy something which no other student body possesses, in that they have a possibility of control in their own hands. This relates them to the beginning of learning in the West ... when students determined their courses and chose their masters. This gave the possibility of having a real working democracy, because democracy is not confined to the ballot box but is something which should permeate into action.

In June 1939, a month after supporting Chermayeff, Bernal spoke at the School of Planning and Research for National Development (an offshoot of the Architectural Association on the other side of Bedford Square) on 'The Influence of Scientific Discovery on Developments in Physical Planning'. 'I am afraid that I am probably one of the least qualified people to talk to a school of planning,' he began. He called for a setting-out of human needs, not just for food and drink but in terms of communication and transport, so as to give physical planning its foundation. He foresaw energy becoming 'a common property as air is or as water has almost completely become', and pointed out that machines were already available to handle the mathematical analysis of the large number of variables that planning had to manipulate. He used an aside to illustrate how developments in war provide a later benefit: 'for example, the domestic use of cast-iron pots and fire-backs were a result of a desperate attempt of the Sussex iron industry to recover some use in the days of James I for their products when cannons were no longer needed'.[20]

By the eve of the war, the magnetism of Bernal's analysis, his vision and his advocacy were creating their own circle of younger architects who were applying his approach and campaigning for it under their own steam. The journal *Focus*, first produced in 1938, was edited by the architect Anthony Cox. For two years it became the radical voice of the younger members of the profession. Cox opened the first number with 'The Training of the Architect', a challenge to the establishment and a call for research, as scientists applied it, to be central to designers' skills and a foundation for their work. Bernal's advocacy of the previous three or four years was starting to be adopted. *Focus*, no. 4 (the final issue before war closed the journal down for good) included 'The Present Position in ARP' and a review by E. J. Carter of *The Social Function of Science* which had been published earlier in 1939.[21]

This book was the culmination of Bernal's attempt to bring science, its discipline and clear-sightedness, into the arena of public discussion. It embraced building, town and country planning and housing. In the chapter 'Science in the Service of Man' he dealt with energy conservation and with new materials, especially lightweight cement materials, an interest he sustained throughout the war and after.

He stressed that to be able to advance technique, social conditions would have to be changed. Fifty years and more later, his prophecies have yet to be fulfilled:

It will soon be possible to break altogether with the tradition of putting stone on stone or brick on brick, unchanged since the time of the Pharaohs, and to move in the direction of rational fabrication.... The totally enclosed, spacious air-conditioned town is rapidly becoming a practical proposition.... A well-designed ventilating system, together with the prescription of the liberation from all forms of dust, smoke, gases or vapours, should make such city air indistinguishable from that of the country ... (with) adequate concentration of urban construction to leave far more space to wild nature.[22]

With demand still increasing today for products that consume rain-forests and for houses that consume greenbelts, Bernal's 'rapidly' has to be qualified by grimmer prophecies. Already, the Baltic beaches and deeps have become choked from inadequate sewerage schemes. Yet Bernal showed himself to be aware even then of these dangers: 'we are throwing away an appreciable proportion of consumable materials and doing so in a way which destroys the amenities of country and town alike. To a large extent this is a problem of social organisation and control.'[23] The emergence of green political pressures in the late eighties shows that there is a fairly straight line connecting Bernal's analyses to the environmental campaigning of today, even though this is still mainly limited to specific issues, such as whales or rainforests, rather than social organization or control. A political leadership that thought it knew what was best on this whole class of issues, and acted on it, might be as frightening as others we have had to endure: unless there were a truly democratic involvement, which Bernal repeatedly stressed as being a necessary first move.

Here again, Bernal dealt trenchantly with the utopianism with which others had tried to tar him:

All Utopias present two repulsive features – a lack of freedom consequent on perfect organisation, and a corresponding lack of effort. To be a citizen of a modern Utopia, the critics feel, is to be well cared for, regulated from birth to death, and never needing to do anything difficult or painful. The Utopian seems, notwithstanding his health, beauty and affability, to partake too much of the robot and the prig. Fairly envisaged, it seems hardly worthwhile sacrificing much in the present if this is all the future has to offer.[24]

The outbreak of war in September 1939 caused a sharp increase of intensity in his many activities for and with architects and planners. His involvement became virtually full-time. 'I am right out of science,' he said.[25] But although this sounds regretful, it was a move of his own making. Shortly after his paper to the RIBA in 1937, the institute had formed an Architectural Science Committee: Bernal had been a member and the RIBA President, Sir Raymond Unwin, an active chairman. By December 1939, an Architectural Science Group had been formed, bringing together the Building Research Station (BRS) and the RIBA's War Executive. The Godfrey Samuel papers in the institute's library show that after the initiating meeting of thirty-six people, including Bernal, five committees were formed which, by a very early stage, had held a total of eighty meetings.

The hectic pace continued. Bill Allen, then a young scientist at the

Building Research Station (later principal of the Architectural Association School), remembers being asked to come to a meeting of the Tots and Quots dining club, as described in Chapter 7. It was held Au Jardin des Gourmets in Soho on 12 June 1940. (The invitation came from Bobby Carter, who by now must surely be the *éminence grise* of this whole story.) This gathering famously led to the twenty-seven scientists present writing a Penguin Special, *Science at War*, in a matter of days and Allen Lane publishing it a week later. Bernal, with Bill Allen and Alec Skempton, a civil engineer, wrote the chapter on the building industry, pillorying the wastage still widespread and calling for the Building Research Station and scientists generally to be closely brought into planning and building technology, and particularly into the study of the use of buildings.

The book scored immediately at two levels: it sold 120,000 copies at sixpence each and, while still in galley proof, was put before a member of the Cabinet, who met with Bernal and Zuckerman as editors. Shortly before Zuckerman died, he remarked in conversation that that meeting of the Tots and Quots had been one of the turning points of the war, in bringing science into the front line. (Bill Allen also remembers a phone call from Carter in October that year. The club had met in Hampstead the night before. One of a stick of bombs had fallen in the garden and shaken the house. The club members resolved never to come together again: too much of the UK scientific effort in the war was by then in their hands.)[26]

During the next few years Bernal continued to pour out interventions on architectural and planning matters, in addition to all his other work. In May 1941 he contributed a long note on 'The Place of Science in Architecture'.[27] In June, he wrote a much longer note on 'Research Organisation in the Building Industry' for the 1940 Council ('a council to promote the planning of social environment').[28] It covered the importance of research, a programme for getting at consumer needs, the industry's capabilities, and the costs implied: it called for the setting up of a building research council. This note was completely redrafted in July, under the names of E. J. Carter, Bernal and Mr Ellis, with the title, 'Science and Building Construction'.[29] After this came a much more detailed and longer memorandum, 'On a Central Body to Coordinate Research in Building'.[30]

In February 1943 he participated in an official RIBA/BRS publication and exhibition 'Rebuilding Britain', which he used to ask for a research body to survey the social, economic and educational fields to see what reforms were most needed and then to try to influence the professions and the public to carry them out. His note, probably written at the height of the war, is revealing for mentioning that the functional requirements that any building has to meet, although most explicit in such as a factory, are 'just as necessary domestically in provision for cooking, cleaning and other work'.

On the suitability of buildings for social life, he argued that 'the degree that they help to create positive attachment and pride in their users' needed studying. Bernal referred to the increasing but still vague appreciation of what good architecture can do for its users. To make that appreciation 'precise and effective, the architectural profession needs to develop an enterprising and modern public relations service'. In many of the postwar planning studies of the 1950s and 1960s, serious social surveys and analysis prefaced the proposals; but it is only recently and mostly on the smaller scale of the neighbourhood about to be facelifted, that citizen participation has been allowed to interpose itself.

After the war, Bernal summarized what had happened during those wartime years in a paper on 'The Organisation of Building Science Research' given to the then Architectural Science Board in March 1946.

> The original members of the Architectural Science Group were far-sighted enough to see that, if the practice of building was not to lag permanently behind all other industrial practices, it would need to be supplemented by research and by the scientific education of architects to enable them to profit to the greatest extent from the results of that research. At that time there could be no immediate prospect of actually carrying out the research, but those discussions on what research was necessary which were carried out in the spare time of men already overburdened with war work, have proved of enormous value. The present organisation and the future programme of research in building undertaken by the Ministry of Works and the Ministry of Health is very largely the direct result of the proposals of the Architectural Science Group.[31]

Throughout the war, architects and others had been daring to sketch out postwar reconstruction plans. By 1945, these could be looked at more confidently. In March, Bernal was asked to chair the Scientific Advisory Committee of the Ministry of Works. 'I am grateful for your taking on this important and urgent task and I hope that you will be able to devote to it a reasonable amount of your time,' wrote the minister, Duncan Sandys.[32]

The need for Bernal to organize his own time must always have been paramount, between these demands and his pure-scientific and other activities. Later in the year, he spoke on the organization of building,

particularly of housing prototypes, which were mushrooming. The
Minister of Health was responsible for housing in this period, and
Bernal said that, as the minister had admitted, 'he threw onto me at
very short notice this terrible task, which had baffled all architects for
years, and which it is certain will get me down also, of studying every
kind of the 1,300 varieties of prefabricated house which have been put
forward, deciding which seem the best, and pulling all the ideas
together'.

> When you come to do that you do find that science is of some little value.
> For instance, if instead of prefabricated houses we were dealing with insects,
> we should find that ordinary people, looking at insects, would say, 'There
> are an awful lot of them; you cannot make sense out of them.' In the course
> of time, however, scientists have found classificatory methods of dealing with
> something like 20,0000,000 species of insects, so that a mere 1,300 prefabri-
> cated houses is a comparatively simple business![33]

Obviously, building was still the same fragmented industry that he had
described in 1936. Bernal, who had dealt with human requirements,
external cladding, improved bricklaying methods and the experimental
building programme, was 'talking as an amateur in building science'.
He saw that the best that could come from appraising the 1,300 choices
would be a Mark I; it would take years, time that was then not spare, to
develop better houses. None the less, some of these Mark I prefabs
have been demolished only in the 1990s after forty years' occupation;
with not much rehabilitation during their lifetime and with sufficient
regard for them by their tenants. Time has shown that their constructional and space-for-living standards were soundly conceived and carried through, both within the 'units' and in their gardens and
neighbourliness; being put down often on inner-city sites after war
damage, they made homes where they were wanted.

In 1945, Bernal would have been quite familiar with the ideas and
details of prefabrication from the early Gordon Square days. *The Modern
House* by F. R. S. Yorke, first published in 1934, had a final chapter on
prefabricated systems, with examples and work-in-progress photographs.[34] By 1944, the fifth edition included a number of new systems.
The book would have been a vade mecum for the architects with whom
he had collaborated in those days (and incidentally it is as refreshing
to study today as it was just after the war).

Initially, twelve groups of fifty traditional houses each were erected;
and these were timed and costed and compared with 21 groups of fifty
prefabricated houses. Living in the houses was also to be studied, to

find out the relative advantages of different forms of heating and cooking. One of the house types in which Bernal took a particular interest was the Riley house, of which an experimental group was built at Wythenshawe in Manchester. This house had a frame of steel stanchions with lattice floor beams. Although he had made a deep study of concrete, steel appealed to Bernal – it should always be used in a way that would allow it to be recycled, he said: a sound attitude to conservation, even if it had been prompted by postwar shortages.

When the eighty Mark IV Riley houses came to be built, other aspects of real life kept breaking in to threaten the smooth organization of production. The go-ahead had been given on the eve of the freeze-up, suppliers were in the midst of power cuts, and skilled tradesmen were hard to find. But finally a letter from the promoters reported to Bernal, 'the tenants were highly delighted'.[35] Bernal himself may well have regarded this report with some scepticism. As he pointed out in a talk he gave in 1946, 'Our only information is based on one sample survey of fifteen housewives. It is probable that we know far more about the domestic habits in the Trobriand Islands than in the London suburbs.' He ended: 'We cannot advance just by telling people about Science, but by getting them to be scientists themselves.'[36]

On one of his wireless talks (he gave eight that year) in the Home Service 'Science Magazine', on 30 September 1945, Bernal discussed what kinds of houses were needed, what they should be built of and how it should be done – which would have to include the full involvement of the building workers. He regarded them as a most fruitful and neglected source of innovation, even though 'building workers often had, in the past, to put up with bad conditions of cold, damp, bad feeding and general discomfort. Such conditions are not only unjust, they prevent good work. . . . The dangers, particularly the health and accident dangers, in the building trades, are also being looked into': he made clear that there needed to be no fear of unemployment following improved productivity.[37]

On building materials, he combined far-sighted and energy-conservation-conscious ideas with the organization of up-to-the-minute research and discussion. Early in 1946 he spoke again to the RIBA, on 'Science in Architecture'. Asked about the possible merits of a large hollow insulated brick, he replied, 'I should not hold with it, because we waste a great deal of coal in making a high-temperature material. If possible, we want to use no coal, or very little, in making any building materials, the reason being that building materials are needed in such vast quantities that they use up an enormous amount of coal.'[38]

In May 1946, arising from a suggestion by Bernal, a symposium was held at the Royal Institution on the Shrinkage and Cracking of Cementive Materials. He had put the idea to the Roads and Building Materials Group of the Society of Chemical Industry which 'seized with both hands' the opportunity of organizing the gathering. Bernal wrote the introduction to the five papers although not present himself (he was in the USA for the Ministry of Works). 'Concrete is nightmare to the fundamental scientist – it has a gross structure, a colloidal structure and several different crystalline phases.'[39] Some years before, James Jeffery had noticed Bernal on a concrete path between the buildings at the Building Research Station. The path was cracked across every ten feet or so – why? The question became grit in the oyster, until the Royal Institution symposium could move towards more wakeful knowledge, relying partly on his own discipline of X-ray crystallography.

Over and above any one subject needing investigation, Bernal evolved an organizational plan for the spectrum of researches that would help building and planning to be more effective. A year after his appointment, Bernal gave a paper to the Architectural Science Board on 'The Organisation of Building Science Research'. As well as studies of building materials, he called for user surveys involving ordinary tenants, so as to assess, for example, how they would respond to central heating; for studies of the most efficient interior plans and their equipment, for work study of traditional and experimental houses on real building sites; for social exploration of what it was like to live in old-established neighbourhoods; for a Human Efficiency Panel to look at training and working conditions for those on site; and for research into building economics and finance.

Substantial results were to come from this plan in the next fifteen years or so. The first were already on the drawing board in Hertfordshire, where the County Architect, C. H. Aslin, brought together a team with a vision that overlapped very largely with Bernal's own. There was a Research and Development Team, the first in a local authority with a major building programme, to study both the form of the county's schools as a whole and the form of their individual components. The designers worked with the teachers. They could also bring their wartime study and experience to bear fresh fruit. Oliver Cox recalls that Zuckerman's studies of soldiers was turned to as the basis for the design of new children's chairs.

The new schools were built at a fierce pace, of steel frames with an external cladding of precast concrete panels: no bricks, no hand-over-hand labouring. After three years, the main inspirer of the team, Stirrat

Johnson-Marshall, became Chief Architect at the Ministry of Education and was to establish a Research and Development Group which provided thoroughgoing research, analysis and model solutions, usually executed in working schools so as to test the new ideas, for more than a generation to come.

Andrew Saint gives the history fully, derived from the recollections of the participants. 'They were the happiest days of my life,' remembered Cleeve Barr, one of the architects who joined the Hertfordshire team in 1947. Saint adds:

> For these young men, eager to get on at last with their careers after the frustration of war, to create instead of destroying, these years set an abiding ideal of what architecture should be all about. The emotion of the period was tremendous. They were, in a manner of speaking, patriots fulfilled.[40]

Bernal's thinking and advocacy had shown the way to that ideal. He had helped to transform the techniques of planning and design by calling for the brief to be based on society's needs; and his unifying approach led to the design of new building types rather than merely components. He was still ahead of many of those at the drawing board. At the end of a talk he gave in 1946, he remarked on the need for democratic control in new towns: '. . . there should be no distinction between the planners and the planned, when both are at the same time the observers and the observed in the scientific experiment.'[41]

Housing production in the public offices altered course also. The London County Council (LCC) took responsibility away from its Valuer (where plans from the thirties had been dusted off in the first postwar rebuilding) and set up a Housing Division under the Chief Architect, with a sociologist, Margaret Willis, in the R&D team led by Oliver Cox. In the late 1950s, Graeme Shankland, a close friend and follower of Bernal, who was one of the leading architect–planners in the LCC, was working on proposals for the council to build its own new town to be sited at Hook in Hampshire. (Previous New Towns had all been central government initiatives.) Shankland led the multidisciplinary team, including sociologists, evolving the brief and designing the structure of the town.

By the time the Hook study was published, in 1961, a decision (wholly at odds with the housing needs of London and the region) had been taken not to implement the proposals. Cox, who had been working beside Shankland on the study, moved to the Ministry of Housing to lead its Research and Development Group where he produced a series of Design Bulletins which were touchstones of good guidance, both technical and sociological, until the mid-sixties. Shankland and Cox later established their own planning and architecture practice, running on the lines that Bernal had foreseen as essential for the recasting of where we live. All the practice's planning and housing projects were preceded by sociological studies, for example, 'La vie dans un grand ensemble', 1971; 'Inner London policies for dispersal and balance', 1977; and 'Third World Urban Housing', 1977.

The work on school building in Hertfordshire, and then the transfer to other public offices of those who were committed to this advanced, client-controlled production, gave rise in 1957 to CLASP (Consortium of Local Authorities' Schools Programme) which brought major economies of scale and ironed out the peaks and troughs of a single county's needs. Its centre of gravity was in Nottinghamshire, where the programme's designers evolved a new steel-framed and concrete-clad system.

The resulting system had to meet not only the urgent demand for new schools but a wholly unrelated dimension of difficulty. Most of the sites were on the edges of towns or villages where coal mining was likely to occur, at times which the Coal Board itself could not forecast: mining gives rise to rolling waves of subsidence which the new schools would have to endure without serious damage.

It was a Pyke-type problem. (Geoffrey Pyke, the third member of Mountbatten's Combined Ops scientific team, with Bernal and Zuckerman, is described in Chapter 7. He believed that the solution to any real problem could be found by stating it in its most extreme and 'impossible' form.) The CLASP solution to withstanding the impact of subsidence had diagonal cross-bracing which was spring-loaded between the uprights of the steel frame. These allowed the settlement of the ground to move progressively through the building without it causing harm.

This system was also a step along a road that Bernal had predicted in 1946 in his paper to the RIBA: that of reduction of the weight of structures by means of the development of new materials. He had written then:

> There are now possibilities of relying on insulation and cutting down the actual weight of material in a house by a factor of at least 10 at present, and I should not be at all surprised if in a measurable time we reach a factor of 50 or 100.[42]

In 1949, Bernal paid his fifth visit to the Soviet Union. One of his major interests was their building activity. He wrote on his return in

Soviet Weekly about the high-speed bricklaying (1,300 bricks per man or woman per shift), about working in all weathers down to 40 below, and about the high earnings in what had already become 'a fully mechanised industry'.[43] He gave a much longer report at the Architectural Association, arranged by the Association of Building Technicians and the Society for Cultural Relations with the USSR, which went in detail into the construction of skyscrapers such as Moscow University and gave instances of the inventiveness of site workers being taken up for widespread use. The dominant neoclassical style was, he found, already coming in for some local criticism: 'the cost of the portico rather ran away with all the money intended for the whole building'.[44]

Both these audiences were probably largely of the converted. But in July 1950 he spoke on the BBC's Third Programme and this talk was printed in the *Listener* (20 July 1950) as 'Building Methods East and West'. In this he compared techniques in the USA, India and the USSR, drawing the lesson that Britain could never equip itself with the social buildings it needed – schools, hospitals and public housing – until it had transformed its antiquated relationships and wasteful building practices. Today, public housing is virtually at a standstill and many hospitals are under threat of closure, whether or not their buildings have gone up since the war: but the building industry's social relations are much as they were in 1950.

The following year a letter arrived from the Ministry of Works.

19th April 1951

Dear Professor Bernal

We saw that the 'Daily Worker' contained the other day a report said to originate from yourself about modern methods of house-building in Leningrad and Moscow. The Minister also spotted the article, and has himself expressed interest.

We are as you know always anxious to hear in as much detail as possible about improved methods of building; and I am wondering whether you could point the way in which we might secure fuller details of the building methods and materials mentioned in the 'Daily Worker'. Any help you can give us will be greatly appreciated.

Yours sincerely

K. Newis

On 20 July a follow-up letter came: 'We have now read with great interest your article in the *Architect's Journal* of July 5th. We would be extremely grateful for any further information, particularly on prefabrication, bricklaying methods and winter working measures.' The report, ten pages of typescript, did come, at the end of September. It appears not to have vouchsafed much beyond what had by then already been published from talks Bernal had already given; but his covering letter hoped 'it will be in a suitable form. I have left out most of the statistics and technical details which however can be provided from the existing literature if it were thought worthwhile to do so.' The letter, enclosing a report on advanced techniques, ends on a note that is an amusing glimpse of the pre-photocopier era: 'I have not included any illustrations or plans as I did not imagine you had any facilities for reproducing them'.

There was a subtext to that letter from the Ministry of Works in 1951: an increasing official interest in numbers (numbers of new houses or flats) and an increasing support for commercial and foreign-based multistorey building systems. These became the state-subsidized way of achieving the housing targets of the late fifties and sixties, as a result of ministerial pressure and the political desire to boost 'housing starts' figures. The industry veered towards prefabrication but with scant attention to people's needs or to the relevance of imported systems to the future occupiers' lives – the opposite of Bernal's teaching. It was a disastrous and unpopular mistake which is still costing millions to alleviate and redress. One who worked with Bernal at the time called the letter 'the kiss of death'.

Probably the high point of showing what was going on in building and planning in the Soviet Union was the exhibition and meeting 'Man Conquers Nature' held at Battersea Town Hall in January 1952. Bernal, F. Le Gros Clark, Dr Sidonie Manton and others had been on a delegation arranged by the Society for Cultural Relations with the USSR. Bernal characterized the canal building and river management as 'for the first time working on the scale of nature itself'. The canals and inland seas would allow extensive deepwater navigation, and the diversion of the Siberian Arctic-bound rivers into the Asian deserts would make them fertile. All the schemes were to be finished by 1957.

> When they are, the additional food and power that they will produce will put the Soviet Union once and for all out of reach of any fluctuations of climate and will provide such riches as will ensure a peaceful transition to communism, where each gives according to his ability and receives according to his need.
>
> We left the town hall buoyed up for a long time ahead. More recently, we have had a lot more to learn, at deeper political levels, as well as learning of

environmental problems such as the fact that the Aral Sea has shrunk to one third of its former size, partly as a result of river diversions.

In the following years, Bernal was gestating another book which was to be as original in its way as *The Social Function of Science*: a book that brought into focus once again his technological far-sightedness, his socialist convictions, the breadth of his sharp-eyed appraisal of current official statistics, and his certainty that resources could start to be planned, from tomorrow, in a better and more humane way. *World without War* was published in 1958, with a 'PS' and a 'PPS' inserted when it was already in the printer's hands. It was his unique contribution to the peace movement, in that it showed the vast potential for restructuring and improving the lot of Third World countries, given the possibility of standing up free of the millstone of the arms race around our necks. Its political forecasting on the long-term scale (up to today) was either wrong or very much in abeyance; nevertheless, it is a book that needs writing again today, when the resources of the world that are going to waste are even vaster than they were then, the plight of populations is almost as severe as when they were daily threatened by possible nuclear conflict, and when we know even more than we did then of what technology can do in building bridges between resources and needs.

In 1962, Bernal gave his own paper at the Architectural Association, on 'Modern Science in Architecture'.[45] He described himself as a 'back number', having spoken there twenty-six years before, when 'I knew nothing about architecture: now I have forgotten all about it'. He had been one of the first to advocate pre-stressed concrete, even though no one was then certain how long it would last. But he looked for still more radical changes in structure: he found the similarities between the structures of certain viruses and the geodesic frameworks of Buckminster Fuller exciting (and he would have been pleased that Carbon 60, recently identified in outer space and since made in the laboratory, has been dubbed Buckminster Fullerine and is thought to have very wide application).

He looked, too, for much cheaper building. The ill-chosen, ill-designed and poorly built prefabrication that was in train at the time he was speaking had not yet shown itself to be illusory. In his 1962 talk, he reviewed his experience of it just after the war:

> We started with a wonderful competition, which was quite the wrong way to start. Anyone could design a house and even get a licence to build it. Then the BRS spent a lot of time literally demolishing the houses and finding out

how difficult it was to do so. Of over a thousand, three hundred survived in the end and are still with us; but those that did not fall to pieces of themselves or were not blown away by the artificial wind that the BRS put on them, and were deemed to be beautiful and suitable in every respect, failed because they cost too much to build. To get anything that could actually be built by prefabrication and did not cost too much was beyond the wit of man, and so we went back to the bricklayer.

His imagination was as unstoppable as ever. 'We really ought to be able to design the whole traffic process as a unit.' Electronic control would improve the effectiveness of traffic by a factor of five, he postulated. A machine to plough up a whole city and start again, sorting and digesting the old bits and rebuilding at its other end, was feasible. 'It might solve a lot of the architect's problems.' (It might – for a little while.)

A little later, in January 1963, he gave a talk to the Student Christian Movement entitled 'Need There be Need?'

It is possible to relieve all human want and misery but this is not done. In the past you could accept it because there were no means to alter it. The difference in morality now is that we have the means to alter it, and if you don't find out what they are and don't do anything about it – you're responsible, though you may not have been responsible before. It's the distinction between ignorance and invincible ignorance: if you didn't know something was a sin, then of course you couldn't be guilty; but if you deliberately took pains not to learn whether it was, that was an ignorance that was itself a form of guilt. Our duty therefore nowadays includes first the understanding and then the changing of the world we live in. I would prefer however to look at it as a purely human consideration. I see in these nearly three thousand million people there are in the world enormous potentialities: for science, for poetry, which is stifled at the moment by sheer poverty. And I want to see mankind realise its full potential.

Shortly after this, Bernal suffered his first stroke and so the above words could very well stand as a closing statement. Fortunately, he still had something to give to planning and development at the world scale and to architecture and building science at home.

In June 1963 there was a discussion at the Royal Society on 'The Mechanical Properties of New Materials' which he organized with three other Fellows.[46] Sixteen papers reviewed glass, polymers, thermoplastics and new metals. Once again, he had brought together people from hitherto exclusively separate disciplines, and after two days of their exchanges he was able to draw together the many-sided contributions

to the nature, the manufacture and the use of the new structural
materials.

> As a result we can all go away and think how these different ideas can be
> combined together in a logical and scientific way to provide us with the basic
> theory and practice of consciously evolved materials, to take the place of
> those admirable materials, bone, wood and stone, that nature has given to
> early man.

Physical fundamentals showed themselves as clearly as ever in his
thinking and writing, whatever the pressing necessities of the political
agenda. Nares Craig's last recollection of Bernal is of him puzzling
about why water flows. My own is of him at the exhibition of the Physics
Society at Alexandra Palace, working on the structure of liquids, the
molecular model made of ball bearings at his elbow. 'What is the
difference between a pile and a heap?' he asked.

In 1966, by which time he was quite ill, he established Socio-Technic
Planning, a small discussion group which included the scientists Blackett and Burhop, Stephen Bodington of the Institute for Workers'
Control as well as Oliver Cox and Graeme Shankland. Bernal prepared
a short paper for the opening meeting, 'Towards a Complete Socio-
technics – Planning Production for Social Benefits'. Having put forward
the idea of computer simulation of the social consequences of technical
innovation, he said:

> It would be useless to apply these ideas to any particular manufacturing
> problem for they might be short-circuited by another competitive project.
> On a larger scale there might even be difficulties in applying them even to a
> whole country, especially a small country which is at the mercy of price and
> market changes beyond its control. But there are also difficulties in the old
> way of thinking, putting a premium on lucky guesses and making heavy losses
> inevitable for the unlucky. In the age of computers it should be possible for
> international organisations, ideally one of the agencies of the United
> Nations, to draw up a matrix of the major features of variation to be expected
> and to indicate the kind of social implications that follow them. There is no
> reason to fear that this will be too complicated for the machine because its
> programme would have to be drawn up on broad lines only. Some of the
> major problems are already visible, for instance, (1) the problem of food and
> population, (2) the problem of transport, (3) the financial aspects to
> development and liquidity. All of these call not so much for actual calculation
> as for good will and readiness to take long term views. It should be clear by
> now that failure to do this is going to lead to a progressively worsening
> situation which may reach a point where breakdown will occur leading to
> famine or war, or both. The plea at present is only that we should look
> forward as well as we can and issue warnings which may be heeded in time.

This sounds very like the warnings that he had sounded, nearly forty years before, in *The World, the Flesh and the Devil.* In the years between, although much had changed, some social fundamentals were still the same. Of the changes, he had made many of them happen by his own insights and by his inspiring of others. When he died, the generation he inspired still had some of that feeling of 'patriots fulfilled'. Today, governments seem deaf to warnings, and the headlines show that 'in time' is obliterated by 'too little, too late'.

Yet it does move . . .

Notes

I would like to thank Oliver Cox for his helpful comments on an earlier draft of this chapter.

1. Maurice Goldsmith, *Sage: A Life of J. D. Bernal,* London, Hutchinson, 1980, p. 27.
2. J. D. Bernal Archive, Cambridge University Library, Box 105: looseleaf notebook.
3. J. D. Bernal, *The World, the Flesh and the Devil,* 1929; second edition, Jonathan Cape, 1970, pp. 27–8.
4. Ibid., pp. 17–18.
5. Ibid. Author's Foreword to the Second Edition, p. 9.
6. Ibid., p. 56.
7. 'A group of economists, scientists and technicians', *Britain without Capitalists,* London, Lawrence and Wishart, 1936.
8. Ibid., p. 140.
9. Ibid., p. 147.
10. Ibid., p. 180.
11. Ibid., p. 174.
12. Ibid., p. 180.
13. Ibid., pp. 181–2.
14. Private letter from Berthold Lubetkin to Francis Aprahamian, 1983.
15. *Circle,* London, Faber and Faber, 1937.
16. Letter from Barbara Hepworth to J. D. Bernal, undated, from 7 The Mall, Parkhill Road, London, NW3.
17. *Journal of the Royal Institute of British Architects,* 26 June 1937, pp. 805–12.
18. Cambridge Scientists' Anti-War Group, *The Protection of the Public from Aerial Attack, being a critical examination of the recommendations put forward by the Air Raid Precautions Dept. of the Home Office,* London, Victor Gollancz, 1937, p. 122.
19. *New Statesman and Nation,* 27 November 1937, Supplement.
20. Lecture, Open Forum meeting, 8 June 1939, in J. D. Bernal Archive, B.3.55.
21. *Focus,* no. 4, Summer 1939.
22. J. D. Bernal, *The Social Function of Science,* London, Routledge, 1939, p. 350 and p. 352.
23. Ibid., p. 370.
24. Ibid., p. 381.
25. Goldsmith, p. 170.
26. William Allen CBE, Bickerdike, Allen Partners, London, recent communication.
27. J. D. Bernal, 'The Place of Science in Architecture', British Architectural Library, RIBA, Manuscripts Collection, SaG/99/4 (Godfrey Samuel papers).
28. J. D. Bernal, 'Research Organisation in the Building Industry', British Architectural Library, RIBA, SaG/99/2.

29. E. J. Carter, J. D. Bernal and Mr Ellis, 'Science and Building Construction', British Architectural Library, RIBA, ScGG/99/3.

30. J. D. Bernal, 'On a Central Body to Coordinate Research in Building', British Architectural Library, RIBA, SaG/99/1.

31. J. D. Bernal, 'The Organisation of Building Science Research', *Journal of the RIBA*, vol. 53 no. 6, pp. 236–40. The paper was delivered on 6 March 1946 (J. D. Bernal Archive, B.3. 102).

32. J. D. Bernal Archive, H.33.1.

33. J. D. Bernal speaking at a conference as chairman of the Scientific Advisory Committee of the Ministry of Works, reported in the *Builder*, 16 November 1945, pp. 400–02 (J. D. Bernal Archive, B.3.89).

34. F. R. S. Yorke, *The Modern House*, Architectural Press, 1934.

35. Letter from Cawood Wharton and Co., Leeds, 3 May 1948, signed by G. W. Walker. J. D. Bernal Archive, H.33.4.

36. Address to the Jubilee Congress of the South-Eastern Union of Scientific Societies, held at Tunbridge Wells, 9–13 July 1946; Report in *Nature*, vol. 158, p. 210 (J. D. Bernal Archive, B.3.118).

37. 'The Housewife', J. D. Bernal Archive, B.5.16.

38. 'Science in Architecture', RIBA lecture, 12 February 1946, reported in the *Builder* (22 February 1946) p. 198, in reply to discussion.

39. Symposium organized by the Roads and Building Materials Group of the Society of Chemical Industry in conjunction with the Ministry of Works, 8 May 1946.

40. Andrew Saint, *Towards a Social Architecture: The Role of School Building in Post-War England*, Yale University Press, 1987, p. 64.

41. 'Is Town Planning a Science?' Lunchtime talk at the Town and Country Planning Association, 12 December 1946, published in the *Surveyor*, January 1947, vol. 105, p. 981.

42. *Journal of the RIBA*, March 1946, pp. 155–8.

43. *Soviet Weekly*, 22 September 1949.

44. *Keystone* (Journal of the Association of Building Technicians), March 1950, p. 57.

45. J. D. Bernal, 'Modern Science in Architecture', *Architectural Association Journal*, November 1962, pp. 156–66.

46. *Proceedings of the Royal Society*, A, vol. 282, 1964, pp. 1–154.

Obiter Dicta

J. D. Bernal

Against Social Darwinism

Technically, Galton's work marked the first crude use of statistics in inheritance studies, and it led to the foundation of the socio-biological science of *eugenics*, which has ever since concerned itself largely with attempting to prove, on genetic grounds, the superior value of upper class stocks and stressing the need to protect them against the careless breeding of the inferior poor. This biological interpretation of mankind, with its emphasis on race and breeding, affected to a greater or lesser degree most progressive thinkers in the social and historical science. It was popularized by historians like Green and novelists like Wells, who were never able to see that by reducing the development of man to a lower level of evolution they were making nonsense of history and social science. Worse results were to follow in practice. This double transportation of social images into biology and back again was to leave a terrible sequel when applied in practice in the twentieth century. More than anything else it sapped the foundations of the older belief, insecurely established in traditional religion, that man belonged to society and that his very individuality could find expression only through society. Imbued with the false biological view of humanity, as a race rather than a community, even the limited sanctions implied in religious morality fell away. Life became a free-for-all where the doctrine of race could be used to justify any degree of class or colonial exploitation: could even be used to prove that white and black men were of different species. Its full horror, however, was reserved for our time, where the excuse of race superiority, fanatically believed in by thousands of the followers of the Nazis, was used to perpetuate under conditions of incredible cruelty and degradation the largest and most senseless massacres in history. . . . Besides glorifying race this perversion of Darwinism also glorified war, for it was in war that the race proved itself and it was in war that the fittest survived. It is true that these ideas were for the most part those of ignorant and fanatical men. . . . Nevertheless, nineteenth- and twentieth-century scientists cannot escape the blame. Their fear of entangling themselves in politics meant that they left the social application of their own ideas to other people, and made no effective protest against the perversion of the products of their own researches.

Science in History (London, 3rd edn, 1965), pp. 786–7

The Nazi attack on science

Naturally, even with the rapidity and brutality of Nazi methods, the spirit of German science could not be destroyed all at once. The attack on it had taken several forms, the first and most dramatic being the expulsion of the Jews from science. The peculiar tragedy of the Jewish people is that whenever they are tolerated sufficiently long to enable them to devote themselves to socially useful tasks they are sooner or later made the scapegoats for all the misfortunes of the countries in which they happen to live. . . . These prejudices, which had hitherto in Germany, as they have in countries such as England, been restrained by common sense and tolerance of the majority inside and outside the professions, were now turned into an official dogma backed up by the whole force of the law and the violence of black and brown shirts brought up on the anti-Semitic and anti-Communist propaganda.

The Social Function of Science (London, 1939), p. 214

The technological revolution of the small manufacturers

The gentlemen and merchants of the seventeenth century had been in fact too successful. The normal course of development of capital and the growth of trade had given them all they wanted. Science was a plaything of which they soon got tired. A new class was emerging, however, that of the small manufacturers, who were taking advantage of the new markets won by the trade wars and of the new demands created by them to push new wares and new ways of making them.

Science in History, 1965

Capitalist science and workers' interests

Even where the profit motive is effective in leading to further scientific research, the application of that research is directed without any consideration of the convenience, or even the health and life, of the workers operating the process. The application of science has vastly improved the methods of industrial production, but it has not made them any more comfortable or less dangerous. Often new scientific industries, such as the match industry, have meant death and disaster to thousands of workers. The discovery of new chemical means for extracting gold cost the lives of more natives in South Africa through silicosis than centuries of inter-tribal warfare. Even when science has been applied for the safety of the workers, the results have been, paradoxically, not less, but more dangerous for them. The classical example of this was the invention of the miners' safety lamp. In 1812 ninety-two men and boys had been killed in an explosion in the Felling Mine in Gateshead-on-Tyne. This and other explosions prompted J. J. Wilkinson to form a society for the study and prevention of mining explosions. Davy was invited to co-operate. The result was the safety lamp. It certainly decreased the danger, at the same time decreasing the illumination. The result, however, did not conduce to the safety of miners. All that happened was that more and more dangerous seams were worked, and explosions continued as they continue to this day.

    J. D. Bernal, 'Science and Education' (in 'A group of economists, scientists and technicians', *Britain without Capitalists*, London, 1936),

p. 432

On the inefficiency of cruelty in farming

Advances in gland physiology and genetics may produce even greater changes in animal husbandry. Up till now these changes have been almost purely commercial in incentive and though they have resulted in much greater yields of such products as eggs and milk, it has been at the expense of increasing the incidence of disease which, in the case of tuberculosis in milk, is passed on to human beings. The conditions of forcing animals are not only unnatural and cruel but inefficient. There is no reason why in a well-ordered economy the well-being of animal populations should not be a primary consideration of husbandry.

The Social Function of Science, pp. 347–8

On the human being inside the white coat

Nearly all scientific publications are written so as to conceal as far as possible the personality and aims of the author. Such things are not considered important in science, and this is one reason why science is so little read or understood outside the ranks of the specialists. In literature and painting we enjoy the work of the artist himself. The character of the scientist's thoughts, his [*sic*] imagination and even his actions are deliberately and carefully concealed.

A scientific paper is a kind of sketch map to the truth. It has the advantage that it indicates how it can be verified, so that by carrying out certain experiments anyone will arrive at the same answer as the author found. What the reader rarely discovers is why or how the scientist did his work, because the logical steps described in his paper are quite different from the steps which he actually took. Sometimes these steps are the result of hunches, and hunches are very often useful as much when they are wrong as when they are right. In a science of science, these real activities would be just as much an object of study as would the formal, logical research presented in the scientific literature.

'Towards a Science of Science', *Science Journal*, 1965

On science and planning

The difference between scientific research and other human occupations is that, as it is looking for the unknown, it cannot decide beforehand what it is going to meet and when it is going to meet it. Luckily for science, the completely unknown is not met so very often. Most actual workers know what they are looking for and many of them find it. It is true that the occasions on which they do not succeed in their object are often the most fruitful in discovery, but still the safest plan to follow is to search as if something were going to be found, and to be prepared to modify one's search if something different is found or if insuperable difficulties arise. It is as feasible for groups of scientists to adopt this course as it is for an individual scientist. . . . The plan of work is laid down at the beginning of the year – such and such a field to be explored, the relation of such and such properties to be examined – but in the plan there is a very large margin left for unexpected difficulties and the time to follow side clues, and there is always a proviso that if anything of extreme importance is found the whole original plan can be scrapped and a new research started. Some kind of plan is in effect necessary even in the laboratories of capitalist countries. A provision has to be made for apparatus of a specified kind, often apparatus which will take years to make and be expected to be in use for years. Suitable men have to be found for assistants, and all this has to be done before the results of the work for which these preparations are made can be known. The drawing up of the plan must be the work of competent people – the director in a very small institute, or the director and the laboratory council in a larger one.

The plans for individual institutes need, further, to be coordinated under the general scheme which will be described later; but this general plan is not an imposition from above on individual institutes, but an integrated sum of the plans of the institutes after taking into consideration the demands for the solution of problems which are presented to them and of their own internal programme of fundamental research.

'Science and Education', pp. 450–51

On communication between scientists

At the same time as the administrative organization, the means of communication between scientists will need to be vastly improved. At present they are unbelievably chaotic. This state of affairs is recognized by most scientists, but the vested interests of learned societies and publishing houses has prevented for years anything effective being done.

There exist something like 36,000 scientific and technical periodicals whose fields all overlap to a greater or lesser degree. They can roughly be classified into the journals which contain original papers in considerable detail, and the magazines which have the more general function of giving reviews and summaries of important papers, and at the same time giving news of their recent scientific advances. Besides this, there exist the abstracts giving briefest accounts of all effective papers.

This system certainly ensures that no piece of scientific work should go unrecorded. It may, if it is at all good, even be recorded five or ten times over. The main criticism, however, is that it is recorded in an entirely unsystematic way. The result is that in field after field it is becoming actually easier to do a piece of work over again than to find out where it has been done.

Actually science has grown so large that a Science Intelligence Service is urgently needed. Its function would be to keep every scientist throughout the world posted with all the information that he needs, not only in his own subject, but in all other subjects which may be relevant to his work, either in providing him with new material or with problems which he may solve.

The written word, however, is not enough. Personal communication and visits are an essential part of scientific communication, particularly with relation to experimental technique. This could also be made possible in an organized way on a scale not at all manageable at the present moment. It should be possible and desirable for all scientists to spend at least three months of the year working in laboratories other than their own, and another month or so in the field in contact with agricultural or industrial workers, or in some remote part of the world. The expenses of scientific travel and hospitality would be part of the regular expenses of scientific institutes, and these exchanges could also be supplemented by congresses deliberately held more for informal conversations than for the reading of papers, which can as well be done at home.

'Science and Education', pp. 451–3

On X-ray photography and the double helix

'Now that we have the electron microscope, viruses can be seen and much of their gross structure distinguished. Even earlier Bernal and Fankuchen had shown by x-rays that the first virus to be isolated by Stanley and by Bawden and Pirie in 1934, that of tobacco mosaic virus (TMV), was a rod-shaped body with an inner regularity. In 1954 this was shown by Watson, Wilkins and Franklin to be a tube constructed of protein molecules into which was woven a thread of nucleic acid. In the apparent spherical viruses the protein molecules formed a regular polyhedral cage for the nucleic acid.

Science in History, p. 665

When the chemical analysis of Chargaff showed the number of purines and pyrimidines were exactly balanced, Crick and Watson put their famous hypothesis of the arrangement being not a *single* helix but a *double* one, the purine of one chain being linked with the pyrimidine in the one twined with it. This was subsequently verified by x-ray analysis by Wilkins and Franklin. . . . The implications of this purely structural discovery were enormous; it was the greatest single discovery in biology.

Science in History, pp. 663–4

James Watson

The role of Bernal's work in the discovery of the double helix

A vital component of TMV was nucleic acid, and so it was the perfect front to mask my continued interest in DNA. . . . TMV had been previously looked at with x-rays by J. D. Bernal and I. Fankuchen. This in itself was scary, since the scope of Bernal's brain was legendary and I could never hope to have his grasp of crystallographic theory. . . . Though the theoretical basis for many of their conclusions was shaky, the take-home lesson was obvious. TMV was constructed of a large number of identical subunits. How the subunits were arrayed they did not know. Moreover, 1939 was too early to come to grips with the fact that the protein and RNA components were likely to be constructed along radically different lines.

James Watson, *The Double Helix*, London 1968, pp. 110–11

Conceivably a few additional x-ray pictures would tell how the protein subunits were arranged. This was particularly true if they were helically stacked. Excitedly I pilfered Bernal's and Fankuchen's paper from the Philosophical Library and brought it up to the lab so that Francis could inspect the TMV x-ray picture. When he saw the blank regions that characterise helical patterns he jumped into action, quickly spilling out several possible helical TMV structures. From this moment on, I knew I could no longer avoid actually understanding the helical theory.

The Double Helix, pp. 112–13

Books by J. D. Bernal

The World, the Flesh and the Devil, London, Kegan Paul, 1929
The Social Function of Science, London, George Routledge, 1939
The Freedom of Necessity, London, Routledge and Kegan Paul, 1949
The Physical Basis of Life, London, Routledge and Kegan Paul, 1951
Marx and Science, London, Lawrence and Wishart, 1952
Science and Industry in the Nineteenth Century, London, Routledge and Kegan Paul, 1953
Panstwowe Wydawnictwo Naukowe Science for Peace, Warsaw, 1953. Collection of essays in Polish
Science and Society, Moscow, Publishing House for Foreign Literature, 1953. Collection of essays in Russian
Science and Society, Budapest, Szikra, 1954. Collection of essays in Hungarian
Science in History, London, C. A. Watts, 1954
World without War, London, Routledge and Kegan Paul, 1958
Prospect of Peace, London, Lawrence and Wishart, 1960
The Origin of Life, London, Weidenfeld and Nicolson, 1967
The Extension of Man, London, Weidenfeld and Nicolson, 1972

J. D. Bernal contributed to 10 scientific books, and wrote 224 papers published in scientific journals. He contributed to 21 non-scientific books, and wrote 392 articles published in non-scientific papers and journals.

Calendar of Events

1901   Born 10 May, County Tipperary, Ireland, first child of Samuel and Elizabeth Bernal; educated (1906–19) in Nenagh and at Hodder College, Stonyhurst College (one term) and Bedford School.

1919   Samuel Bernal dies; Mathematics scholarship to Emmanuel College, Cambridge.

1922   Marries Eileen Sprague.

1923   Graduates; commences research under Sir William Bragg at the Davy Faraday Laboratory, using X-ray techniques to investigate the arrangements of atoms in molecules; joins 1917 Club; 'The Analytic Theory of the 230 Space Groups', unpublished MS.

1924   'The Structure of Graphite', *Proc. Royal Society*, London; first Bernal chart.

1926   Son, Michael Bernal born; actively supports the General Strike.

1927   Appointed Lecturer in Structural Crystallography, University of Cambridge.

1928   Visits scientific research centres in Germany and Holland, reporting to Sir William Bragg.

1929   Helps organize conference of X-ray crystallographers at the Royal Institution; 'X-rays and Crystal Structure', *Encyclopaedia Britannica*; *The World, the Flesh and the Devil*, London.

1930   Son Egan Bernal born; begins work on amino acids.

1931   Begins work on sterols; attends History of Science and Technology Congress, London, addressed by Bukharin; first visit to Soviet Union; Tots and Quots started by Solly

Zuckerman; 'The Crystal Structure of the Natural Amino
Acids and Related Compounds', *Z. fur Krist.*, Leipzig.

1932   Attends Theoretical Biology Club, organized by J. H.
Woodger; Dorothy Crowfoot joins laboratory; 'The Crystal
Structure of Vitamin D and related Compounds', *Nature*,

1933   Lectures at Leningrad and Moscow universities.

1934   Appointed Assistant Director of Research in Crystallography, University of Cambridge; takes first usable X-ray
diffraction photograph of a protein crystal; meets Paul
Langevin and Frédéric Joliot-Curie; helps form Anglo-
French Scientific Committee; becomes first president of
Cambridge Scientists' Anti-War Group; helps revive Cambridge branch of the AScW; 'X-ray Photographs of Crystalline Pepsin', with Dorothy Crowfoot, *Nature.*

1935   Organizes first Congress for Academic Freedom, Oxford.
Gives evidence to the Royal Commission on the Private
Manufacture and Trade in Arms.

1936   Attends Paris conference called by the Comité de Vigilance des Intellectuels Antifascistes; British delegates form
For Intellectual Liberty.

1937   Son Martin Bernal born to Margaret Gardiner; elected
Fellow of the Royal Society; appointed to Chair of Physics
at Birkbeck College, University of London. Criticizes
government proposals for air-raid precautions at meeting
of MPs.

1938   'Science and National Defence' memorandum, copies sent
to Leslie Hore-Belisha, Liddell Hart and Sir Arthur Salter;
'A Speculation on Muscle in *Perspectives in Biochemistry*',
Cambridge, presented to Sir Frederick Gowland Hopkins.

1939   *The Social Function of Science*, London; attends conference
at All Souls, Oxford, organized by Sir Arthur Salter, with
Sir John Anderson, Lord Privy Seal with responsibility for
Civil Defence; appointed to Civil Defence Research Committee; lectures in US, June to September, full-time work
for Ministry of Home Security.

1940   *Science in War*, London; appointed to Scientific Advisory
Committee, Ministry of Supply; Birkbeck College hit by
incendiary bomb.

1941   Attached to Bomber Command; not allowed to go to
Moscow.

1942	Appointed Scientific Adviser to Lord Mountbatten, Combined Operations; 'Science and the War Effort', *Nature*.
1943	Visits Cairo and Tripoli; sails with Churchill and Mountbatten to attend Allies' Quadrant Conference in Quebec.
1944	Visits Normandy, D-Day Plus 1; visits Mountbatten in Ceylon, visits Arakan front.
1945	Research restarts at Davy Faraday Laboratory; appointed chairman, Ministry of Works Scientific Advisory Committee; awarded Royal Medal of the Royal Society; involved in creation of UNESCO.
1946	Inaugural meeting of World Federation of Scientific Workers; drafts the constitution for this.
1947	Awarded US Medal of Freedom, lecture tour in USA; becomes President of Association of Scientific Workers.
1948	Biomolecular Research Laboratory opens at Birkbeck; appointed member Building Research Board; helps organize Royal Society conference on scientific information; attends first conference of International Union of Crystallography, at Harvard; attends Wrocław Cultural Conference.
1949	World Peace Council founded; elected vice-president; refused visa by USA.
1949–1965	His work for the World Peace Council and scientific conferences take him frequently to Paris, Moscow, Prague, and Stockholm, and during this period he visits most European countries attending conferences and lecturing.
1950	World Peace Council planned for Sheffield moves to Warsaw after many delegates are denied admission to UK; Picasso draws mural on wall of his London flat; meets Nehru with Joliot-Curies in India, asks Nehru to release the Telengana prisoners.
1951	Meets Lukács in Hungary; *The Physical Basis of Life*, London.
1952	Becomes President of Marx Memorial Library; refused entry to France; *Marx and Science*, London.
1953	Daughter, Jane Bernal, born to Margot Heinemann; *Science and Industry in the Nineteenth Century*, London.
1954	Awarded prize for the Strengthening of Peace among the Peoples, interview with Khrushchev in Moscow; visits India and China, meets Mao Zedong and Nehru in Peking, and Nehru in Delhi.
1955	Congress of French Association for the Advancement of Science, Caen; attempt made to deport him while lecturing.

1956 First proposal for a separate department of Crystallography at Birkbeck; course of lectures at Lomonosov University, Moscow; speaks at a meeting in London protesting against the Soviet invasion of Hungary.

1957 Attends first congress of the International Union of Crystallography, Montreal; symposium on 'Origin of Life on Earth', and special conference on 'Science in History', Moscow.

1958 Frédéric Joliot-Curie dies; Bernal is elected chairman of the presidential committee of the World Peace Council; International Conference on Scientific Information, Washington; *World without War*, London.

1959 Awarded Grotius medal; World without War exhibition at Caxton Hall; visits China, has interview with Ho Chi Minh.

1960 Visits Ghana for Commission on University Education, meets Nkrumah; Colloquium on Structure of Liquids, Cern; *Prospect of Peace*, London.

1961 Private conference on disarmament with Canon Collins, CND, and Homer Jack; visits India.

1962 Lectures in South America. World Congress for General Disarmament and Peace, Moscow; Pugwash meeting, London; International Symposium on Higher Scientific and Technical Education, Moscow.

1963 Attends UN conference on 'Application of Science and Technology for the Benefit of Less Developed Areas', Geneva; visits UNESCO department of Natural Sciences, Paris; in hospital for investigation; lecture tour of USA; first stroke, July; appointed Professor of Crystallography, Birkbeck College.

1964 'After Twenty-Five Years', in M. Goldsmith and A. Mackay, eds., *The Science of Science*, London.

1965 World Peace Council, Congress, Helsinki, JDB resigns chairmanship of the Presidential Committee; made a Fellow of Emmanuel College, Cambridge; second stroke.

1967 *The Origin of Life*, London.

1968 Retires from Birkbeck College; Bernal Peace Library opens at Marx House.

1969 Elected Fellow of Birkbeck College.

1971 Dies 15 September.

1972 *The Extension of Man*, London.

Notes on Editors and Contributors

RITCHIE CALDER, 1906–82, was a Scottish journalist and educationalist. Specializing in the spread of scientific knowledge to lay readers, he wrote numerous books, including *Men Against the Desert* (1951), *Men Against the Jungle* (1954), *Living with the Atom* (1962), and *The Evolution of the Machine* (1966). He was made a life peer in 1966.

CHRIS FREEMAN, Emeritus Professor at the University of Sussex, was the first director of the Science Policy Research Unit there, from 1965 to 1981. After working in market research and adult education, he has recently concentrated on problems of technical change and unemployment. He edited *Economics of Hope* (Pinter, 1992), and co-authored *Work for All or Mass Unemployment?* (Pinter, 1994) and *Economics of Industrial Innovation* (Pinter, 1997).

ERIC HOBSBAWM, CH, FBA, Emeritus Professor of Economics and Social History, Birkbeck College, was for many years Bernal's colleague at Birkbeck. He was born in 1917 and educated in Vienna, Berlin, London and Cambridge. Since retiring from Birkbeck he has taught at the New School for Social Research in New York. He has written many books, including *The Age of Revolution, The Age of Capital, The Age of Empire,* and *The Age of Extremes,* covering the years from 1789 to 1991.

ROY JOHNSTON, PhD, graduated from Trinity College, Dublin, in 1951, with a double moderatorship in physics and mathematics. He did research at the Paris École Polytechnique and Dublin Institute of Advanced Studies on characterizing sub-nuclear cosmic ray particles. Later he switched into industrial process technology and computer-based system modelling, with a techno-economic dimension. He is concerned with the mismatch between scientific capability in Ireland

and the uptake of the results of scientific research in Ireland. From 1970 to 1976 he ran a weekly column on science and technology in the *Irish Times.* Some of his current work is accessible via his website at http://www.iol.ie/rjtechne

PETER MASON, who graduated from London University during World War Two, worked on quartz crystals for radar sets needed in the invasion of Normandy. Later he changed to research on physical problems involved in building and roads, rubber and plastics. Interest in biological molecules took him to the Commonwealth Scientific Industrial Research Organization in Australia to study proteins in wool. In 1966 he was appointed to the foundation chair of physics at Macquarrie University. He has broadcast with the Australian Broadcasting Corporation's Science Unit, and has written books including *Genesis to Jupiter, The Light Fantastic,* and *Blood and Iron* (Penguin, 1984).

IVOR MONTAGU, 1904–84, studied biology at Cambridge, doing his thesis on small mammals. He became a film editor, director and critic. He filmed in Spain during the Civil War, in London, and Hollywood. He was a friend of Eisenstein and Charles Chaplin. He formed the Association of Ciné Technicians, and together with Sydney Bernstein founded the first film society in London. For many years he was president of the English Table Tennis Association. He was on the staff of the *Daily Worker* and worked with J. D. Bernal for the World Peace Council. He wrote an autobiography, *The Younger Son,* and many books on the cinema.

HILARY ROSE is Visiting Research Professor of Sociology at City University and Professor Emerita of Social Policy at the University of Bradford. She has written *Love, Power and Knowledge* (Polity Press, 1994) and is currently working on *Feminism Confronts the Human Genome Project.*

STEVEN ROSE is a neuroscientist and Professor of Biology at the Open University. He has written *Lifelines* (Penguin, 1997). Together, Hilary Rose and Steven Rose wrote *Science and Society* (Penguin, 1969) and edited *The Political Economy of Science* and *The Radicalisation of Science* (Macmillan, 1976). They are currently editing *Coming to Life,* a multi-disciplinary critique of evolutionary psychology.

FRED STEWARD, PhD, is senior lecturer at Aston Business School, Birmingham, leading the innovation research programme. A former member of the *Marxism Today* editorial board, he has been concerned with environmental and consumer issues. He is currently conducting research projects on technical innovation, sponsored by the European Commission.

ANN SYNGE, 1916–97, the daughter of Adrian Stephen, psychiatrist brother of Virginia Woolf, studied medicine at Cambridge and Dublin. She married the Nobel prizewinner R. L. M. Synge. She taught, and translated scientific books from Russian, including Oparin's *Origin of Life*.

PETER TRENT was a lecturer, and later the Faculty Tutor in Bernal's Physics Department at Birkbeck during all of Bernal's time there. For the last decade he has been a member of the Oxford group working on the construction of the Sudbury Neutrino Observatory in Canada, an experiment which will start recording data in 1999.

CHRIS WHITTAKER is an architect and town planner. He was with the London County Council and the Ministry of Housing, later the Department of the Environment, working on area improvement. He directed the Barnsbury Environmental Study. In practice, he worked with a large number of tenants' groups and residents' associations. He has written *Environmental Powers* (Architectural Press, 1976), and co-authored *Solar Houses in Europe* (Pergamon Press, for EEC 1981).

FRANCIS APRAHAMIAN, 1917–91, was Desmond Bernal's research assistant at Birkbeck College, and worked extensively with him on *Science in History, World without War*, and many other books. He was largely responsible for the posthumous publication of *The Extension of Man*. He was then appointed Science Editor at the Open University, where his responsibilities extended to the production of material for the Women's Studies course.

BRENDA SWANN, PhD, worked with Desmond Bernal as his secretary and research assistant on *The Social Function of Science* from 1938 to 1940. She was the organizing secretary of the Association of Scientific Workers from 1940 to 1945. After studying economic history at the London School of Economics, where she obtained her PhD, she worked on the registration of common land for the Commons Society. From 1970 to 1976 she worked at the Public Record Office, writing guides to the records newly opened by the Labour government.

Index

Adorno, Theodor 153
'After Twenty-Five Years' (Bernal) 137, 153
All Quiet on the Western Front (Remarque) xiii
Allen, Bill 280–81
Allen, Clifford 53
America by Design (Noble) 145
Amery, Leo 191
Anderson, Perry
  transient radicals 139
Anderson, Sir John (later Viscount Waverly) 164–5, 173, 216
Angell, Norman 40
Anglo-French Society of Science 171
Appleyard, Rollo 19
Aprahamian, Francis xxi, 132, 233, 252, 268
Architectural Association 290
Architectural Science Board 282
architecture
  Bernal addresses RIBA 275–6, 284, 287
  MARS 273–4
  physical planning 278
  postwar reconstruction and housing 282–9
  prefabrication 290–91
  RIBA's Architectural Science Group 280
  social conditions 279–80
  The Social Function of Science 279–80
  urban needs 270–72
  wartime safety 276–8
Armytage 129
Arons, Dr Leo xvi
'Art and the Scientist' (Bernal) 274

Ashby, Sir Eric 250
Aslin, C.H. 285
Association of Building Technicians 288
Association of Scientific Workers xxiii, 135, 140, 151
Association of Special Libraries and Information Bureaux (ASLIB) xxii
Astbury, William Thomas 88–9
astronomy
  Birr telescope 18–20
Attlee, Clement
  against World Peace Conference 221–2
Auger, Pierre 170

Back to Methuselah (Shaw) 138
Bacon, Francis 136
Bagnold, Brigadier R.A. 179
Baker, J.R. 117, 146
Barnes, Sylvia 44, 46
Barr, Cleeve 286
Bauhaus School 271
Bayly, Launcelot 8
Beaupre, Beautemps 180
Bedford School 7–10, 38, 79
Bell, Daniel 111
Bell, Martin 113
Bergerac, Cyrano de 265
Bernal, Catherine (grandmother) 1
Bernal, Eileen (*née* Sprague, wife) xxi
  Labour political activity 51
  marries Bernal 13–14, 30
  undergraduate days 46
Bernal, Elizabeth (*née* Miller, mother)
  dismisses Clongowes 21

Bernal, Elizabeth (*cont.*)
marriage and motherhood 2–4, 6, 9
Bernal, Fiona (sister) 3, 6
Bernal, Geraldine (sister) 3, 9, 23, 79
Bernal, Godfrey (brother) 6
Bernal, Jane (daughter) 133, 231
Bernal, John Desmond
personal life: family and birth 1–3; family property 20; social context of family 7–8; school years 3–10, 38, 78; career options in Ireland 28–32; childhood interest in science 6, 7, 18–20, 126–7; inspiration to others ix–x, 86–7; loss of religious faith 12–13, 41, 45; marries Eileen 13–14; sexuality 132–5; unacceptable to Trinity 31; wide range of activities xxii–iv; first stroke after punishing schedule 229–31, 238–9, 250, 252; ill-health 99, 231–3; death 99
architecture: addresses RIBA 275–6, 284, 287; postwar reconstruction and housing 282–9
politics: becomes a Communist 42–4, 46–7; building internationalism 136–7; Cold War commitment to peace and socialism 72–5; conversion to socialism 37–41; eulogy for Stalin 151–2; influence on Labour Party 153–4; involvement in party politics 51–4; Irish politics 5–6, 22–8, 32–4, 45; leads World Peace Council 227–30; nationalism versus socialism 41–2; optimism 54–7, 58–60; peace movement xi, 216–25; radical mobilization 137–42; reaction to Lysenko affair 150–51; reconciling nationalism with socialism 38–43; 'red science' x–xx, 49–51; vice-president of World Peace Council 223; visit in 1949 to Soviet Union 288; warns about atomic weapons 217
science: appointed to Birkbeck 92, 237; biochemical substances 87–94; breadth of knowledge 268–9; crystallography 84–7, 92–9, 235; Engels's science xvi–xvii; faith in the nature of science 256–9; formation of Crystallography Department 247–52; provisionalism 256; role at Birkbeck 238–42, 242–7; sparks wartime controversy over direction of science 169–70; retirement from Birkbeck 252; Snow's assessment of science career 100; vision of the future 128–31
wartime activities: research on effect of weapons 215, 216; recruited 160, 163–5, 191; at Princes Risborough 165–8; planning Normandy landings 178–85, 194–5; D-Day 196–211; effect of WWII 69–70; Habbakuk project 185–8; inspects D-Day beaches 188–9; with Mountbatten 176–8, 189–90; Mountbatten's memories of 191–5; parting with Zuckerman 188; scientific approach to bombing Germany 173–6; visits France in wartime 170–71
see also titles of works in bold
Bernal, John (grandfather) 1
Bernal, Kevin O'Carroll (brother) 3
school days 6–7, 9–10
takes the family farm 30
Bernal, Martin (son) xxi, 132–3
Bernal, Ralph 1
Bernal, Samuel (father)
death 11, 30
fatherhood 4, 6
property 20
youth and marriage 1–2
Bernal Peace Library 233
Bernal prize 136
Beveridge, William 135
Biquard, Pierre 214, 223, 226
Birkbeck College 91
after Bernal 253
annual Bernal lecture 233
Bernal's position 235
bitter civil war over proposed change of focus 249–50

Birkbeck College (*cont.*)
  formation of Crystallography
    Department 247–52
  history and tradition 235–7
  laboratory working conditions 92–4
  new building 99
  Oblomov Room 244
  physics department under Bernal's
    leadership 242–7
  political and social openness
    239–40
Blackett, Patrick M.S. (later Baron)
    ix–x, 92, 153, 175, 292
  Birkbeck 237, 241
  French Committee 171
  Intellectual Liberty 215
  memorial meeting 252
Blanco-White, Justine 275–6
Bodington, Stephen 292
Bohm, D.J. 247
Bohr, Niels 84
The Bolshevik Theory (Postgate) 44
Booth, A. D. 97, 98, 246
Booth, Kathleen 98
Boulier, Abbé 220
Boyle, Robert 33
Bragg, Sir William (Henry) xxii, 14,
    80
  crystallography 83–4
  'The Imperfect Crystallisation of
    Common Things' 88–9
  inspired by Bernal 86
Bragg, Sir (William) Lawrence 14, 83,
    246
Brailsford, Henry Noel
  War of Steel and Gold 40
Brecht, Bertolt
  The Life of Galileo 258–9, 265
Britain
  Air-Raid Precautions measures
    160–63
  bureaucratic management of
    science 120–21
  Cold War restrictions 220–21
  and Egypt 229
  eugenics 135–6
  postwar science 153
  'red science' era x–xx
  spending on science R&D 105–8,
    110–16, 139–42
  UN Security Council 217
Britain without Capitalists (multi-
    author) 142

Bernal's chapter on architecture
    269–72
British Association for the
    Advancement of Science
  social function of science 140–41
  suspends Bernal 221
British Broadcasting Corporation 288
British Peace Committee 221
  Labour Party blacklists 225
British Peace Conference 222
Bronowski, Jacob 165
Brooke (Cambridge friend) 43
Brougham, Lord Henry 235
**'Building Methods East and West'
    (Bernal)** 288
Building Research Station 280–81
Bukharin, Nikolai Ivanovich xvii, 60,
    214
  influence on Bernal 143
  tried and executed 143
Burhop, Eric 142, 153, 292
Buzzard, Sir Farquhar 165

Calder, Ritchie 146
Cambridge Scientists' Anti-War
    Group xi, 64, 141
  air-raid safety 276–7
  criticize WWII civil defence 160–63
  practical weapons studies 214
Cambridge University
  Bernal's appointment 85
  Bernal's student years 11–15, 79–80
  the Labour Club 213
  political education 40, 42–4
  see also crystallography
Cambridge University Socialist Society
    39
Campaign for Nuclear Disarmament
    151
capitalism
  bias in technical change 114
  and fascism 69–70
  incompatible with scientific
    development 153
  management of R&D 107–8,
    114–16, 116–25, 144–5, 266
  reorganization 71–2
  role of scientist 65, 68
  small manufacturers 296
  stabilization in Great Depression 69
Caraffi, A.J. 237
Carlisle, C.H. 88, 93, 241, 244, 246
Carter, Bobby 277, 281

Carter, E.J. 275–6, 279, 281
Castro, Jose de 223
Catholicism
  Bernal's early social perceptions 5
  Bernal's loss of faith 12–13, 41, 45
  Irish education 20
  world-view 39
Caudwell, Christopher xviii
Cecil, Lord Robert 164
Chamberlain, Neville 216
The Chemistry of the Candle (Faraday)
  126
Chermayeff, Serge 278
Cherwell, Lord (Frederick
  Lindemann) 169–70, 187
  scientific approach to bombing
    Germany 173–6
China
  Bernal's hope for 55
  clocks 258
  schism in socialist world 73
  UN Security Council 217
The Church and Science (Windle) 41
Churchill, Winston S.
  boats in the bath 184–5, 186,
    194–5
  doesn't want Scientific Advisers
    187, 193
  orders piers for French beaches
    181, 183
  Quebec Conference 176
Civil Defence Research Committee
  168, 216
The Claims of Labour (Pickering) 40
Clark, F. Le Gros 289
Clark, Lord Kenneth 172
class
  Bernal family 21–2, 25
  early experience 24
  not consciously rational 47–8
Clongowes School 21
Cockcroft, Sir John Douglas 171
Cold War
  beginnings 217–18
  Bernal's commitment to peace and
    socialism 72–5
  effect of Lysenko affair 149–51
  restrictions on freedom 220–21
  Stalinism and science 146
Cole, G.D.H. 42
  guild socialism 44
St Columba's School 21
Communism

Bernal's adoption 42–4, 43–4, 46
  fragmentation of world movement
    73–4
  versus Freudianism 60, 62
  natural science 145
  optimism for the world 54–7
  and psychology 46–8
  uncomfortable experience of
    intellectual within 60–61, 63
Communist Party of Great Britain
  Bernal's involvement in party
    politics 52–4
  Bernal's membership 228
  economic crisis of 1930s reaffirms
    Bernal's support 60–64
  effect of Lysenko affair 149–51
  and Labour 52
  loyalty to USSR 149
  Postgate, Raymond 44
  programme for Soviet Britain 68
concrete 284–5
Conservative Party of Britain 153–4
Consortium of Local Authorities'
  Schools Programme 287
Cotton, Madame 223
Cox, Anthony 279
Cox, Oliver 285–7, 292
Craig, Nares 292
Crawford, Virginia 46
Creative Revolution (Paul) 43
Crick, Francis ix, 149, 244
'A Criticism of the Manifesto of the
  Federation of Progressive
  Societies and Individuals'
  (Bernal) 65
Crowfoot, Dorothy see Hodgkin
Crowther, J.G. 123, 171, 220
  Fifty Years with Science 169
  Science at War (with Whiddington)
    167
crystallography
  Bernal's work 80, 84–7
  biochemical 88–92
  Birkbeck becomes centre 244–5
  formation of Department at
    Birkbeck 247–52
  historical perspective 80–84
  influence on Crick and Watson 304
  space group thesis 80
  X-rays xxii
Czechoslovakia
  attack on Prague Spring 74

Dalton, John 80
Darlington, Cyril 171
Darwin, C.G. 83
Davy Faraday Laboratory 88, 93
De Gaulle, Charles 226
De Valera, Eamon
    young Bernal's admiration for 22,
    26
democracy
    applying Western model 72
    forms versus power 65–6
    giving authority to philosophers/
        scientists 130
    global outlook 69–70
    pre-Cold War humanist balance
        71–2
    Soviet Union 70
dialectical materialism
    economic determinants of science
        143–4
    Marx and Science 155
    role of individual 66–8
    science and Marxism xvi–xix
'Dialectical Materialism' (Bernal)
    143–4
The Dialectics of Nature (Engels) xvi–
    xvii, 143
Dickinson, Henry
    Communism 43
    political influence 11, 13, 24, 27,
        37–8
    scepticism 48
    socialist movement at Cambridge
        39
Dienes, Paul 237
Dirac, Paul 171
Dobb, Ivor 213
Dobb, Maurice 42, 43
The Double Helix (Watson) 152
Dutt, R. Palme 64

Economic Consequences of the Peace
    (Keynes) 40
economics
    determinism 55–7
    effect of Great Crash 69
    research and development 102–3
    see also capitalism; Communism;
        Marxism
Eddington, A.J. 12, 49
Edgeworth, R.L. 33
education
    Bernal's belief in 38

Ireland 20–22
Egerton, Sir Alfred 171
Egypt
    wants British out 229
Ehrenberg, Werner 93, 94, 246
    Professor of Physics 250, 251
Ehrenburg, Ilya 219
Einstein, Albert xvi, 139, 217, 256,
    266
Eisenhower, Dwight D. 221
electron microscopy
    viruses 300
Ellis, Mr 281
Engels, Friedrich
    dialectical materialism 213
    The Dialectics of Nature xvi–xvii, 143
    science xv
Engels Society xxiii, 149
England
    compared to Ireland 38
eugenics 135–6, 141
Ewald, P.P. 82, 83, 84

Fabianism 43–4
'The Failure of the Scientist' (Bernal)
    50
Fankuchen, Isadore 90–92, 241, 300,
    301
Faraday, Michael 262
    The Chemistry of the Candle 126
Farrington, Benjamin 138, 139, 143
fascism 54
    eugenics 135–6, 141
    French opposition 214
    part of democratic capitalism
        69–70
    role of scientists 130
    worries scientists xiii–xiv
Felling Mine 297
Fergusson, Bernard 184
Ferri, Enrico
    positive science 49
    Socialism and Positive Science 40
Fifty Years with Science (Crowther) 169
Fisher, R.A. 165
Focus (journal) 279
Foley, Galway 24
Forest Products Research Station 165
Forster, E.M. 215
Fowler, R.H. 86–7, 98
    work on water 185
France
    French wartime scientists 170–71

France (*cont.*)
  influence of scientists on Bernal 214
  UN Security Council 217
Frankfurt School 153
Franklin, Rosalind 93, 95, 134, 239, 244, 247, 300
Freedland, Jonathan 136
Freedom and Necessity (Bernal) 70
Freeman, Chris 144, 154
French Association for the Advancement of Science 178
French Committee 171
French family 20
Freud, Sigmund
  Bernal's undergraduate interest 46–7
  The Interpretation of Dreams 46
  The Psychopathology of Everyday Life 46
Freudianism 46–8
  versus Communism 60, 62
  repudiated by Bernal 67
  social theory 56
Friedrich (crystallographer) 83
From Apes to Warlords (Zuckerman) 163–4, 177
'The Frustration of Science' (Bernal) 67
Fry, Margaret 215
Fry, Roger 12
Fuller, Buckminster 290
Furberg (Birkbeck colleague) 244

Gabo, Naum 274
Galileo Galilei 258
Galton, Sir Francis 295
Gardiner, Margaret xxi, 132–3, 215, 274
Garwood, Dr F. 173
General Strike of 1926 52–3
genetics
  Crick and Watson 149
  DNA 301
  eugenics 135–6, 141
  the Lysenko affair 147–51
  Social Darwinism 295–6
  Watson on Bernal 301
George VI, King 177
Germany
  civil objectives for R&D 115–16
  Nazi persecution of the Jews 295–6
The Golden Notebook (Lessing) 133

Goldsmith, Maurice 132, 154
Gollancz Publishers 215
Goodeve, Charles 187
Goodge, Professor 42
Goossens, Eugene 12
Gould, Stephen Jay xix
Grace, J.H. 79
Greaves, C.D. 27
Greece
  slaves hold back industry 260
Green, Alice Stopford
  The Making of Ireland and Her Undoing 25
Gregory, Richard A. xiii, 141
Gropius, Walter 271
Grote, George 235
'The Growth of Organized Research' (Bernal) 57–8
Grubb, Howard 19
Grubb, Thomas 19

Habbakuk project 178, 185–8, 192–4
Habermas, Jürgen 153
Halban (refugee scientist) 170
Haldane, J.B.S.
  dialectical materialism xvi
  on effects of bomb blasts 167–8
  as Huxley character x
  influence 86
  preface to Engels 143
  reaction to Lysenko affair 149–50
  Royal Society and British Association 140
  sterols 88
Haldane, Lord (Richard) 51
Hamilton, Cicely
  Marriage as Trade 46
Hamilton, Sir William Rowan 8, 33
Hanaghan, Jonty 13
Hardiman, T.P. 31
Harrison, Jane 12
Harrison, John 258
Haüy, René-Just 81
Heinemann, Margot xxi, 231
Hepworth, Barbara 215, 274–5
Hermann, Carl 80
Hessen, B. 144, 214
  'The Social and Economic Roots of Newton's *Principia*' xvii, 143
Hill, A.V. xii, 171
Hill, Bradford 165–6

A History of Embryology (Needham) xvii
History of the Russian Revolution
    (Trotsky) 43
History of Trade Unionism (Webb and
    Webb) 40
Hitler, Adolf 65
    worries scientists xiii, xiv
Ho Chi Minh 228
Hodder Place 6–7
Hodgkin, Dorothy Crowfoot ix, 88
    Cambridge to Oxford 90, 93
    Oxford department 247
    radical mobilization 142
    works on insulin 92
Hodgkin, Thomas 235
Hodgkinson, Lovell 8–9
Hopkins, Sir Frederick Gowland 141,
    257
Hore-Belisha, Leslie
    science for national defence 163–4
Horkheimer, Max 153
housing 270
Huc, Abbé 180
Hughes, A. F. W. 215
Hughes-Hallet, Admiral John 183,
    194
Hungary
    schism in socialist world 73
Hussey, Captain T.A. 184, 191
Hutchinson, Arthur H. 14, 79, 80
    and Bernal's appointment 85
Hutt, Allen 13, 26, 43
Huxley, Aldous x, 59
    Intellectual Liberty 215
Huxley, Julian xxiii, 141
    organization of scientific research
        169
    reviews *Science in War* 172
    scientific research 104–5, 107
    Scientific Research and Social Need xii
Huygens, Christiaan 81, 258

idealism 57
'The Imperfect Crystallisation of
    Common Things' (Bragg) 88–9
imperialism 229
Independent Labour Party 40
industrial technology
    career options in Ireland 29–32
    young Bernal's influences 28–9
Infield, Leopold 223
'The Influence of Scientific Discovery

    on Developments in Physical
        Planning' (Bernal, lecture) 278–9
Institute for Workers' Control 292
Intellectual Liberty (IL) 214–16
Intellectuals for Peace
    Wrocław (1948) conference 218–20
    Paris (1949) meeting 219–21
    Stockholm (1950) meeting 221
    Warsaw (1951) meeting 221
International Conference on the
    History of Science 214
International Peace Campaign 215
International Union of
    Crystallography 230–31
The Interpretation of Dreams (Freud)
    impact on Bernal 46
Ireland
    'adolescent nationalism' of young
        Bernal 5–6, 22–8
    Bernal's disengagement 45
    brain drain 32–4
    career options 28–32
    compared to England 38
    Easter Rising 9
    nationalism versus socialism 41–2
    social perceptions of 5
Irish Republican Army 22

Jackson, T.A. 27
Jackson, W. Gordon 248
James, R.W. 83
Japan
    research and development 112,
        115–16
Jarrett, Father Bede 41
Jeffery, James W. 93, 94, 96, 246
    cracks in concrete 285
Jewkes 104
Jews and Judaism
    persecution by Nazis 295–6
Joad, C.E.M. 237
Johnson-Marshall, Stirrat 285–6
Johnston, Joseph 31
Johnstone Stoney, George 33
Joliot-Curie, Frédéric xv, 170, 171,
    214
    compared to Bernal 227–8
    death 226–7
    Intellectuals for Peace 219
    president of World Peace Council
        223
    refused entry into Britain 222
Joliot-Curie, Irène xv, 226

Joly, John 31
Joyce, James 20

Kaldor, Mary 116
Kaldor, Nicholas 102
Kendrew, Sir John C. ix, 189, 252
Kepler, Johannes 265
Keynes, John Maynard 109
    Economic Consequences of the Peace 40
Keynesianism 102
    heresy in economics 119
Khrushchev, Nikita 228
    denounces Stalin 147, 151
Kidd, Ronald 215
King, Admiral 187, 193
Klug, Aaron ix, 93, 94, 95, 247
Knipping (crystallographer) 83
Kölman (Soviet scientist) 261
Kowarski (refugee scientist) 170
Krutitsky, Nikolai 222

Labour Party
    Bernal in party politics 51–3
    Bernal's influence on Wilson 74
    blacklists peace organizations 225
    disciplines Zilliacus 220
    local elections of 1920 40
    MacDonald's turmoil 63
    open to technoeconomist
        argument 153–4
    pessimism about 44
    Poplar councillors 51–2
Lane, Allen 172
Langevin, Paul xv, xxiii, 170, 212, 214
Laski, Harold 130, 238
Laue, Max von 82–3
Laugier, Henri 170
Le Guin, Ursula 265
Left Book Club 64, 215
Lenin, Vladimir Illych
    cult of personality 48
    Materialism and Empiriocriticism xvi
Lessing, Doris
    The Golden Notebook 133
**'Lessons of the War for Science'
    (Bernal)** 97
Levine (Birkbeck colleague) 93
Levins, R. 148
Levy, Hyman 140, 143, 215
    reaction to Lysenko affair 149, 150
Lewison, Jeremy 274
Lewontin, R.C. 148
Liddell Hart, Sir Basil 163–4, 215

The Life of Galileo (Brecht) 258–9, 265
Lindemann, Prof. Frederick *see*
    Cherwell
Linnaean Society of Caen 178, 180
Lloyd George, David 27
Lockwood, Sir John
    opposition to Bernal at Birkbeck
        248–9, 250–51
London Mechanics' Institution 92,
    235
London School of Economics 236
London University 92
    see also Birkbeck College
Longchambon, H. 170
Lubetkin, Berthold 272–3
    Planned ARP (with Skinner and
        others) 277
Lucas, James 25
Lucas Aerospace 266
Luxemburg, Rosa 125, 130
Lysenko, Trofim Denisovich 136
    effect on Western debate 116, 124,
        131, 149–51
    genetics and proletarian science
        147–51
    Soviet Biology 147

MacDonald, Ramsay 63
Mackay, Alan xxiv, 93, 212
Mackenzie, Dean 187
Macmillan, Harold xiii
The Making of Ireland and Her Undoing
    (Green) 25
'Man Conquers Nature' (exhibition,
    Battersea Town Hall) 289–90
Mandel, Ernest 153
Manton, Dr Sidonie 289
Mao Dzedong
    reads Bernal's book 228
Marcuse, Herbert 153
Marriage as Trade (Hamilton) 46
Marshall, Father 41
Martin, Kingsley 215
Martin, Leslie 274
Marx and Science (Bernal) 144
    dialectical materialism 155
    nature and knowledge as processes
        73
Marxism
    Bernal's conversion 11–12, 37–41
    Bernal's dissatisfaction with 46–7
    changing views of science 135–7
    dialectical materialism xvii–xix

Marxism (*cont.*)
  economic determinism 55–7
  economic theory of historical
    change xvii, 69
  and Freudianism 67
  happy marriage with science
    138–42
  impact on natural science x–xi
  and Irish patriotism 11, 22
  politicization of scientists xv–xx
  research and development of
    science xvi–xx, 102–3
  role of individual 66–8
  unity and diversity xviii
materialism 57
Materialism and Empiriocriticism
    (Lenin) xvi
McLaughlin, Dr T.A. 29–30
Medical Research Council 167, 168
Megaw, Helen 86, 246
  inorganic crystallography 96
Melchett, Lord (Alfred Mortiz Mond)
    169
Mellows, Liam 27
The Melting Pot (Zangwill) 24
Mendel, Gregor 148, 149
Metadier (of French Admiralty) 171
Michurin, Ivan Vladimirovich 149
military technology 144
  Bernal and Zuckerman visit
    wartime France 170–71
  biochemical weapons 144
  Cold War scientific mobilization
    146
  CSAWG studies 214
  effects of bomb blasts 167–8
  poison gases and CSAWG criticism
    of gas masks 162–3
  research and development 115–16
Mill, James 235
Miller, Bessie *see* Bernal, Elizabeth
Miller, Jack (uncle) 3
Miller, Laetitia (aunt) 2, 3–4
Miller, Reverend William
    (grandfather) 2
mining 297
Mitscherlich, Eilhard 81
Modern Architecture Research Group
    (MARS) 273–4, 277–8
The Modern House (Yorke) 283
'Modern Science in Architecture'
    (Bernal, lecture) 290
Molotov, Vyacheslav Mikhailovich 218

Montagu, Ivor 53
Moore, Henry 215
Morgan, General Frederick 183
Morgan, Thomas 148, 149
Mosley, Oswald 136
Mountbatten, Lord Louis
  Department of Wild Ideas x, 177–8,
    181–8, 191–5
  memories of Bernal 191–5
  working with Bernal xxii, 176–8
Moynan, Carlo 20, 28
Mulberry project 177, 182–5, 194
Muskie, Edmund 187

National Association of Local
    Government Officers (NALGO)
    163
National Peace Council (NPC) 215
National Science Foundation 119–20
National Union of Scientific Workers
    xxiii, 140
  Bernal's pessimism 50–51
  see also Association of Scientific
    Workers
nationalism
  Bernal's conflict 41
nature
  as a process 73
Nature (journal)
  social implications of science xiii
'Need There be Need?' (Bernal,
    lecture) 291
Needham, Dorothy 86
Needham, Joseph 86
  on Bernal's attractiveness 133
  Chinese clocks 258
  dialectical materialism xvi, xvii
  A History of Embryology xvii
  Royal Society and British
    Association 140
Nehru, Jawaharlal 228
Neruda, Pablo 232
Nettleton, H.R. 241
Nicholson, Ben 274
1917 Club 52
Noble, David
  America by Design 145
Noel-Baker, Philip 215
Nolan, J.J. 31
nuclear weapons xiv, 71
  Bernal's warnings about atomic
    bomb 217

nuclear weapons (*cont.*)
Stockholm Intellectuals for Peace statement 221
Nuffield Foundation 246

'**On the Analytic Theory of Point Systems' (Bernal)** 80
Oparin, Alexander Ivanovich 223
'**The Organisation of Building Science Research' (Bernal)** 282, 285
Organization for Economic Cooperation and Development (OECD) 106, 136–7
Otto, Frei 269
Outline of History (Wells) 40
Owen, Robert 235

Parker, Mr 18, 25
Parsons, Sir Charles 19, 33
Paul, Eden and Cedar
Creative Revolution 43
Pavitt 116
peace movement xi, 73
Bernal's pre-war experiences 213–16
and Bernal's role in national defence 164
building internationalism 136–7
Cambridge Scientists' Anti-War Group xi, 64, 141–2
CND 151
left scientists 151–2
Penguin Books 172
Perrin, Jean xv, 170
Perutz, Max F.
Cambridge 90, 92, 93, 247
Habbakuk project 186, 187, 193
Nobel Prize ix, 186
Phillips, Magda xxiii, 64
Philpot, John 90
physics 57
'**Physics of Air Raids' (Bernal, RI Lecture)** 167
Picasso, Pablo 139, 218, 222, 246
Pirie, Norman W. 91, 257, 300
research on effect of weapons 215
Place, Francis 235
Planck, Max 263
Plato 130
Poland
stand against Russia 42
Polanyi, Michael

'Republic of Science' 117
Society for Freedom in Science 146
Polevoi, Boris 126
Pollitt, Harry 297
pollution 144
The Popes and Science (Walsh) 41
Porat, Marc 111
Portugal
wartime neutrality 188, 193–4
Postgate, Raymond
The Bolshevik Theory 44
unconvincing 46
Potter, Stephen 237
poverty 129, 263
Prenant, M. 143
Price, Derek de Solla 154
Princes Risborough 165
'**The Problem of the Metallic State' (Bernal)** 86
The Protection of the Public from Aerial Attack (multi-author) 215
Protestantism
Bernal's early social perceptions 5
psychoanalysis *see* Freudianism
'**Psychoanalysis and Marxism' (Bernal)** 67
psychology
Bernal's interest at Cambridge 45–8
mass manipulation 48
'**Psychology and Communism' (Bernal)** 47–8
The Psychopathology of Everyday Life (Freud) 46
Pyke, Geoffrey x, 185–8, 191, 192–4, 287
pykerete *see* Habbakuk project

Quebec Conference 176, 184, 194

racism 154
anti-Semitism 136
eugenics 135–6, 141
fascism worries scientists xiii
and genetics 295–6
Redmond, John 22
relativism 49
religion
see also Catholicism
Remarque, Erich Maria
All Quiet on the Western Front xiii
'Republic of Science' (Polanyi) 117

Riggs-Miller, Mrs (aunt) 1–2, 5
Rimmel, Anita xxi, 239, 246, 252
Robertson, J. C. 235
Robeson, Paul 220
Roentgen, Wilhelm Konrad von
    81–2
'Roman de Rou' (Norman poem)
    180, 181
Roosevelt, Franklin D. 176, 187,
    217
Rose, Hilary and Steven
    Science and Society 153–4
Rosse, Earl of
    telescope 18–20
Rothschild Report 120
Royal Institute of British Architects
    Architectural Science Group 280
    Bernal addresses 275–6, 284, 287
Royal Institution
    Bernal speaks on 'Lessons of the
        War for Science' 97
    Bernal takes a place 14
Royal Society 230
    discussion of new materials 291–2
    exchange of knowledge 93–4
    give Bernal Royal Medal 217
    impregnable to the left 140
Royal Society of Chemistry 93
Royal Statistical Society 93
Royden, Maude
    Women and the Sovereign State 46
Ruhemann, Martin 123
Russell, Bertrand 12, 125, 130
    on Marxist scientists xv
Russell, Dora 12
Russia (pre-Soviet)
    Bernal's support for x, 24
    cult of personality 48
    effect of Revolution on young
        Bernal 11–13
    hope for the revolutionary
        experiment 39, 54
Rutherford, Lord Ernest xii–xiii, 165
Ryazanov, D.B. xvi

Saint, Andrew 286
    Towards a Social Architecture 271
Salter, Sir Arthur 164–5, 216
Samuel, Godfrey 280
Sandys, Duncan 282
Sayre, Anne 134, 239–40
Scandinavia
    eugenics 135–6

Schreiner, Olive
    Women and Labour 46
Schumpeter, Joseph 103
science
    academic research versus
        implementation 122–3
    administrative and legislative
        activity 119–20
    Bernal sparks wartime controversy
        over direction of science
        169–70
    Bernal's school days 78–9
    bias in technical change towards
        capital goods 114
    biochemical structures 87–94
    Cold War restrictions 220–21
    computers 98
    corporate feelings of scientists
        57–8
    crystallography 80–86, 88–92
    education and information 113
    the elite and craftsmanship 259–61
    faith in the nature of science
        256–9
    government impatience 187
    the Lysenko affair 147–9
    meeting humanity's physical needs
        262–7, 263
    for military 115–16, 122, 146
    misuse of 256, 296
    organizing 130–31
    planning with the unknown 298
    as a political arena 50–51
    postwar Britain 153–5
    radicalization of the 1930s 138–42
    'red science' era x–xx
    scientists with capitalists and
        workers 65
    as a social subsystem 101
    socialist planning for science and
        technology 116–25
    space research 99, 127–8, 268–9
    state commitment to R&D 103–8,
        110–16
    structure of liquids 96–7, 98
    technocracy 57–8
    vision of the future 128–31, 268–9
    wartime defence 163–4
    X-rays 78, 81–4, 89–92
    Science and Society (Rose and Rose)
        153–4
    'Science and the Humanities' (Bernal)
        97

Science at War (Crowther and Whiddington, eds) 167
architecture 281
born out of Tots and Quots meeting 172
Science for Peace 151
Science in a Developing World (Bernal)
brain drain problem 32
'Science in Architecture' (Bernal, lecture) 284
Science in History (Bernal) 73, 152
childhood influences 19, 20
different editions mark changes in science 255–6
Ireland 33
Lysenko affair 151
Science of Science Foundation 154
The Science of Science (Goldsmith and Mackay, eds) 34, 257
Science Policy Foundation 154
Scientific Research and Social Need (Huxley) xii
'The Scientist and the World Today' (Bernal)
scientists and intellectuals
authority for philosopher/scientists 130
Bernal versus today's perspective on science 137
capitalism and scientists 65, 68
communication 303
and Communism 60–61, 63
declining importance of individuals in knowledge 63–4
how scientists work 298
learned societies and university departments 93–4
political engagement 139
radicalization in the 1930s 137–42
Scotland 32–3
The Search (Snow) xxiv
Segal, Walter 270–71
sexuality
Bernal's interest in psychology 45–6
homosexuality 134
Shankland, Graeme 286, 292
Shaw, G.B.
Back to Methuselah 137–8
Shostokovich, Dmitri 222
Siday, R.E. 241

Sinn Féin
Bernal's 'adolescent nationalism' 22–8
nationalism versus socialism 41–2
Skempton, Alec 281
Skinner, Freddie
Planned ARP (with Lubetkin and others) 277–8
slavery
holds back industry 260
Snow, C.P.
on Bernal the scientist 100
on Bernal's appointment to lectureship 85
estimating politics of scientists xi–xii
faith 257
Intellectual Liberty 215
memorial meeting 252
portrays Bernal in *The Search* x, xxiv
'Science and Government' lecture 175
two cultures 212
'The Social and Economic Roots of Newton's *Principia*' (Hessen) xvii, 143
Social Darwinism 40, 295–6
Social Democratic Party (Germany)
science and Marxism xv–xvi
The Social Function of Science (Bernal) xi, 68
architecture 279–80
development of academic Bernalism 154–5
effect on British Marxism 136
Flexitime 120–21
influences and context 142–7
objectives of book 101–3
perfectibility of organizational systems 130
recommended commitment to research 110–16
research and development 103–8
socialist planning for science and technology 116–25
social sciences
Bernal's first works 58–60
science as social subsystem 101
socialism xxiii
Bernal's faith in science 266
bias in technical change towards capital goods 114

socialism (*cont.*)
Fabian 137–8
guild socialism challenged by Communism 43–4
planning for science and technology 116–25
Socialism and Positive Science (Ferri) 40
Society for Cultural Relations with the USSR 66, 288, 289
Society for Freedom in Science 146
Society for the Social Studies of Science 136
Society of Visiting Scientists 171
Socio-Technic Planning 292
Soddy, Professor 50–51
South Africa
chemical gold mining 297
Soviet Biology (Lysenko) 147
Soviet Union *see also* Russia
authoritarianism 124–5
Bernal's late criticism 74
Bernal's visits 63, 288
encouragement for scientists 139
grimly overcoming capitalism 66
influence of scientists on Bernal 214
military development 115–16
model of democracy 70
planning for science and technology 122–5
proletarian science and Lysenko 147–51
science R&D 105–7, 110
shines against 1930s fascism 139
UN Security Council 217–18
Spanish Civil War 68, 161
Sprague, Eileen *see* Bernal, Eileen
Stalin, Joseph
Bernal's obituaries for 151–2
Khrushchev denounces 147, 151
two camps view 136
Stalinism
alienates scientists xix
Lysenko affair 148–9
Stanley, Wendell 300
Stapledon, Olaf 265
Steno, Nicolaus 81
Stonyhurst College 6, 7
Stopes, Marie 12, 135
Stradling, Reginald 165, 173
'The Structure of Liquids' (Bernal) 96
'The Structure of Proteins' (Bernal, RI lecture) 241

'The Structure of Solids as a Link between Physics and Chemistry' (Bernal, lecture) 241
Suffolk, Earl of 170
Swann, Brenda xxi
Synge, Ann xxi
Synge, R.L.M. 86
gas masks 163
research on effect of weapons 215

Tansley, A.J.
Society for Freedom in Science 146
Tawney, R.H. 215
technocracy xii, 57–9
rejected by Bernal 62
technology
academic research versus implementation 122–3
military 122, 144, 146
socialist planning 116–25
Tecton (architects) 277–8
Teller, Edward 266
Titmuss, Richard 135
Tizard, Sir Henry 171, 175, 176, 191
Topolski, Feliks 218
Tots and Quots club
architecture 281
leads to *Science in War* 168–9, 171, 172
'Towards a Complete Socio-technics – Planning Production for Social Benefits' (Bernal, lecture) 292
Towards a Social Architecture (Saint) 271
Trinity College 31
Trotsky, Leon 153
History of the Russian Revolution 43
Turing, Alan 134
Twamley, Broughton 8, 12

Union of Democratic Control (UDC) 39
United Nations
seeds of Cold War 217–20
United States
Cold War restrictions on freedom 219–20, 247
eugenics 135–6
give Bernal Order of Freedom with Bronze Palms 217
military development 115–16
spending on science R&D 105–7

University of London
takes in Birkbeck 236, 237
Unwin, Sir Raymond 275, 280

Vavilov, Nicolai 148
'The Vectorial Geometry of Space
Lattices' (Bernal) 79–80
viruses
electron microscopy 300
The Visible College (Werskey) xxiv

Waddington, C.H. xvii, 171, 275
research on effect of weapons
215
Walton, E.T.S. 33
War of Steel and Gold (Brailsford) 40
Watson, James 134, 149, 300
on Bernal's work 301
The Double Helix 152
Webb, Beatrice 43, 124, 137
History of Trade Unionism (with S.
Webb) 40
Webb, Sidney 43, 124, 137
Birkbeck College 236
History of Trade Unionism (with B.
Webb) 40
Weismann, August 148, 149
Wells, H.G. 130, 137, 265, 271
Outline of History 40
Werskey, P. Gary xxiii, 133, 138
The Visible College xxiv
'What is an Intellectual?' (Bernal) 139
Whiddingston, Prof. R.
Science at War (with Crowther) 167
Wilkins, Maurice ix, 134, 142, 300
Wilkinson, J.J. 297
Willis, Margaret 286
Wilson, Harold 154
influence of Bernal 74
Wilson, Woodrow 11
Wittgenstein, Ludwig xix
women
at Birkbeck 239–40
education 38
lack of in science 260
working in scientists' brotherhood
134
Women and Labour (Schreiner) 46
Women and the Sovereign State
(Royden) 46
Women's International Democratic
Federation 219
Wood, Alexander 14

Woolf, Leonard 215
Wooster, Nora 162, 215
Wooster, W.A. (Peter) 162, 215
The World, the Flesh and the Devil
(Bernal) 58–60, 102, 125–31
far future possibilities 269
(Bernal) (*cont.*)
optimism 145
in perspective 137
World Federation of Scientific
Workers x, 97, 151, 230
World Peace Conference 221–5
World Peace Council x, 222–5
Bernal leads 227–30
Helsinki and Bernal's last
conference 231–3
independence for colonial
countries 229
Warsaw resolutions 224
World War I
Bernal's childhood experience of
8
World War II
architecture and safety 276–8
Bernal's D-Day diaries 196–211
CSAWG criticize civil defence
measures 160–63
as defence against fascism 161
effect on Bernal's politics 69–70
end of the Phoney War 171–2
Germany bombs Britain 173–4
World without War (Bernal) 73, 145,
152
childhood influences 20
Mao dislikes 228
Third World countries 290
written in a busy life 225
Wrocław Liaison Committee 219

Yates, Frank 165
Yorke, F.R.S.
The Modern House 283

Zangwill, Israel
The Melting Pot 24
Zavadovsky, B.M. xvii
Zhdanov, Andrei Alexandrovich 148
Zilliacus, Kenni 220
Zuckerman, Solly (later Baron) x,
153
From Apes to Warlords 163–4, 177
children's school chairs 285

Zuckerman, Solly (*cont.*)
    French Committee 171
    memorial meeting 252
    monkeys in the trenches 167–8
    puts together *Science in War* 172
    science for wartime defence 163–4

    sees Habbukuk as nonsense 185,
      187, 188
    studies effects of bombing 173,
      174–5
    Tots and Quots club 168–71
    wartime post 170–71, 176, 177, 191

Printed in the United States
by Baker & Taylor Publisher Services